Recombinant Poxviruses

Edited by

Matthew M. Binns, Ph.D.
Head of the Virology Section
Department of Infectious Diseases
Animal Health Trust
Newmarket, England

and

Geoffrey L. Smith, Ph.D.
Reader in Bacteriology
Sir William Dunn School of Pathology
University of Oxford
Oxford, England

CRC Press
Boca Raton Ann Arbor London Tokyo

Library of Congress Cataloging-in-Publication Data

Recombinant poxviruses / editors, Matthew M. Binns, Geoffrey L. Smith.
p. cm.
Includes bibliographical references and index.
ISBN 0-8493-6179-6
1. Recombinant poxviruses. 2. Viral vaccines. 3. Synthetic vaccines. I. Binns, Matthew M., 1954– . II. Smith, Geoffrey L., 1955–
[DNLM: 1. Poxviridae--genetics. 2. Poxviridae--immunology. 3. Recombination, Genetic--physiology. 4. Vaccines, Synthetic--immunology. 5. Vaccines, Synthetic--therapeutic use. QW 165.5.P6 R311]
QR412.R2 1992
576'.6484--dc20
DNLC/DLC
for Library of Congress 91-47917
CIP

Direct all inquiries to CRC Press, Inc., 2000 Corporate Blvd., N.W., Boca Raton, Florida, 33431.

International Standard Book Number 0-8493-6179-6

Library of Congress Card Number 91-47917

Printed in the United States of America 1 2 3 4 5 6 7 8 9 0

Printed on acid-free paper

PREFACE

Poxviruses, particularly vaccinia virus, have played a unique role in the development of virology, immunology, and vaccines. In 1798, infection with cowpox virus was shown to confer protection against the antigenically related pathogen smallpox. This discovery founded the science of immunology and eventually led to the elimination of smallpox, "the greatest scourge of the human species." Today vaccinia virus continues to be at the forefront of immunological research as an expression vector used to study antigen recognition by the immune system. Vaccinia itself possesses a formidable array of defenses against the immune system, having specific mechanisms to block the function of complement and interferon, suppress inflammation, block peptide presentation to cytotoxic T cells, and downregulate antibody responses. Edward Jenner is rightly called the Father of Immunology.

Vaccinia virus was the first animal virus to be physically purified, chemically characterized, and accurately titrated. It was the animal virus in which virus-encoded enzymes were discovered in 1967. This revolutionary finding was followed by the demonstration of enzymes in many other viruses, including reverse transcriptase in retroviruses. A quarter of a century later, new enzymes are still being discovered in vaccinia virus and are providing further insights into the fundamental biochemical processes of transcription and DNA replication.

The use of cowpox virus to prevent smallpox established the principle of vaccination. Indeed, our words vaccine, vaccinia, and vaccination all derive from the latin *vacca* (cow) from whence the virus came. Cowpox virus was the first human vaccine and the intensive use of the modern day smallpox vaccine, vaccinia, resulted in the global eradication of that disease. In 1980, the World Health Organization confirmed the demise of smallpox and has recently proposed the destruction of all remaining smallpox virus stocks and cloned virus DNAs after the nucleotide sequence of the virus genome has been determined in 1993. This will complete a chapter of medical history.

Paradoxically, despite the eradication of smallpox, there has been an explosion of interest in poxviruses, as this book testifies. This interest has stemmed in part from the application of molecular genetics to clone and express foreign genes from recombinant poxviruses. These viruses have multiple applications, but the most talked about, least used, and most controversial is the potential to reuse vaccinia virus as a live vaccine against pathogens other than smallpox. In this respect, vaccinia virus scores another first as the virus with which the concept of constructing genetically engineered live vaccines was established. This principle has subsequently been applied to many other microorganisms, both viruses and bacteria.

The aim of this book is to bring together in one volume comprehensive reviews about recombinant poxviruses with an emphasis on the potential of these viruses as new vaccines. A wide range of issues involved in producing new genetically engineered live vaccines are considered, such as efficacy, safety, stability, cost, host range, immune response, immunization route, use of multivalent vaccines, and need for revaccination. The opening chapter describes the origin of vaccinia virus, its use to eradicate smallpox, and the pathogenesis of poxvirus infections. Thereafter the molecular biology of poxviruses, the methods of contructing vaccinia virus recombinants, and the many ways in which they may be used are described. The use of poxvirus recombinants as live vaccines, the immune responses induced by these viruses, and the results of field trials with vaccinia-rabies virus recombinants are then considered. Subsequently, the important issues of the safety and immunogenicity of vaccinia virus are described, and lastly, two chapters report the progress that has been made developing *avipoxviruses* and *parapoxviruses* as candidate recombinant vaccines.

THE EDITORS

Matthew Binns is Head of the Virology Section, Department of Infectious Diseases, Animal Health Trust, Newmarket, England.

Dr. Binns graduated from Hatfield Polytechnic in 1977 with a B.Sc.(Hons.) degree in Applied Biology. He then moved to the University of Kent, Canterbury and obtained a Ph.D. degree in Microbiology in 1980. After doing post-doctoral work at Washington University School of Medicine, St. Louis, Missouri, and the University Medical School, Geneva, Switzerland, studying pathogenic bacteria, he was appointed Senior Scientific Officer, Institute for Animal Health, Houghton Laboratory, U.K. in 1983. While there he developed an interest in using fowlpoxvirus as a species-specific vaccine vector for poultry. In 1990 he assumed his present position.

Dr. Binns is a member of The Society for General Microbiology and The American Society of Microbiology and is serving on the Editorial Board of the *British Veterinary Journal*. His current work is funded by The Equine Virology Research Foundation, The Horserace Betting Levy Board, and by private industry. He has published approximately 50 papers and continues to use poxviruses as vectors to aid in the study of a number of equine infectious diseases.

Geoffrey L. Smith is Reader in Bacteriology, Sir William Dunn School of Pathology, University of Oxford.

Dr. Smith graduated from the University of Leeds in 1977 with a B.Sc. degree in Microbiology and Biochemistry. He then moved to the National Institute for Medical Research, London and in 1981 obtained a Ph.D. degree in Virology from the Council for National Academic Awards. This was followed by post-doctoral work at the National Institute for Allergy and Infectious Diseases, Bethesda, Maryland, where, together with Michael Mackett in Bernard Moss's laboratory, techniques were developed for the use of vaccinia virus as an expression vector. In 1985 he returned to the U.K. to a lectureship in Virology in the Department of Pathology, University of Cambridge. From here he moved in 1989 to a Readership in the Sir William Dunn School of Pathology, University of Oxford. In 1988 Dr. Smith was awarded the Jenner Fellowship from the Lister Institute for Preventive Medicine, and in 1992 he received the Fleming Award from The Society for General Microbiology.

Dr. Smith is a member of The Society for General Microbiology, The American Society for Microbiology, The British Society for Immunology, and Christ's College Cambridge, and is a Fellow of Wadham College, Oxford. His present work on vaccinia virus is funded by the Medical Research Council, The Wellcome Trust, and The Lister Institute for Preventive Medicine. He has published approximately 80 papers and continues fundamental research on the molecular biology and pathogenesis of vaccinia virus and other cytoplasmic DNA viruses.

CONTRIBUTORS

Marion E. Andrew, Ph.D.
Principal Research Scientist
CSIRO Australian Animal Health Laboratory
Geelong, Australia

Marc Artois, D.V.M., Ph.D.
Centre National d'Etudes Vétérinaires et Alimentaires
Laboratoire National d'Etudes sur la Rage et la Pathologie des Animaux Sauvages
Ministère de l'Agriculture
Malzéville, France

Michel Aubert, Ph.D.
Director
Centre National d'Etudes Vétérinaires et Alimentaires
Laboratoire National d'Etudes sur la Rage et la Pathologie des Animaux Sauvages
Ministère de l'Agriculture
Malzéville, France

Jean Blancou, D.V.M., Ph.D.
Director General
Office International des Epizootics
Paris, France

Michael E. G. Boursnell, Ph.D.
Immunology Limited
Cambridge, England

David B. Boyle, Ph.D.
Principal Research Scientist
CSIRO Australian Animal Health Laboratory
Geelong, Australia

Bernard Brochier, D.V.M., Ph.D.
Department of Virology and Immunology
Faculty of Veterinary Medicine
University of Liege
Liege, Belgium

R. Mark L. Buller, Ph.D.
Head
Poxvirus Pathogenesis Group
Laboratory of Viral Diseases
National Institute of Allergy and Infectious Diseases
National Institutes of Health
Bethesda, Maryland

Gilles Chappuis, D.V.M.
Head of Development
Rhône-Mérieux
Laboratoire IFFA
Lyon, France

Barbara E. H. Coupar, Ph.D.
Principal Research Scientist
CSIRO Australian Animal Health Laboratory
Geelong, Australia

William I. Cox, Ph.D.
Senior Scientist
Virogenetics Corporation
Troy, New York

Philippe Desmettre, D.V.M.
Director of Research and Development
Rhône-Mérieux
Laboratoire IFFA
Lyon, France

Frank Fenner, M.D.
Emeritus Professor
John Curtin School of Medical Research
Australian National University
Canberra, Australia

Marie-Paule Kieny, Ph.D.
Assistant Scientific Director
Department of Virology and Immunology
Transgene S.A.
Strasbourg, France

Bernard Languet
Engineer
Rhône-Mérieux
Laboratoire IFFA
Lyon, France

Jean-Pierre Lecocq, Ph.D.
President
Transgene S.A.
Strasbourg, France

David J. Lyttle, Ph.D.
Research Fellow
Health Research Council of New Zealand
Virus Research Unit
University of Otago
Dunedin, New Zealand

Michael Mackett, Ph.D.
Senior Research Scientist
Department of Molecular Biology
Paterson Institute for Cancer Research
Christie Research Centre
Christie Hospital NHS Trust
Manchester, England

Bernard Moss, M.D., Ph.D.
Chief
Laboratory of Viral Diseases
National Insitute of Allergy and Infectious Diseases
National Institutes of Health
Bethesda, Maryland

Gregory J. Palumbo, Ph.D.
Laboratory of Viral Diseases
National Institute of Allergy and Infectious Diseases
National Institutes of Health
Bethesda, Maryland

Enzo Paoletti, Ph.D.
Director of Research
Virogenetics Corporation
Troy, New York

Paul-Pierre Pastoret, D.V.M., Ph.D.
Professor
Department of Virology and Immunology
Faculty of Veterinary Medicine
University of Liege
Liege, Belgium

Anthony J. Robinson, Ph.D.
Director
Health Research Council of New Zealand
Virus Research Unit
University of Otago
Dunedin, New Zealand

Geoffrey L. Smith, Ph.D.
Reader in Bacteriology
Sir William Dunn School of Pathology
University of Oxford
Oxford, England

James Tartaglia, Ph.D.
Assistant Director of Research
Virogenetics Corporation
Troy, New York

TABLE OF CONTENTS

Chapter 1

VACCINIA VIRUS AS A VACCINE, AND POXVIRUS PATHOGENESIS

Frank Fenner

TABLE OF CONTENTS

I. THE FAMILY *POXVIRIDAE*

Because of the severity and widespread occurrence of smallpox and the introduction of inoculation with vaccine virus to prevent it in the early years of the 19th century, the poxviruses were among the first viruses to be studied in detail in the laboratory. The large size of poxvirus virions led to their early visualization with a microscope and made possible their purification by differential centrifugation and thus a study of their chemical composition. The occurrence of diseases in several species of animals characterized by skin lesions from which virus particles of similar size and appearance could be obtained led to early proposals for grouping such viruses as a "genus".[1,2] Following the establishment of a Poxvirus Subcommittee by the Sixth International Congress for Microbiology in 1953, Fenner and Burnet[3] produced a short description and classification of the poxviruses of vertebrates. In 1966 virus classification acquired formal status with the establishment of the International Committee for the Nomenclature of Viruses (later the International Committee for the Taxonomy of Viruses). The current classification recognizes the family *Poxviridae,* two subfamilies, *Chordopoxvirinae* and *Entomopoxvirinae,* and eight genera within the subfamily *Chordopoxvirinae* (Table 1).

For historical reasons and because of its favorable growth characteristics, the vast majority of laboratory studies with poxviruses have been done with vaccinia virus, the prototype virus of the genus *Orthopoxvirus*. The biology of this genus has been reviewed in a recent book,[4] and various aspects of the biology of vaccinia virus relevant to the use of recombinant poxviruses are described in the first seven chapters of this book. The other two chapters describe the use of poxviruses of two other genera, *Avipoxvirus* and *Parapoxvirus,* as host-specific recombinant vectors.

II. VACCINIA VIRUS AS A SEPARATE SPECIES OF *ORTHOPOXVIRUS*

The genus *Orthopoxvirus* contains ten species (Table 2), of which five, cowpox, ectromelia, monkeypox, vaccinia, and variola viruses, will be mentioned again in this chapter. Because of its long history of use as a vaccine and in the laboratory, there are a large number of strains of vaccinia virus that differ somewhat from each other in their properties. Nevertheless, because of the constellation of biological properties that they exhibit and their characteristic DNA maps (Figure 1), all of these strains clearly belong to a distinct species of *Orthopoxvirus*. Biological characteristics of vaccinia virus that are useful for species diagnosis within the genus *Orthopoxvirus* are the broad host range (shared only with cowpox and monkeypox viruses), the rapid growth of large pocks on the chorioallantoic membrane, and the high ceiling temperature for growth on the chorioallantoic membrane (Table 3). In

TABLE 1
Classification of Poxviruses of Vertebrates

Family: *Poxviridae;* Subfamily: *Chordopoxvirinae*

Genus	Prototype virus
Avipoxvirus	Fowlpox virus
Capripoxvirus	Sheeppox virus
Orthopoxvirus	Vaccinia virus
Leporipoxvirus	Myxoma virus
Molluscipoxvirus	Molluscum contagiosum virus
Parapoxvirus	Pseudocowpox virus
Suipoxvirus	Swinepox virus
Yatapoxvirus	Yaba poxvirus

addition, vaccinia virus grows more rapidly on the chorioallantoic membrane and in cultured cells than other poxviruses.

III. VACCINATION AGAINST SMALLPOX

A. HISTORY OF SMALLPOX VACCINE

The history of Jenner's discovery and promotion of the use of cowpox virus for protection against smallpox has been well described in a book by Baxby.[5] Although others had inoculated children with material from cowpox lesions to protect them from smallpox before 1796, Jenner's demonstration of its efficacy by challenge inoculation with variolous material, his recommendation for maintenance of the virus by arm-to-arm inoculation of children, his recognition of "spurious" cowpox (now recognized as pseudocowpox), and his vigorous promotion of the practice justify full credit for the discovery of vaccination being given to him.

B. SMALLPOX VACCINE PRODUCTION

Over the years since its introduction in 1798, the method of production of smallpox vaccine evolved from serial passage in humans to a quality-controlled freeze-dried vaccine, but even through the smallpox eradication campaign, which was successfully concluded in 1980, smallpox vaccine production retained features that would not have been tolerated in a vaccine developed since 1950, notably the unavoidable bacterial and potential viral contamination of vaccine lymph prepared in animal skin.

1. Early Methods

Jenner developed the idea of maintaining a source of virus for vaccination by arm-to-arm passage in children, but every now and then this failed, and in some parts of Europe rewards were offered for the discovery of cases of cows with cowpox or horses with "horsepox", to obtain a new source of vaccine.

TABLE 2
Species of the Genus *Orthopoxvirus*[a]

Species	Host range in laboratory animals	Animals found naturally infected	Geographic range of natural infections
California vole poxvirus	? Broad	Voles	Western U.S.
Camelpox virus	Narrow	Camel	Africa and Asia
Cowpox virus	Broad	Carnivores, cats, cow, elephant, gerbils, man, okapi, rats	Europe and U.S.S.R.
Ectromelia virus	Narrow	Laboratory mouse, ? voles	Europe
Monkeypox virus	Broad	Anteater, great apes, man, monkeys, squirrels	Western and central Africa
Raccoon poxvirus[b]	? Broad	Raccoon	Eastern U.S.
Tatera poxvirus	Narrow	*Tatera kempi* (a gerbil)	Western Africa
Uasin Gishu poxvirus	Medium	Horse	Kenya, Zambia
Vaccinia virus[b]	Broad	Buffalo, cow, man, pig, rabbit	India (buffalopox[c])
Variola virus	Narrow	Man (disease now eradicated)	Formerly worldwide

[a] Distinguishable by biological characteristics and restriction endonuclease mapping (see Figure 1A).
[b] Used for recombinant vaccines.
[c] Probably derived from smallpox vaccine.

Based on Fenner, F., Wittek, R., and Dumbell, K. R., *The Orthopoxviruses,* Academic Press, San Diego, 1989, 9. With permission.

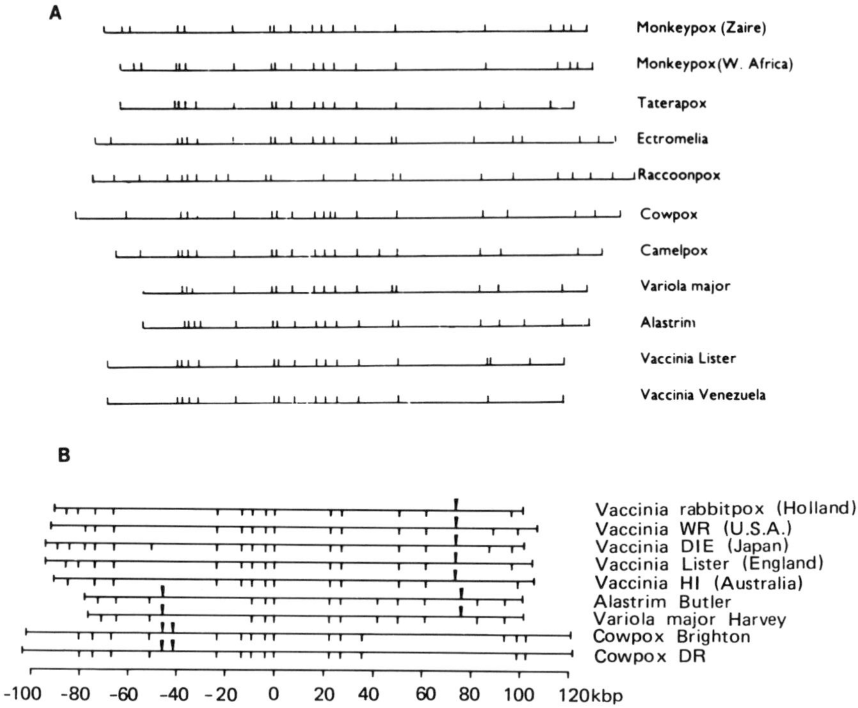

FIGURE 1. (A) Orthopoxvirus DNA *Hin*dIII maps — physical arrangements of fragments of the genomic DNA of one or two isolates of each species except Uasin Gishu poxvirus and the vole orthopoxvirus, the DNAs of which have not been examined. The consecutive order of digest fragments was determined by cross-hybridizations with cloned DNA fragments of orthopoxviruses with known *Hin*dIII maps. From top: Monkeypox (Zaire) = monkeypox virus, Congo-8 strain; Monkeypox (West Africa) = monkeypox virus, Copenhagen strain; Taterapox = tatera poxvirus, Dahomey strain; Ectromelia = ectromelia virus, Hampstead strain; Raccoonpox = raccoon poxvirus, Maryland strain; Cowpox = cowpox virus, Brighton strain; Camelpox = camelpox virus, Somalia strain; Variola major = variola virus, Harvey strain; Alastrim = variola virus, Butler strain; Vaccinia Lister = Vaccinia virus, Lister strain; Vaccinia Venezuela = vaccinia virus, Venezuela strain. (Based on Esposito, J. J. and Knight, J. C., *Virology,* 143, 230, 1985. With permission.) (B) Comparison of the cleavage sites after digestion with *Xho* I and *Sma* I of DNA from several strains of vaccinia virus and DNA from the two species of virus from which it has been suspected to have been derived, cowpox and variola viruses. (From Mackett, M. and Archard, L. C., *J. Gen. Virol.,* 45, 683, 1979. With permission.)

Passage in calves was used as a source of vaccine in Italy from 1840 and in France from 1864, stimulated by the fear of the transmission of vaccinial syphilis from human sources of vaccine, but calf-derived vaccine was not used in England until 1881 and arm-to-arm vaccination was popular there until it was banned in 1898.

By the end of the 19th century there were a multitude of small backyard "vaccine parks" in most countries of Europe, each supplying calf-skin vaccine

TABLE 3
Biological Characteristics of Vaccinia, Cowpox, and Variola Viruses[a]

	Virus		
Characteristic	**Variola**	**Vaccinia**	**Cowpox**
Pock on chorioallantoic membrane	Small opaque white	Strains vary; large opaque white or ulcerated	Large, bright red
Ceiling temperature	37.5°C-38.5°C	41°C	40°C
Skin lesion after intradermal inoculation of rabbit	Nil or small nodule, nontransmissible	Strains vary; indurated, sometimes hemorrhagic nodule	Large, indurated hemorrhagic nodule

[a] Unequivocal species diagnosis can be made by the examination of electropherograms produced after restriction endonuclease digestion of viral DNAs.

From Fenner, F., Wittek, R., and Dumbell, K. R., *The Orthopoxviruses,* Academic Press, San Diego, 1989, 144. With permission.

to a small clientele. Later vaccine production was taken over by factory-type laboratories and glycerine was introduced as a suspending material which inhibited bacterial contamination. In the mid-1920s various countries introduced regulations to cover the production and quality of vaccines to be used in humans and standards for the purity and potency of smallpox vaccine were developed.

2. Development of Freeze-Dried Vaccine

Glycerinated vaccine lymph was reasonably stable at refrigerator temperatures, but it was quite unsatisfactory in tropical countries, where there was an inadequate public health infrastructure and few refrigeration facilities. From Jenner's day, air-dried vaccine had been used for despatch of virus to distant places. Otten[6] improved on this by drying vaccine *in vacuo,* but the material was often heavily contaminated with bacteria and it was difficult to reconstitute. The solution to the problem, embarked upon independently in France, the U.S., and England, was the introduction of freeze-drying. The method that was finally adopted worldwide was based on the work of Collier.[7,8]

3. Vaccine Production in the Intensified Smallpox Eradication Programme

Global eradication of smallpox was adopted as a goal by the World Health Organization in 1959, but no earmarked funding was provided and little was achieved until the Intensified Smallpox Eradication Programme was initiated in 1967. One of the first actions of the Intensified Programme was to survey smallpox vaccine production facilities throughout the world.[9] In 1967 information was obtained from 59 producers in 44 countries. Most of these producers used calves, 12 sheep, and 6 water buffalo for vaccine production; 3 producers were making vaccine in cultured cells and 3 were making vaccine in developing chick embryos. Twelve different strains of vaccinia virus were used. The major changes in production methods since the 1920s were that the animals were kept in more hygienic conditions, high titer seed virus was applied by scarification rather than by inoculation, and in most cases vaccine was freeze-dried. However, tests of the vaccine produced in that year (1967) showed that less than half of the samples came up to the standards of potency and stability of freeze-dried vaccine that had been laid down by the World Health Organization (WHO) in 1959. A program was therefore established by the WHO for upgrading vaccine production and ensuring that WHO standards for vaccine quality were met for all vaccine used in the Intensified Smallpox Eradication Programme. By 1973 most vaccine producers, in developing as well as developed countries, were producing satisfactory vaccine.

C. VACCINATION TECHNIQUES

Scarification — Jenner inoculated human subjects with vaccine virus by making a light scratch of the skin through a drop of vaccine, which introduced virus into the epidermal cells. With variations in the depth of the "insertion" and the number of sites (from one to four, usually over the deltoid muscle), this method remained in vogue for the next 150 years. In 1927 Leake[10] introduced a "multiple pressure" method that introduced vaccine into the epidermis with less chance of subdermal inoculation, and this method was widely adopted.

Jet injection — As the worldwide campaign for smallpox eradication got under way, tests were begun for a more "up-to-date" method of delivery of vaccine on a mass scale and the jet injector, introduced in the 1950s for mass immunization by subcutaneous inoculation, was adapted for intradermal delivery of smallpox vaccine. Jet injectors were used extensively in the eradication campaigns in West Africa and Brazil, but the need to assemble large numbers of vaccinees in one place and difficulties of maintenance of the instruments militated against their use elsewhere.

Bifurcated needle — The major advance in vaccination technique during the global eradication campaign was the development of the bifurcated needle (Figure 2). The technique could be learned rapidly even by illiterate persons and it could be applied to small numbers of vaccinees in rural areas. An ingenious container was developed for carrying and dispensing sterile needles and boiling them after each day of work, so that they could be used again the next day.

D. INTERPRETATION OF THE RESULTS OF VACCINATION

1. Primary Vaccination

In susceptible individuals, smallpox vaccination produced a pustule at the inoculation site and frequently produced swelling and tenderness of the

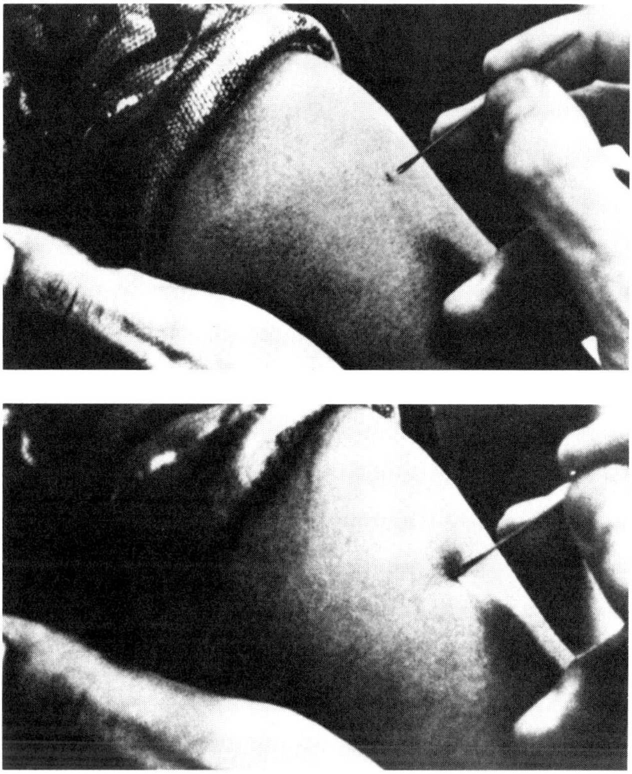

FIGURE 2. Vaccination with the bifurcated needle. The requisite amount of reconstituted vaccine is held between the prongs of the needle and vaccination is done by multiple puncture: 15 strokes at right angles to the skin over the deltoid muscle, in an area about 5 mm in diameter. (From Fenner, F., Henderson, D. A., Arita, I., Jezek, Z., and Ladnyi, I. D., *Smallpox and Its Eradication,* World Health Organization, Geneva, 1988, 570. With permission.)

draining lymph node. The course of the reaction is illustrated in Figure 3. A papule appeared at the vaccination site on the third day after vaccination and within 2 or 3 days this became vesicular, to constitute the umbilicated and loculated ''Jennerian vesicle''. The vesicle soon became pustular due to the entry of polymorphonuclear cells and the surrounding area became erythematous and indurated. The area of erythema reached a maximum between the 8th and the 12th day, and at this time the draining lymph nodes were enlarged and tender and the patient was often feverish. The pustule dried from the center out to become a dry brown or black scab which fell off about 3 weeks after vaccination, to leave a typical pitted scar.

Observation of the lesion enabled the vaccinator to decide whether the virus had multiplied and the patient thus rendered immune to smallpox, and in the longer term, the vaccination status of an individual or a population

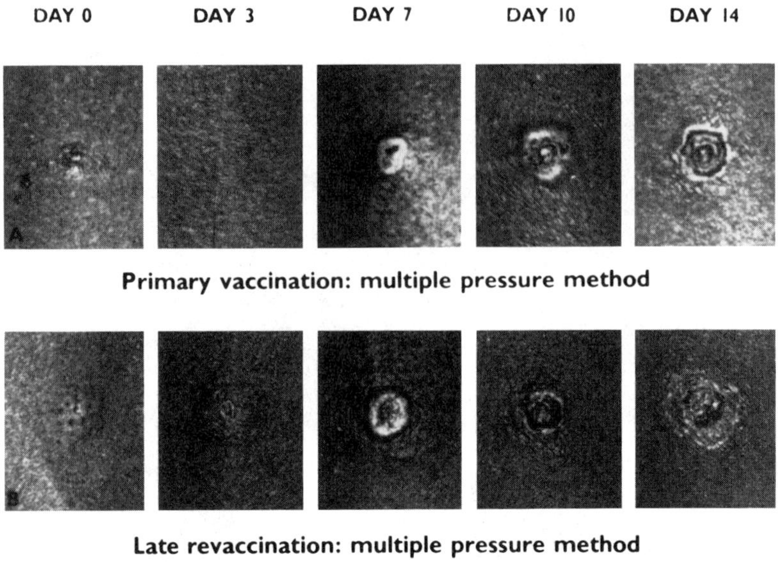

FIGURE 3. Skin reactions after primary vaccination and revaccination several years after primary vaccination, by the multiple pressure method. (From Fenner, F., Henderson, D. A., Arita, I., Jezek, Z., and Ladnyi, I. D., *Smallpox and Its Eradication,* World Health Organization, Geneva, 1988, 297. With permission.)

could be determined by visual examination for vaccination scars, dispensing with the need for a serological survey.

2. Revaccination

The response to revaccination depended on the potency of the vaccine and the interval since primary vaccination. Except when an interval of a decade or more had elapsed, the lesion produced by revaccination always progressed more rapidly than after primary vaccination. Sometimes there was no reaction at all, a result that was usually due to the use of vaccine of low potency and which was impossible to interpret correctly. Otherwise the reaction to revaccination could be maximal at any time between the second and the eighth day.

E. COMPLICATIONS OF SMALLPOX VACCINATION

Not only did a successful vaccination always produce a local lesion and some constitutional disturbance, but as Wilson[11] has said: "Smallpox vaccine has probably been followed by more complications and been responsible for more deaths than any other vaccine." The complications fell into two groups: those affecting the skin (Figure 4) and those affecting the central nervous system (CNS). The first category included two types of disease which should have been avoidable, because the predisposing conditions (eczema and im-

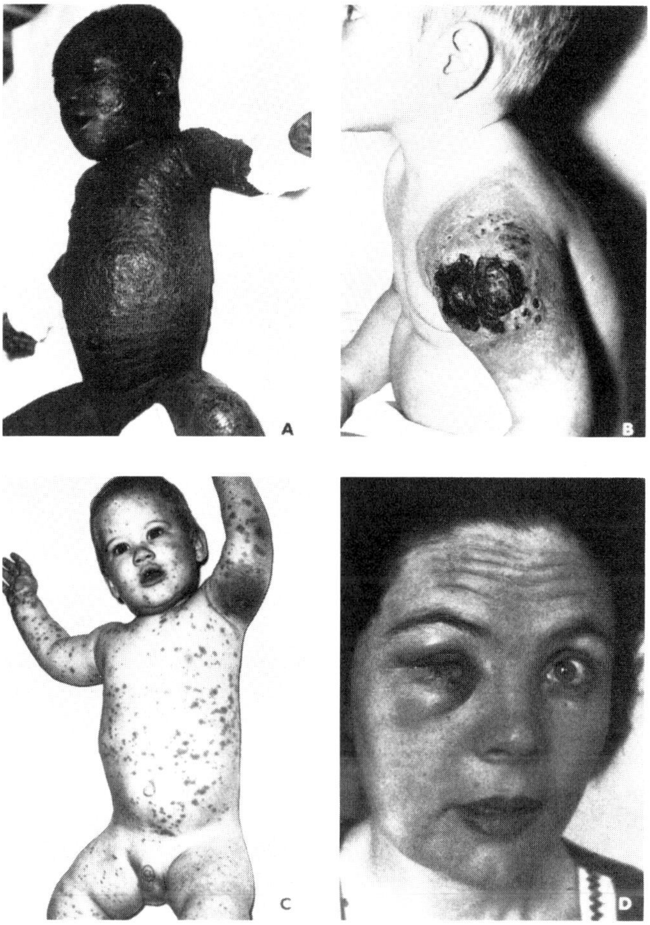

FIGURE 4. Cutaneous complications of smallpox vaccination. (A) Eczema vaccinatum in the unvaccinated contact of a vaccinated sibling; (B) progressive vaccinia, which was fatal, in a child with congenital immunodeficiency; (C) generalized vaccinia ten days after vaccination; this had a benign course, with no residual scarring; (D) accidental ocular infection with vaccinia virus. (From Fenner, F., Henderson, D. A., Arita, I., Jezek, Z., and Ladnyi, I. D., *Smallpox and Its Eradication,* World Health Organization, Geneva, 1988, 298. With permission.)

munological deficiency diseases) were contraindications to vaccination. The most common was eczema vaccinatum (Figure 4A), which also sometimes occurred in eczematous family contacts of vaccinated subjects (Table 4). Progressive vaccinia (Figure 4B) was found only in subjects with a serious impairment of their cell-mediated immune system of either genetic or acquired origin (immunosuppressive drugs, lymphoreticular malignancy). Details of the nature and frequency of these complications are given in Fenner et al.[9] The best available data were obtained in a year-long survey conducted in the U.S. in 1968, summarized in Table 4.

TABLE 4
Frequency of Complications of Smallpox Vaccination in the U.S. in 1968, as Determined by a National Survey

Complication	Reported cases (deaths) per million vaccinations		
	Primary vaccination	Revaccination	Unvaccinated contacts
	5,594,000[a]	8,547,000[a]	—
Encephalopathy (age <1 year)	6.5 (4.9)	—	—
Encaphalitis (age >1 year)	2.2 (0.2)	—	—
Progressive vaccinia	0.9 (0.4)	0.7 (0.2)	—
Eczema vaccinatum	10.4	0.9	60 cases (1 death)
Generalized vaccinia	23.4	1.2	2 cases
Accidental infection	25.4	0.8	44 cases
Other complications	11.8	1.0	8 cases
Total	74.7	4.7	114 cases

[a] Numbers in group.

Reprinted by permission of the New England Journal of Medicine. (281, 1201, 1969.)

Because it was unpredictable, the most important complication was CNS disease, which included encephalopathy, a rare but serious disease in infants, and postvaccinial encephalitis, a syndrome in older persons associated with demyelination, with a case fatality rate of about 30%. The reported incidence of cases varied considerably in different countries (Table 5). These differences may have been due in part to inconsistencies in reporting, but the strain of vaccinia virus used for vaccination also played a role. This was demonstrated when the Lister strain, which had long been used in the U.K., was introduced in Austria, Belgium, the Netherlands, and Switzerland in the 1960s, and the incidence of post-vaccinial encaphalitis fell dramatically.[9] CNS disease also occurred in smallpox, in which it was substantially more common than after vaccination, but was overshadowed by the severity of the other symptoms. The complications of vaccination with vaccinia virus remain a concern for scientists and public health officials interested in immunization based on the use of vaccinia virus as a vector for genes from other infectious agents.

F. SOURCES OF VACCINIA VIRUS STRAINS USED FOR SMALLPOX VACCINE

The virus originally used by Jenner[12] to protect human subjects against smallpox was obtained from the pustule on the hand of a milkmaid infected with cowpox. Initially supplies were maintained by serial arm-to-arm passage in children, but periodically new strains were obtained from cows or from lesions of horsepox.

In 1939 Downie[13] showed that cowpox virus differed in many biological

TABLE 5
Postinfection Encephalitis Associated with Smallpox and Primary Vaccination[a]

	Cases per million		
Survey	**Encephalopathy age 2 years**	**Encephalomyelitis age 2 years**	**Vaccinia strain**
	Smallpox		
Variola major	2000		
Variola minor	500		
	Primary Vaccination		
Hamburg (1939–1958)	93	449	? Bern
Bavaria (1945–1953)	56	121	?[b]
Holland (1930–1943)	50	348	?[b]
Holland (1924–1928)	39	232	?[b]
U.K. (1951–1960)	15[c]	15	Lister
U.S. (10 states, 1968)	42[c]	9	New York City Board of Health
U.S. (National, 1968)	7[c]	2	New York City Board of Health

[a] Postvaccinial encephalitis was very rare after revaccination.
[b] Not the Lister or New York City Board of Health strain.
[c] Age less than 1 year.

From Fenner, F., Henderson, D. A., Arita, I., Jezek, Z., and Ladnyi, I. D., *Smallpox and Its Eradication,* World Health Organization, Geneva, 1988, 307. With permission.

properties from the material then being used for smallpox vaccination (vaccinia virus), a finding that found support from studies of DNA mapping (see Figure 1). No strain of virus used for vaccine production since 1967, and probably very much earlier, has been characterized as cowpox virus; all strains studied have had the biological characteristics of vaccinia virus.

If in Jenner's time smallpox vaccine virus was indeed derived from cowpox virus, when were such strains replaced by vaccinia virus? We can only speculate, but there are some hints. The first depends on the very prominent regular-shaped eosinophilic A-type inclusion bodies produced by cowpox virus but not by vaccinia virus. If cowpox virus was used for vaccine production in the 1890s, when pathologists first described poxvirus inclusion bodies,[14,15] they would have recognized the A-type inclusions. However, the only type of inclusion body then described in tissues infected with variola virus and "vaccine virus" was the irregular B-type Guarnieri body. The second clue depends on the fact that cowpox virus does not occur naturally in North America. Vaccine strains were repeatedly sent to the U.S. from Europe, especially England. They included a strain of "nonhumanized" cowpox virus sent from France in 1870. However, the strain long used for vaccination in that country (New York City Board of Health strain) is said to be a lineal derivative of a strain first used for vaccine production in the U.S. in 1876. The New York City Board of Health strain is classical vaccinia virus, not cowpox virus, which suggests that vaccinia virus was being used for vaccination at least as early as 1876.

IV. THE ORIGIN OF VACCINIA VIRUS

All strains of smallpox vaccine in use during the last 50 years that have been adequately studied consist of vaccinia virus and not cowpox virus. The origin of vaccinia virus has long been disputed, and no certain answer is possible. Five hypotheses have been advanced:

1. Vaccinia virus was derived from variola virus by transfer and adaptation by passage in cows.
2. By continuous passage in human skin through variolation in the 18th and early 19th centuries, variola virus became attenuated and altered to become vaccinia virus.
3. Vaccinia is a hybrid between cowpox and variola viruses, occurring in the early years of the 19th century when cowpox virus was used to vaccinate patients in smallpox hospitals.
4. Vaccinia virus is a "fossil" and represents the maintenance in the laboratory of a virus of a domestic or wild animal that has otherwise become extinct.
5. Vaccinia virus was derived from cowpox virus by repeated passage on the skins of cows, sheep, and other animals.

A. POSSIBLE DERIVATION FROM VARIOLA VIRUS BY PASSAGE IN COWS

In his published paper,[12] although not in earlier manuscripts of it,[16] Jenner used the term "variolae vaccinae" (smallpox of the cow). The notion that cowpox was really smallpox in bovines was popular among scientists in the late 19th century and even later and led to the frequent introduction of variola virus from patients into the inoculum used for the preparation of smallpox vaccine. In consequence, several strains in use for vaccine production in the 1970s are said to have been "derived" from variola virus.[17] It is impossible to determine what happened in these cases. Variola virus has a very narrow host range and rarely produces even a transient nodule in cows. On the other hand, since orthopoxviruses are notoriously resistant to environmental conditions, laboratory contamination was an ever present risk when such manipulations were carried out (as they usually were) in institutions where vaccine was being produced. Indeed, Kelsch et al.[18] showed that in such circumstances cows scarified with sterile glycerine often developed a few vaccinia pustules.

Eventually, the careful experiments of Herrlich et al.,[19] carried out over several years in premises in which neither vaccinia nor cowpox virus had ever been used, were entirely negative. These workers could not achieve the ''transformation'' of variola virus into vaccinia virus. This conclusion is supported by the DNA maps of the two viruses; there is no way in which a variola virus DNA molecule could be readily converted into the quite different DNA molecule of vaccinia virus (see Figure 1).

B. SERIAL PASSAGE OF VARIOLA VIRUS IN HUMANS, BY VARIOLATION

In an attempt to debunk Jenner, Razzell[20] has postulated that vaccinia virus was derived from variola virus by continued intradermal passage in humans. However, as with the suggested derivation of vaccinia virus from cowpox virus, DNA mapping of variola and vaccinia viruses provides no support for this hypothesis.

C. HYBRIDIZATION BETWEEN COWPOX AND VARIOLA VIRUSES

In the early days of vaccination there would have been ample opportunities for hybridization to occur between cowpox virus and variola virus.[5] Hybridization between these viruses occurs readily in the laboratory[21] and variants showing a wide variety of biological characteristics occur, although none of those examined resembled vaccinia virus in all characteristics. A final decision on the likelihood that vaccinia virus originated in this way would be assisted by restriction endonuclease mapping of the DNA of selected hybrids. So far this has not been done. One argument against such an origin is that all strains of vaccinia that have been examined have very similar DNA maps and biological characteristics. Such homogeneity would be unlikely to occur in hybrids between variola and cowpox viruses produced at different times and places.

D. VACCINIA VIRUS IS A "FOSSIL" VIRUS

Apart from the recent observations on buffalopox in India,[4] no animal reservoir of vaccinia virus has been discovered. Considering its broad host range, it would be reasonable to postulate that if there were an animal reservoir, there would be several alternative ''natural hosts'', as with the other two viruses that have a broad host range, cowpox and monkeypox viruses. Baxby's[5] suggestion that vaccinia virus may have been derived from horsepox, a disease that now appears to be extinct, does not necessarily solve the problem, since it now seems more likely that both horsepox and cowpox were due to the infection of horses and cows respectively with ''cowpox'' virus from a wild rodent source.

E. DERIVATION FROM COWPOX VIRUS BY MUTATION

In some ways the most appealing suggestion is that vaccinia virus is the result of adaptation of cowpox virus to growth in the human and bovine skin. Cowpox virus has the largest genome of all orthopoxviruses and deletion mutants occur with high frequency. Many such mutants resemble dermal vaccinia virus in two related properties: the type of pock produced on the chorioallantoic membrane and the type of lesion produced in the rabbit skin. Most retain the capacity to produce A-type inclusion bodies, but Amano and Tagaya[22] have reported the recovery of mutants of cowpox virus that failed to produce A-type inclusion bodies, although they behaved like wild-type cowpox virus in pock production and pathogenicity for the rabbit skin. Further, they and Patel et al.[23] have shown that there is a close relationship between the 160K protein of cowpox virus A-type inclusion bodies and the 92K LS antigen of vaccinia virus. Clearly, mutants of cowpox virus could be found that resembled vaccinia virus in all three of these biological characteristics.

However, there are problems with this explanation, too. If vaccinia virus was derived from deletion-transposition mutations in cowpox virus, what was the origin of the rabbitpox strains, that produce somewhat hemorrhagic pocks and large lesions in the rabbit skin, but have the characteristic DNA map of vaccinia virus? Also why did the many strains of cowpox/horsepox, in use in different parts of Europe during the first half of the 19th century, all give rise to viruses with DNA maps as similar to each other, and as different from cowpox virus, as those illustrated in Figure 1B? The origin of vaccinia virus remains a mystery.

V. VARIANTS OF VACCINIA VIRUS

A. DIFFERENCES BETWEEN STRAINS USED FOR VACCINES

Strains of vaccinia virus in use for the production of smallpox vaccine at the beginning of the Intensified Smallpox Eradication Programme in 1967 varied considerably in their virulence for man, as determined by the severity of the local skin lesion and associated constitutional symptoms. Some strains were too severe in their effects for use in mass vaccination campaigns; others produced too mild a reaction to promote the level of immunity required in smallpox-endemic countries. There was also a correlation between the frequency of postvaccinial encephalomyelitis and the virulence of the strain of virus employed. After 1967 many producers changed to the Lister strain.[9]

B. NEUROVACCINIA AND DERMAL VACCINIA STRAINS

Two terms occur in older works on vaccinia virus that need some explanation: ''dermal vaccinia'' and ''neurovaccinia''. Early workers usually maintained vaccinia virus by passage through calves, sheep, or rabbits, the animals being inoculated by scarification. When the skin lesions reached a sufficient size the infected skin area was scraped, the material thus obtained being called ''dermal vaccinia'' or ''dermovaccine''. Virus passaged in this way usually produces white pocks on the chorioallantoic membrane and erythematous, nonulcerating nodules when inoculated intradermally in rabbits.

In the period 1920 to 1940 a great deal of experimental work was carried out with vaccinia virus that had been maintained by intracerebral inoculation of rabbits, sometimes with occasional testicular passage.[24,25] Such strains produced encephalitis in rabbits, and they were called "neurovaccinia" virus. One important aspect of neurovaccinia virus was that some strains could establish an enzootic disease in rabbit colonies, often with substantial mortality.[26] These strains were called "rabbitpox virus". Neurovaccinia strains produced ulcerated, slightly hemorrhagic pocks on the chorioallantoic membrane and large indurated lesions with a purple center after intradermal inoculation in rabbit and are more virulent than dermal vaccine strains. Restriction endonuclease digestion reveals that both neurovaccinia (rabbitpox) and dermovaccinia strains have very similar DNA maps.[27]

C. DIFFERENCES BETWEEN LABORATORY STRAINS OF VACCINIA VIRUS

Two systematic studies have been made of the biological characteristics of laboratory strains of vaccinia virus.[28,29] Both found that some of the strains tested produced pocks of different morphology when grown on the chorioallantoic membrane. A variety of other differences in biological behavior were found, involving production of hemagglutinin, heat resistance of the virion, and pathogenicity in rabbits and mice[28] and plaque morphology and virulence for monkeys.[29] The traditional division of strains into "dermovaccinia" and "neurovaccinia" broadly differentiated viruses of lower and higher virulence. Neurovaccinia strains produced hemorrhagic pocks on the chorioallantoic membrane and large indurated skin lesions with a purple center after the intradermal inoculation of rabbits. White pock mutants of neurovaccinia strains produced lesions like dermal vaccinia virus, i.e., white nonhemorrhagic pocks, on the chorioallantoic membrane and small pink nodules in the rabbit skin.

VI. PATHOGENESIS OF POXVIRUS INFECTIONS

Poxvirus pathogenesis has recently been reviewed by Fenner et al.[4] and Buller and Palumbo,[30] the latter with particular emphasis on the correlation of molecular biology and pathology. Among the poxvirus infections that are considered in this book, some produce localized pustules or nodules and others produce a generalized rash or localized lesions in some hosts and a generalized rash in others (Table 6). Analysis of the pathogenesis of generalized poxvirus infections has made use of three model orthopoxvirus infections: mousepox[31] (caused by ectromelia virus), rabbitpox[32] (caused by a neurovaccinia virus), and monkeypox.[33]

A. PORTALS OF ENTRY

Portals of entry of poxviruses include minute abrasions of the skin or oral

TABLE 6
Animal Species in Which Different Species of Poxvirus Produce Either Localized Skin Lesions or a Generalized Rash

Virus	Natural host	Localized pustular skin lesion	Generalized pustular rash
Orthopoxvirus			
Cowpox virus	? Wild rodents	Man, cat, cow	Felidae (many species), rat, elephant, rhinoceros, okapi
Ectromelia virus	? Voles, laboratory mouse	*Mus pahari*	Laboratory mouse
Vaccinia virus (dermal)	?	Man, cow, rabbit	Man (rare), water buffalo
Neurovaccinia virus[a]	?	—	Rabbit
Variola virus	Man	—	Man, monkeys
Avipoxvirus			
Fowlpox virus	Birds	Chicken	Chicken (sometimes)
Canarypox virus	? Wild birds	—	Canary
Capripoxvirus			
Sheeppox virus	Sheep	—	Sheep
Parapoxvirus			
Orf virus	Sheep	Sheep	—

[a] Including rabbitpox virus.

mucous membrane, inoculation (e.g., vaccination or mechanical infection by arthropods), and infection via the respiratory and sometimes the intestinal tract.

1. The Skin

Infection via minute abrasions in the skin occurs in cowpox in cows, cats, and other animals and in vaccinia virus infections of calves sucking dams (cows or water buffaloes) with lesions on their teats and on the hands of humans who milk such animals. The usual result is a localized lesion of the lips or hands, respectively, but systemic infection with a generalized rash may occur in water buffalo calves. Fowlpox is often spread within flocks by beak pecking, which causes lesions on the face and comb, and between flocks by mosquitoes, which act as "flying pins".

Ectromelia virus infection of mice usually occurs via minute lesions of the skin, producing a localized "primary lesion" which is followed by systemic spread, which produces a generalized rash in mice which do not die from acute hepatitis. Roberts[34] found that after scarification the primary epidermal lesions consisted of single infected cells in the Malpighian layer. Dermal lesions were groups of two or three fibroblasts or histiocytes which had often migrated some distance from the site of scarification. Subsequently the infection spread contiguously in the epidermis, aided by spread to more distant sites from infected dermal cells. The skin lesion became edematous about the fourth day, probably as a result of cutaneous anaphylaxis. Epidermal hyperplasia occurred in an annulus of normal cells just beyond the advancing margin of infected cells. Spread from the skin lesion to the lymphatics was probably due to movement of infected histiocytes.

2. The Respiratory Tract

Experimentally, infection by the inhalation of virus aerosols has been demonstrated with ectromelia virus in mice, with rabbitpox virus in rabbits, and with variola virus in monkeys. In natural infections respiratory infection was the principal mode of infection in smallpox and is probably important in outbreaks of fowlpox and sheeppox.

Details of the early events in respiratory infection of mice with ectromelia virus were studied by Roberts.[35] After inhalation of ectromelia virus in an aerosol that contained particles of various sizes, infection of both upper and lower respiratory tracts occurred. The first evidence of viral replication in the lung was seen in alveolar macrophages and in single mucosal cells of small bronchioles. Two days later the number of infected alveolar macrophages had increased greatly and the mucosal foci had increased in size. By the third day, endothelial cells of the adjacent lymphatics were infected, and occasional infected macrophages were found in the pulmonary lymph nodes. The spleen was also infected at this time. Entry of virus also occurred via cells of the olfactory mucosa and macrophages associated with it. From the third day onwards the course of the disease was dominated by viral replication in the lymph nodes, spleen, and liver (see below).

Epidemiological observations during an outbreak of mousepox at the U.S. National Institutes of Health in 1979 suggested that the respiratory route was not of major importance.[36] However, it does seem to be important in outbreaks of rabbitpox among laboratory rabbits; susceptible rabbits housed in the same room as rabbits suffering from rabbitpox could be infected over distances of up to 12 ft.[37]

3. The Oral Route

Oral infection may involve entry via small abrasions of the skin of the lips or the oral mucous membrane, or possibly via the epithelial cells of the small intestine. The oral route may play a role in natural infection in mousepox when cannibalism occurs and in lambs and calves sucking teats of sheep, cows, or buffaloes infected with parapoxivurses, cowpox virus, or buffalopox (vaccinia) virus. Oral infection also seems a likely source of some cases of cowpox virus infection of domestic cats (a not uncommon sporadic infection in Britain[38]) and in the infection of large species of Felidae, as in the outbreak in the Moscow Zoo, which was caused by consumption of white rats infected with cowpox virus.[39]

B. THE SPREAD OF INFECTION THROUGH THE BODY

In mousepox, genetically susceptible mice that do not die from acute viral hepatitis develop a generalized pustular rash. The pattern of the naturally occurring disease can be reproduced by the footpad inoculation of mice with a small dose of virus.[40] If the appearance of the primary skin lesion was taken as the end of the incubation period, it was evidence that during this symptom-free period there was a stepwise progression of virus through the body: infection, replication, and release, usually accompanied by cell necrosis, occurring first at the site of inoculation and then in the regional lymph node, whence virus and/or virus-infected cells entered the bloodstream. Using fluorescent antibody staining to identify virus-infected cells, Mims[41] showed that the phagocytic cells were the first infected in the liver and spleen. When infection had breached the macrophage barrier the virus replicated to high titer in the parenchymal cells of the liver and in the lymphocytes of the spleen, organs which usually showed semiconfluent necrosis. The bone marrow is also an important early site of viral replication.[42] A day or so after infection of these organs, large amounts of virus were liberated into the bloodstream and during this secondary viremia focal infection of the skin, kidneys, lungs, intestines, and other organs occurred. There was again an interval during which the virus replicated to reach a high titer before visible changes were produced, so that a period of about two or three days usually elapsed between the appearance of the primary lesion and the focal lesions of the secondary rash (Figure 5). Mims[41] demonstrated by fluorescent antibody staining that

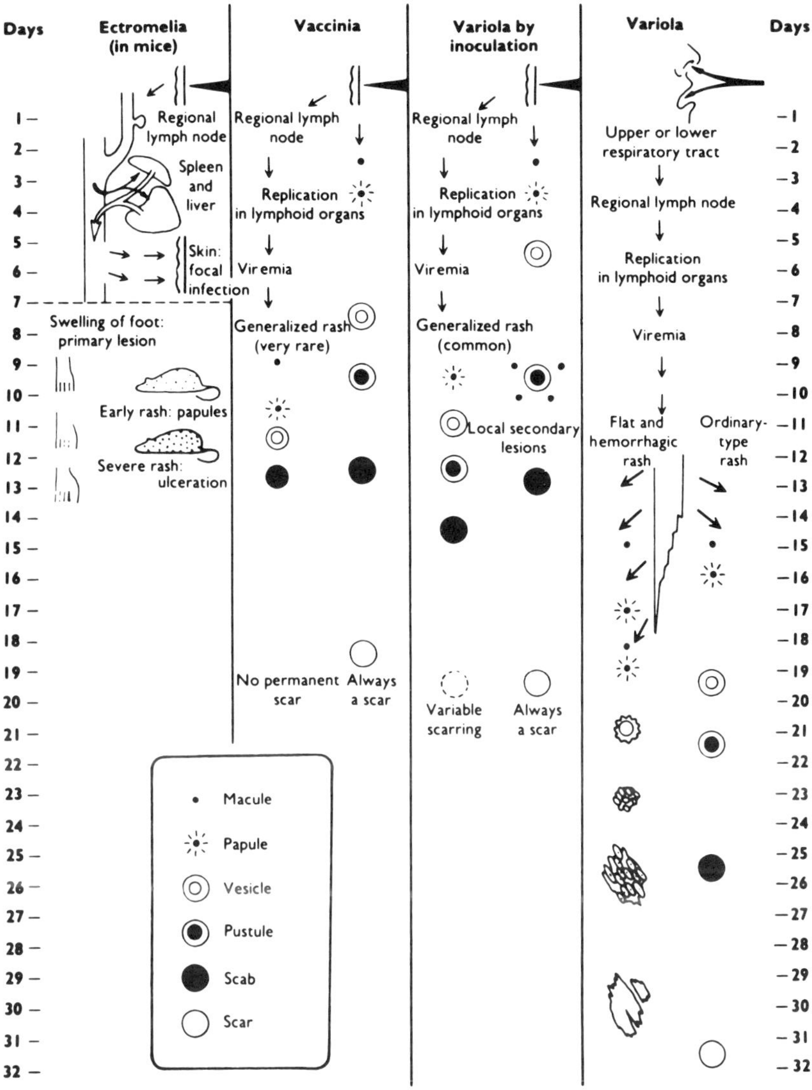

FIGURE 5. The spread of virus around the body and the evolution and healing of skin lesions in a model system (mousepox, after Fenner[31]) and in vaccinia, inoculation smallpox, and "natural" smallpox in man. (From Fenner, F., Henderson, D. A., Arita, I., Jezek, Z., and Ladnyi, I. D., *Smallpox and Its Eradication,* World Health Organization, Geneva, 1988, 125. With permission.)

circulating lymphocytes and monocytes contained viral antigen; only small amounts of virus could be recovered from plasma or serum. However, mousepox appears to be unusual among the generalized poxvirus infections in that the spleen, liver, and bone marrow are early target organs for viral replication. In rabbitpox produced by respiratory infection,[32,37] virus spread through the body in a stepwise manner and could always be isolated from the lungs. Very high titers also occurred in the gonads and the adrenal glands. The incubation period, from infection until the onset of fever, was 4 to 6 days. A rash usually appeared on about the sixth day, although often rabbits died before the rash became evident (''pock-less'' rabbitpox).[43]

In spite of the early replication of rabbitpox virus in the lung, rabbits were not infectious until nasal and conjunctival discharges appeared, and environmental contamination appeared to be maximal 6 to 7 days after infection. Examination by fluorescent antibody staining revealed that after aerosol infection, primary lesions developed at two distinct sites in the respiratory tract: the bronchioles and the alveoli. As well as being spread to the bloodstream via the lymphatics, rabbitpox virus could spread directly from the bronchiolar or alveolar lesions to adjacent blood vessels. Viremia was mainly leukocyte associated. Animals that died early, before the rash appeared, had viremias which increased exponentially to reach very high titers at the time of death; those that died later had much lower titers.[37]

1. The Role of Enveloped Virions

In cell cultures enveloped forms of vaccinia virus are more important than nonenveloped forms in the dissemination of virions over a distance,[44] as shown by the ''anticomet'' test. Payne[45] showed that in infections of mice with vaccinia virus, the capacity to spread to distant organs in the mouse was correlated with the production of enveloped virions (Table 7). Enveloped virions also occur in infections of mice with cowpox virus[46] and ectromelia virus.[47]

2. Some Aspects of Systemic Infection

a. Generalized Rash

The most characteristic feature of generalized poxvirus infections is a generalized rash, the individual lesions being termed pocks. The appearance and development of the pocks differ somewhat in different animals, because of different structures of their skin. Thus the lesions of the rash in the thin skin of the mouse differ histologically from those in the thicker skin of man (smallpox) or sheep (sheeppox).

In a proportion of animals of susceptible species or strains, highly virulent strains of certain orthopoxviruses may cause acute death before a rash develops, e.g., ectromelia virus in mice, rabbitpox virus in rabbits, or early type hemorrhagic smallpox in man. Such humans or animals exhibit profound prostration and other signs of severe toxemia.

TABLE 7
Relationship of *In Vitro* Virus Release to *In Vitro* and *In Vivo* Parameters of Virus Dissemination

	RK 13 cells			Mouse virulence[a]	
	Virus production				
Vaccinia strain	Enveloped virions[d]	Ratio: Nonenveloped to enveloped	Formation of "comets"[b]	Case-fatality rate (%)	Brain titer[c]
South Africa	5.6	250	–	0	—
Cape Town	5.7	100	–	0	—
Venezuela	6.0	160	–	0	—
Lister	6.3	300	–	0	—
Tashkent	6.6	160	–	0	—
WR	6.6	300	–	85	4.7
Lederle 7N	6.7	250	–	0	—
HI-White	6.8	50	–	0	—
Dairen	7.0	16	–	0	2.7
Lafontaine	7.6	40	±	0	3.6
IHD-White	8.0	12	+	25	3.3
Gallardo	8.3	5	+	25	3.3
IHD-J	8.3	6	+	70	4.6

[a] After inoculation with 10^6 plaque-forming units intranasally.
[b] In cell monolayers with liquid overlay (see Figure 9).
[c] Titer in $\log_{10}$ units per gram of brain.
[d] Titer in $\log_{10}$ units per milliliter.

Based on Payne, L. G., *J. Gen. Virol.*, 50, 89, 1980. With permission.

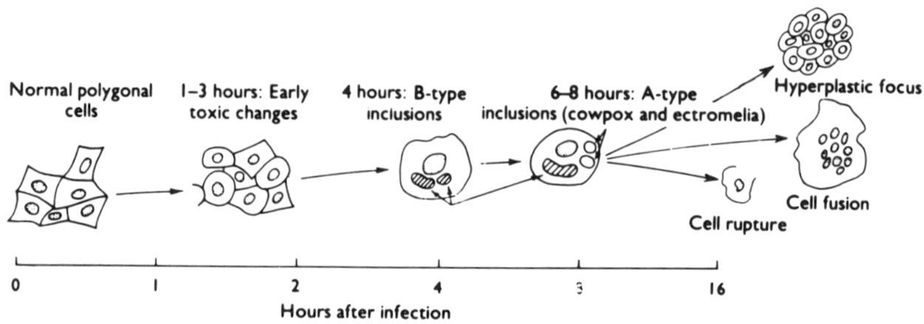

FIGURE 6. Cellular changes seen at various times after infection of a monolayer of cultured cells. (From Fenner, F., Henderson, D. A., Arita, I., Jezek, Z., and Ladnyi, I. D., *Smallpox and Its Eradication,* World Health Organization, Geneva, 1988, 92. With permission.)

b. Toxemia

All clinical observers have commented on the "toxic" appearance of patients suffering from severe attacks of smallpox, and similar toxemia is observed in some cases of mousepox, rabbitpox, and sheeppox. No adequate physiological explanation has yet been found for this toxemia. Viral antigen was readily demonstrable in the plasma of patients with severe smallpox[48] and rabbits with myxomatosis.[49] Perhaps the formation of immune complexes between such antigen and immunoglobulin M (IgM) antibodies, and the associated activation of complement, initiates a series of physiological effects that produced the toxic symptoms.

Boulter et al.[50] examined various physiological parameters in rabbitpox. The only consistent physiological changes observed in sick rabbits was extreme hypotension, leading to a shock-like syndrome, decreased urinary output, and a rise in blood urea and plasma potassium levels. Death seemed to be due to lethal concentrations of potassium ion, which occurred possibly as a consequence of the severe hypotension.

VII. HISTOPATHOLOGICAL CHANGES IN POXVIRUS INFECTIONS

A. CHANGES IN CULTURED CELLS

Within 1 or 2 hr of infection with most species of orthopoxvirus, toxic changes may occur in the infected cells, which in monolayer cultures become rounded and retract from each other (Figure 6). Very soon after infection, new virus-coded antigens are found in the plasma membrane,[51,52] and by the fourth hour there is cytological evidence of viral replication; basophilic areas appear in the cytoplasm, the B-type inclusions,[53] or viral "factories".[54] Eventually gross changes occur in the cells. Depending upon the particular virus-host cell combination there may be an aggregation into hyperplastic foci, adjacent cells may fuse to form large karyocytes, or cell necrosis and rupture may occur, with release of the cell-associated virions. Late in infection,

A-type inclusions (see below) develop in cells infected with cowpox, ectromelia, or raccoonpox viruses, although Lees and Stephen[55] found that cultured hepatocytes infected with ectromelia virus did not become rounded up, and as in the liver of infected mice, they developed B-type but not A-type inclusion bodies.

Some of the earliest changes, and the most significant in relation to the immune response, are the virus-induced alterations in the plasma membranes of infected cells. Some virus-coded antigens are expressed on the surface of the cell within two hours of infection;[52,56] other polypeptides which develop late in infection, including the hemagglutinin and several other envelope glycoproteins, are also incorporated into the plasma membranes of infected cells.[57] Some of these viral antigens promote cell fusion and thus cell-to-cell spread of virions.

B. INCLUSION BODIES

Depending on the species of orthopoxvirus, one or two types of inclusion bodies may be found in infected cells. Irregular, weakly staining inclusion bodies were first described in cells infected with variola and vaccinia viruses by Guarnieri[14] and used to be called "Guarnieri bodies". Much more prominent, eosinophilic inclusion bodies (Figure 7) are found in cells infected with ectromelia virus,[58] cowpox virus,[13] raccoon poxvirus,[23] and fowlpox virus.[59] In a systematic study, Kato et al.[53] proposed that the Guarnieri-type inclusion bodies should be called "B-type" and the Marchal-Downie-Bollinger inclusion bodies "A-type". B-type inclusion bodies are found in all poxvirus-infected cells and are the sites of poxvirus replication.

With most strains of cowpox and ectromelia virus, A-type inclusion bodies can be seen by electron microscopy to contain large numbers of mature virions. Such strains of virus are designated V^+ (Figure 7B).[60,61] However, with some strains of these viruses the A-type inclusion bodies are devoid of virions (V^-, Figure 7C), or rarely, the virions are clustered around the periphery (V^i, Figure 7D).[62] Some strains of cowpox virus fail to produce A-type inclusion bodies, but produce instead a protein that reacts in gel diffusion tests with the LS antigen of vaccinia virus.[22]

Using a V^- strain of cowpox virus, Patel et al.[23] showed that the inclusion bodies consisted of a single species of a 160K protein which is synthesized in large amounts late in the replication cycle. Antibodies to the cowpox virus 160K protein react specifically with a 130K protein found in cells infected with ectromelia virus and a 155K protein found in cells infected with raccoon poxvirus (in each of which it is the A-type inclusion body protein) and a 94K protein found in cells infected with vaccinia virus (the LS antigen). Conservation of these proteins in all species of orthopoxvirus tested has also been demonstrated at the gene level. Thus, vaccinia virus DNA contains a gene whose 5′ end is almost identical with that of the cowpox virus gene encoding the 160K protein. Furthermore, DNA-DNA hybridizations showed that the

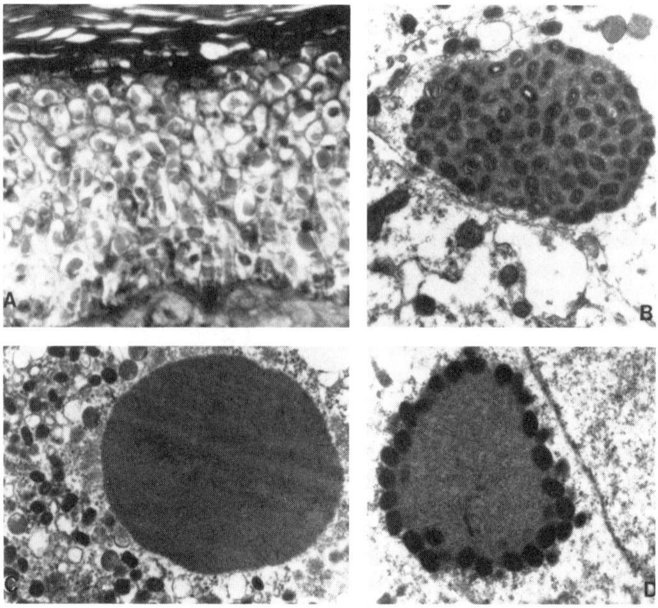

FIGURE 7. A-type inclusion bodies in the cytoplasm of cells infected with ectromelia virus or cowpox virus. (From Fenner, F., Wittek, R., and Dumbell, K. R., *The Orthopoxviruses*, Academic Press, San Diego, 1989, 100. With permission.) (A) Section of skin of mouse after infection with ectromelia virus, stained with Mann's stain. (B,C,D) Electron micrographs of inclusion bodies of three types (see Baxby[61]). (B) V^+ inclusion bodies, with numerous occluded virions; (C) V^- inclusion bodies which completely lack occluded virions; (D) V^i inclusion bodies with virions attached to the periphery. (From Baxby, D., *Zentralbl. Veterinaermed. Reihe B*, 22, 472, 1975. With permission.)

DNAs of other orthopoxviruses, but not of tanapox virus or myxoma virus, contain sequences that hybridize to a probe consisting of the coding region of the cowpox virus gene.[112,113]

The factor necessary for inclusion of virions in the A-type inclusion body (occlusion) has been designated ''VO'' and is present on the surface of the virion, into which it is incorporated a short time before occlusion takes place.[63] In cells doubly infected with V^+ and V^- strains of cowpox virus, the dominant V^+ strain provides the VO factor for integrating V^- virions into the inclusion body.[60,64] Vaccinia virus, which does not produce A-type inclusion bodies, expresses the VO factor and can supply the deficiency for occlusion of V^- cowpox virus strains in doubly-infected cells.

Occlusion within inclusion bodies affects the resistance of infectivity of poxviruses that occur in this form, since the matrix protein provides protection and ensures that the ''infectious unit'' consists of a large number of viral particles. However, single virions which are released from cells as enveloped particles are also produced in cowpox virus-infected cells[45] and probably play a role in the spread of virions from one cell to another, and around the body. Even with viruses that usually produce them, A-type inclusion bodies are not formed in some tissues, such as the liver in ectromelia virus-infected mice.

C. POCKS ON THE CHORIOALLANTOIC MEMBRANE

The potentialities of the chorioallantoic membrane of the developing chick embryo for growing viruses was first demonstrated by Woodruff and Goodpasture,[65] using fowlpox virus. The fact that dilute suspensions of vaccinia virus would produce individual lesions (pocks) on the chorioallantoic membrane was first demonstrated by Keogh,[66] who noted the differences between pocks produced by dermal strains of vaccinia virus and neurovaccinia virus and used the pock counting method to assay viral suspensions and neutralizing antibodies. This method of cultivation has remained an important technique for studying orthopoxivruses, the lesions produced by different species of orthopoxvirus that infect humans being sufficiently distinctive for pock morphology to have been used as a diagnostic method throughout the Intensified Smallpox Eradication Programme. Most species of orthopoxvirus produce white, nonulcerated pocks that vary in diameter from about 0.2 to 3 mm; the pocks produced by other species (cowpox, monkeypox, and neurovaccinia viruses, including rabbitpox virus) are ulcerated or hemorrhagic. Avipoxviruses produce proliferative lesions without necrosis, capripoxviruses produce lesions similar to those of vaccinia virus, and parapoxviruses do not produce pocks on the chorioallantoic membrane.[67]

1. Histology of Pocks

The pocks produced by all species of orthopoxviruses begin as areas of proliferation of the ectodermal cells, which is caused by a diffusible protein that is produced very early after infection (vaccinia growth factor[68] or more generally, "poxvirus growth factor"). Infection of cells spreads centrifugally, proliferation preceding infection. In parallel with the proliferation of the ectoderm there are inflammatory changes and edema in the mesoderm. Within 24 h, eosinophil leukocytes are conspicuous, and when extensive necrosis occurs, as in the pocks produced by vaccinia virus, the mesoderm becomes packed with cells of various kinds including leukocytes, many of them necrotic. B-type inclusion bodies are not readily demonstrable, but A-type inclusion bodies are prominent in cells of the pocks produced by cowpox and ectromelia viruses. The variations seen in lesions produced by different species of poxvirus depend upon several factors: the extent to which proliferation occurs, the stage at which cell necrosis begins and the extent of the necrotic change, the invasion of the mesoderm with leukocytes, and the extent of leakage of erythrocytes from small vessels in the mesoderm. The entoderm is usually slightly hyperplastic. In lesions produced by cowpox virus there is a conspicuous accumulation of erthrocytes just beneath the thickened ectoderm. In pocks produced by neurovaccinia virus, on the other hand, the hemorrhagic appearance of the pock is due to the presence of erythrocytes in the surface tissues of the ulcerated pock.

Pocks produced by ectromelia virus are smaller, those produced by fowlpox virus are larger but less well-defined, but in both, almost every cell in the ectoderm contains an A-type inclusion body.[67]

2. Ceiling Temperature of Viral Growth

When carrying out experiments on the growth of ectromelia virus on the chorioallantoic membrane, Burnet and Lush[69] noticed that ectromelia virus (not then recognized to be an orthopoxvirus) did not produce pocks at 39.5°C, but did if the temperature was lowered to 37°C. This idea was developed by Bedson and Dumbell,[70] who introduced the concept of the ''ceiling temperature'' as the highest temperature at which pock formation by orthopoxviruses would occur. The ceiling temperature has proved to be a useful method for distinguishing between species of orthopoxvirus. Cultured cells can also be used for measuring the ceiling temperature.[71]

3. Pocks Produced by White Pock Mutants

In addition to the opaque white pocks produced by most species of orthopoxvirus, the species that produce hemorrhagic pocks (cowpox, monkeypox, and neurovaccinia viruses, including rabbitpox virus) always produce a small proportion of ''white'' (nonhemorrhagic) pocks; the variants that produce them are designated white pock mutants. These were first observed with cowpox virus. Histologically, the main feature distinguishing the pocks produced by the white pock mutants of cowpox virus from those produced by wild-type cowpox virus is the intense infiltration of the mesodermal tissues with leukocytes. All white pock mutants of cowpox virus that have been studied have deletions of variable lengths of DNA from the right hand end of the genome. By analysis of the DNAs of these mutants and wild-type cowpox virus, Pickup et al.[72] identified and sequenced a gene that is essential for expression of the red pock phenotype; it codes for a 38 K protein resembling serine protease inhibitors. Studies of wild-type cowpox virus and a white pock mutant which lacked only the 38 K protein suggests that the difference between the pocks produced by the two viruses could be attributed to the massive influx of leukocytes into the mutant lesion.[73] Perhaps the 38 K protein inhibits the generation of chemotactic molecules responsible for the migration of heterophils into the lesion.

D. SKIN LESIONS IN ORTHOPOXVIRUS INFECTIONS

The histology of the skin lesions in smallpox has been described by Fenner et al.[9] (Figure 8). Essentially the same features are seen in the skin lesions in vaccinia[74] and monkeypox in man.[75] However, Greene[76] noted that although there was an extensive rash in many cases of rabbitpox, with individual lesions passing through the stages of papule, pustule, and crust, the lesions occurred

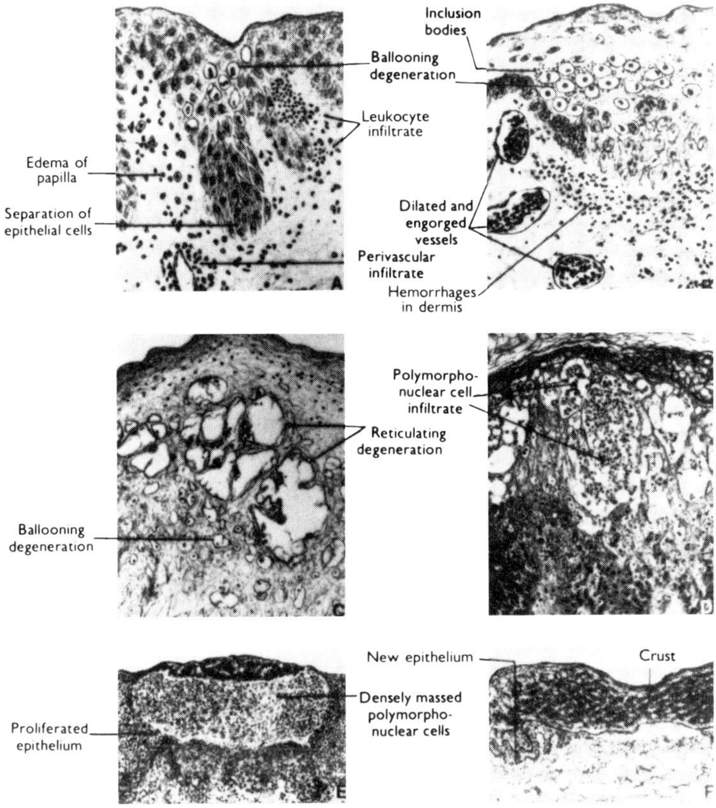

FIGURE 8. Stages in the development and evolution of the skin lesion of smallpox. The Jennerian pustule after vaccination follows essentially the same course. (A) The earliest change was edema of the dermis leading to the separation of epithelial cells of the papillae and lymphocytic infiltration in the dermis, especially around the small vessels. Ballooning degeneration was seen in a few cells in the lower Malpighian layer. (B) These changes progressed and the small vessels became dilated and engorged. Inclusion bodies were visible adjacent to cells showing ballooning degeneration. In early hemorrhagic-type smallpox, illustrated here, there was pronounced bleeding into the dermis. (C) As the pathological process progressed, the epithelial cells broke down by reticulating degeneration to produce a multilocular vesicle. (D) The vesicle formed by coalescence of the smaller cavities became infiltrated with polymorphonuclear leukocytes to produce a pustule, around which were cells containing inclusion bodies. (E) The fully developed pustule became packed with polymorphonuclear leukocytes and the epithelium on either side of the pustule proliferated. (F) Eventually the pustule became a crust, beneath which new epithelium grew in to repair the surface. Such lesions, in which the sebaceous glands were not involved, healed without leaving a pockmark. (From Fenner, F., Henderson, D. A., Arita, I., Jezek, Z., and Ladnyi, I. D., *Smallpox and Its Eradication,* World Health Organization, Geneva, 1988, 134. With permission.)

in the dermis and no epithelial changes were seen. In mousepox, the first histological evidence of the rash consists of localized areas of hyperplastic cells with dark-staining nuclei and occasionally with inclusion bodies in the epidermis.[40] These areas of proliferation and edema increase in size until they become macroscopically visible as pale, slightly raised macules. Large intracytoplasmic A-type inclusion bodies are present in most of the epithelial cells in the lesion. No vesiculation or pustulation occurs in the thin epidermis of the mouse, and within 24 hours the lesion becomes necrotic and then ulcerated, with an adhering scab.

VIII. THE IMMUNE RESPONSE IN POXVIRUS INFECTIONS

The immune response plays an important role in protection from infection or reinfection and in the process of recovery from infection.

A. PROTECTION AGAINST REINFECTION

Responses to revaccination — Depending primarily on the interval since a previous vaccination, the responses obtained on revaccination with vaccinia virus range from a complete absence of reactivity, through an accelerated reaction, sometimes with vesiculation (see Figure 3) to a "major response", indistinguishable from that found in primary vaccination. Other things being equal, skin site plays a role in the severity of the response; revaccination on the ventral surface of the forearm more frequently produced positive reactions than revaccination over the deltoid muscle.[9] Even more striking are the finger lesions sustained by workers in vaccine production laboratories, who often suffered from vaccinial whitlows in spite of the fact that they were revaccinated annually.[77] Such cases often exhibited enlargement of the epitrochlear and axillary lymph nodes and sometimes the reinfection progressed like a primary vaccinial reaction, with maximum vesiculation on the ninth or tenth day.

Heterologous protection — In general, vaccination within the previous 5 years protected persons exposed to smallpox against clinical disease, and statistically some protection was evident for over 30 years. Experiments in laboratory animals reveal that infection with any one orthopoxvirus produces substantial protection against disease produced by any other orthopoxvirus.[4] Although serological tests often provide a clue to the identity of a newly discovered poxvirus, cross protection in a suitable experimental animal was until the availability of DNA restriction mapping the ultimate biological test for the allocation of the virus to the appropriate genus.

B. HUMORAL RESPONSES IN ORTHOPOXVIRUS INFECTIONS

The genome of an orthopoxvirus codes for a very large number of different polypeptides, each of which is potentially antigenic; hence a large number of

TABLE 8
Comparison of the Protective Activity of Three Types of Antiserum Administered Passively to Rabbits Before Challenge Infection with Rabbitpox Virus

Antiserum			Response to challenge	
Against	**Source**	**Neutralization titer**	**Fever**	**Death**
Inactivated vaccinia virus	Horse	800,000[a]	12/12	5/12
	Sheep	500,000[a]	5/5	4/5
Live vaccinia virus	Sheep	150,000[b]	4/5	0/5
	Rabbit	19,000[b]	8/10	0/10
Live rabbitpox virus	Rabbit	32,000[b]	0/5	0/5
None	—	—	18/18	15/18

[a] Measured against nonenveloped virions; no neutralizing antibodies to enveloped virions.
[b] Measured against nonenveloped virions; neutralizing antibodies to enveloped virions also present.

From Boulter, E. A., et al., *Am. J. Epidem.*, 94, 612, 1971. With permission.

different antibodies are produced during orthopoxvirus infections. These may be short-lived IgM or persistent immunoglobulin G (IgG). Many, but not all of these antibodies are also elicited after the inoculation of large doses of inactive virions or viral antigens. Antibodies can be measured by any of the usual techniques. Only the neutralization test, which has some relevance to pathogenesis, will be discussed.

1. Neutralization Test

Although titers are highest with the homologous virus,[78,79] the antigens which evoke neutralizing antibodies after infection with orthopoxviruses show a great deal of overlap. Neutralization tests are therefore useful for allocating poxviruses to the appropriate genus, but not for distinguishing between species within a genus. Traditionally, neutralization tests have been carried out by mixing serum with suspensions of virions obtained by disrupting infected cells. Since such virions lack the envelope antigens found in naturally released virions, such tests do not measure antibodies to envelope antigens,[80] which are important in pathogenesis (Table 8).

Antibodies that neutralize the infectivity of enveloped virions can be assayed by the "anticomet" test[81] (Figure 9). Antibodies that neutralize the infectivity of envelope antigens, added after viral adsorption, prevent the formation of the "comets", but not the development of plaques. Antibodies to nonenveloped virions have no such effect, but prevent plaque production in orthodox neutralization tests, in which virus-serum mixtures are added to the monolayer.

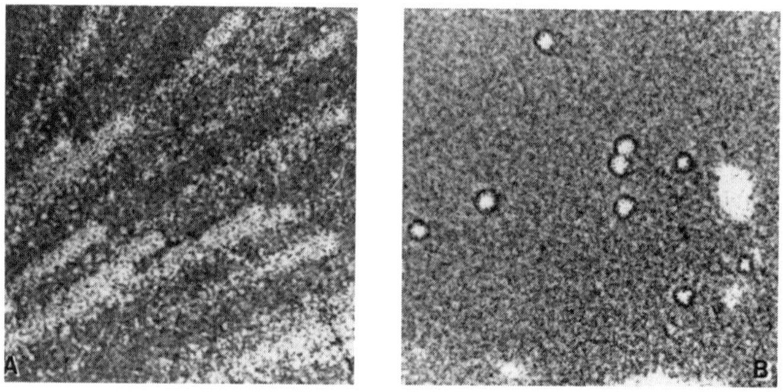

FIGURE 9. The "anticomet" test. (A) Strains of virus which produce substantial numbers of extracellular enveloped virions produce "comet tails" from the initial plaques when suitable dilutions are inoculated in monolayers with a liquid overlay medium; (B) addition to the overlay medium of antibody to viral envelopes or enveloped virions, or of antiserum from an animal that has been infected with vaccinia virus, prevents the production of the "comet tails", but not of the initial plaques. Antiserum to surface tubular elements or inactivated virions, although neutralizing the infectivity of nonenveloped virions in a conventional neutralization test, does not inhibit "comet" formation. (From Appleyard, G., Hapel, A. J., and Boulter, E. A., *J. Gen. Virol.*, 13, 11, 1971. With permission.)

2. The Humoral Response in Relation to Pathogenesis

Antibodies of many different specificities are generated during infection with a virus as complex as an orthopoxvirus. Most of these are probably irrelevant, as far as pathogenesis, recovery, and protection are concerned. The relevant antibodies belong to three classes: (1) antibodies that neutralize viral infectivity, of which there are two subclasses, directed respectively against nonenveloped and enveloped virions,[82] (2) those that, with complement, lyse virus-infected cells, and (3) antibodies that combine with circulating antigens to produce immune complexes.

It has long been believed that specific antibodies generated by the humoral immune response played important roles in both protection against orthopoxvirus infections and recovery from established infections. The protective effect of antibodies is most clearly demonstrated by passive immunization; their putative role in recovery was based mainly on temporal relationships observed during the course of established infections.

a. Passive Immunity

Passive immunization, either by the transmission of antibodies from mother to offspring or by the inoculation of antisera, provides a means of studying the influence of antibodies in the disease process uncomplicated by cell-mediated immunity. The effectiveness and the limitations of passive immunization in generalized orthopoxvirus infections are well illustrated in experiments on mousepox.[83] Active immunity after recovery from ectromelia virus infection usually inhibited viral replication in the inoculated foot and always prevented generalization of the disease. Active immunization with vaccinia virus had little effect on viral replication in the foot, but greatly diminished generalization, although a transient rash sometimes occurred. Passive immunization was much less effective in modifying the course of the disease. Antivaccinial serum had almost no effect and several mice died with acute hepatitis at the same time as the controls. Antiectromelia serum was more effective, but did not influence viral replication in the inoculated foot, and a low level of replication in the spleen and skin was found in most animals. In passively immunized mice the antibody level rose about 2 days later than it did in the vaccinated mice.

These experiments show that active immunization, whether heterologous or homologous, provided much greater protection than the administration of preformed antibodies, even when the antibody titers in the passively immunized mice were higher than those in the actively immunized animals. The reason for the difference is that active immunization with infectious virus provokes the complete range of cell-mediated and humoral immune responses; passive immunization provides only the antibodies present in the convalescent animals from which the sera were obtained. Further, passive immunization with the homologous antibody provided much greater protection than did heterologous antibody, even though the sera crossreacted extensively in neutralization tests.

Experiments on passive immunization in rabbitpox[84] confirmed the protective value of potent homologous antiserum, which protected rabbits from death even when treatment was delayed until overt disease had developed. Subsequent experiments[80] (see Table 8) showed that antisera produced after infection with live virus had a much greater protective effect than antiserum produced by immunization with inactivated virus, even when the neutralizing titer, measured by orthodox neutralization tests, was much lower.

b. Neutralizing Antibodies and Protective Immunity

From the point of view of protection against natural infection, the antibody that neutralizes enveloped virions, which can be detected by the anticomet test (see Figure 9), is more important than the antibody that neutralizes nonenveloped virions.[81] The relevant features of the experiments by Boulter, Appleyard, and colleagues[80,81,85] and later experiments by Payne[45] can be summarized as follows:

1. Sera of rabbits which had recovered from infection with rabbitpox virus contained antibodies that neutralized enveloped virions. Sera from rabbits inoculated with inactivated virions lacked such antibodies.
2. Sera from both groups of rabbits neutralized nonenveloped virions.
3. Crossabsorption of the two types of sera with concentrated enveloped or nonenveloped virions selectively removed neutralizing antibodies of the appropriate specificity.

4. Antibody to isolated envelopes was able to neutralize the infectivity of enveloped virions and to protect mice against the spread of infection.
5. Rabbits immunized with inactivated rabbitpox virus (i.e., with inactivated nonenveloped virions) had high levels of neutralizing antibodies to nonenveloped and none to enveloped virions, but showed only partial immunity on challenge inoculation, whereas rabbits immunized with enveloped virions were fully protected.

C. CELL-MEDIATED IMMUNE RESPONSES

Although observers had long been concerned with the frequent lack of correlation between circulating antibodies and recovery from infections, ''cellular immunity'', invoked in such situations to explain recovery from disease, had no precise meaning until the complex immunospecific responses of T cells were recognized during the 1970s, distinct and independent of the antibody-producing B cells. The T-cell responses, other than those involved in modulation of the humoral response, constitute cell-mediated immunity. The mechanisms by which T lymphocytes exercise antiviral functions are complex; they may involve both the direct effects of T cells on cells with virus-modified surface membranes and the effects of lymphokines.

In vitro experiments with cells infected with ectromelia virus[86,87] showed that some of the cell surface changes relevant to T-cell-mediated lysis occurred before viral DNA replication had begun, i.e., they were early synthetic functions coded for by the input DNA. Cytotoxic T-cell lysis is due to a direct interaction between the appropriate T cells and cells infected with vaccinia virus.[88] Recognition by T cells depended on the presence on the membranes of infected cells of both virus-specified antigens and compatible major histocompatibility gene products. The early lytic effect ensures that cells are lysed before progeny virions are assembled and thus accounts for the efficiency of cell-mediated immunity in promoting recovery in established infections with orthopoxviruses.

1. Cell-Mediated Immunity in Relation to Pathogenesis

Mice are particularly suitable for studying the role of various components of the immune response in orthopoxvirus infections, since so many genetically defined mouse lines are available. Mousepox, which proved so useful for studying other aspects of pathogenesis, has been used to provide information on the relative importance of humoral and cell-mediated immunity and has been extensively exploited for this purpose.[89–93]

Mechanisms controlling viral growth in the major visceral target organs (liver and spleen) become operative 4 to 6 days after primary infection by the natural route, which in the experimental studies was simulated by subcutaneous inoculation into the footpad. Cell-mediated immune responses occur soon after infection: virus-specific cytotoxic T cells are detectable 4 days after infection and reach peak levels in the spleen 1 to 2 days later, while delayed hypersensitivity is detectable by the footpad test 5 to 6 days after infection. In contrast, significant neutralizing antibody is not detectable in the circulation until the eighth day.

Mice pretreated with antithymocyte serum, which acts specifically on T lymphocytes, die from otherwise sublethal doses of virus, because of uncontrolled viral growth in target organs. Such mice have impaired cell-mediated responses, but their neutralizing antibody levels are normal, interferon levels in the spleen are elevated, and the innate resistance in target organs is unchanged.

Very large doses of interferon or immune serum transferred to preinfected recipients are relatively ineffective against the established infection in target organs, although high levels of interferon and high antibody titers can be demonstrated in the sera of the recipients. On the other hand, immune spleen cells harvested 6 days after active immunization of the donor transfer specific and highly efficient antiviral mechanisms which rapidly eliminate infection from the target organs of the recipients, in whose serum neither antibody nor interferon is detectable. The active cells in the immune population can be identified as cytotoxic T cells. Mononuclear phagocytes of immune T-cell recipients, labeled with tritiated thymidine before T-cell transfer, appear in foci of infection in the liver after T-cell transfer, and prior irradiation of immune T-cell recipients in a regimen designed to reduce blood monocyte levels significantly reduces the antiviral efficiency of the transferred cells.

Recognition of the diversity of T cells and the availability of monoclonal antibodies against various subclasses of T cell has made it possible to dissect the role of the subclasses with greater precision. *In vitro* experiments suggest that the generation of cytotoxic T cells ($CD8^+$) is strictly dependent on functions provided by helper T cells ($CD4^+$).[94] However, Buller et al.[95] showed that genetically resistant C57BL/6 mice that had been depleted of $CD4^+$ cells by inoculation with an appropriate monoclonal antibody developed an optimum *in vivo* virus-specific cytotoxic T-cell response and recovered from infection with ectromelia virus, which was lethal in nude mice ($CD4^-$, $CD8^-$). Experiments on depletion of $CD8^+$ cells, although not fully satisfactory, suggested that these cells played an essential part in the recovery process.

Taken together, these findings support the idea that blood-borne cytotoxic T cells with immunologic specificity for virus-induced antigenic changes in infected cell surface membranes enter infectious foci and retard viral spread by lysing infected cells before the maturation and assembly of progeny virions. This T-cell activity attracts blood monocytes which contribute to the elimination of infection by phagocytosis and intracellular destruction of virus. Macrophage activation and locally produced interferon may increase the efficiency of virus control and elimination, but are less important than T cells.

Cytotoxic T cells also appear to be important in expressing the genetic resistance of C57BL mice to ectromelia virus infection. Using a highly sensitive assay system, O'Neill and Brenan[96] showed that cytotoxic T cells could be detected in the popliteal lymph nodes of C57BL mice 3 days after inoculation of a large dose of ectromelia virus into the footpad, compared with 5 days in genetically susceptible BALB/b mice. The delay in the recruitment and proliferation of these cells could well account for the more rapid transmission of virus from the draining lymph node to the liver and spleen in susceptible strains of mice. Experiments with retrovirus immunosuppression also support the view that the cytotoxic T-cell response is much more important than humoral immunity in protecting mice against mousepox. Nevertheless, antisera from animals that have recovered from poxvirus infections do have an effect on the progression of infection with the homologous virus, probably in controlling the extent of viremia.

D. IMMUNOLOGICAL DEFICIENCY STATES IN MAN

The effects of immunological deficiency states in human subjects who have been vaccinated with vaccinia virus provides information on the relative importance of cell-mediated and humoral responses in determining recovery in orthopoxvirus infections. Fulginiti et al.[97] described a number of cases of progressive vaccinia in immunologically deficient infants and children, and Kempe[98] has summarized his extensive experience with these conditions, in relation to vaccination. In children with immunologic defects in cell-mediated immunity, vaccinia virus replicated without restriction, resulting in a continuously progressive primary lesion, persistent viremia, and widespread secondary viral infection of many organs. This response was particularly severe in patients with thymic alymphoplasia. In patients with thymic dysplasia and partially or completely intact immunoglobulin-synthesizing capacity (Nezelof's syndrome), the progression of the primary disease was sometimes slower and less persistent, but a fatal outcome was usual.[99]

Freed et al.[100] analyzed various kinds of human immunological defects and found that cell-mediated immunity was clearly of major importance in controlling vaccinial infection, since individuals with defects in cell-mediated immunity but intact antibody production suffered from progressive vaccinia. Those with defects in antibody production but a satisfactory capacity to mount a cell-mediated immune response usually reacted normally to vaccination. In some circumstances, especially when the immunological defect was acquired and not genetic, the administration of vaccinia-immune globulin would allow cell-mediated immunity to recover sufficiently to control the vaccinial infection. Also, in some cases of Bruton's syndrome, a partially deficient cell-mediated immune mechanism might have been overwhelmed, but could be restored to effectiveness by the administration of vaccinia-immune globulin.

Delayed hypersensitivity reactions could not be evoked in patients with progressive vaccinia, nor could the peripheral blood lymphocytes of such patients be stimulated to mitosis by exposure to inactivated vaccinia virus.[97] Although neutralizing antibody was sometimes present in the serum, its presence did not prevent the development of progressive vaccinia, if cell-mediated immunity was defective.[101] Kempe[98] records having seen 15 boys with very low levels of gammaglobulin who had been routinely vaccinated in infancy without complications. All had histories of enormous and hyperactive delayed hypersensitivity responses to vaccinial antigens.

IX. NONSPECIFIC FACTORS INVOLVED IN HOST DEFENSE

A wide range of physiological conditions of host animals may affect their response to poxvirus infections. These may operate via immunological mechanisms or one or more of a variety of nonimmunological factors may be involved. Buller and Palumbo[30] have discussed in some detail the role of nonspecific factors in the response to poxvirus infections, notably interferons, complement, and natural killer (NK) cells. A few other nonspecific factors affecting poxvirus infections merit brief mention.

A. BODY TEMPERATURE

Body temperature has a dramatic effect on the severity of the leporipoxvirus disease, myxomatosis,[102] and mice housed at 2°C are about 100 times more susceptible to mousepox than those maintained at 20°C.[103]

B. AGE

Among unvaccinated persons, smallpox produced its highest mortality in the very young and the aged and its lowest in the age-group 5 to 20 years.[9] A similar result was observed in mousepox,[104] in which both suckling and aged mice (over 1 year old) were much more susceptible than 6-week-old mice. Viral invasion of the internal organs occurred much earlier in suckling mice and lethal changes were produced before the immune response developed. However, invasion of the liver and spleen occurred at about the same time in 6- to 8- and 56-week-old mice, but replication usually progressed inexorably to produce a lethal outcome in the older mice, but was controlled in the young adult mice. Myxomatosis was much more severe in rabbits less than 3 weeks old than in adult animals,[105] and among rabbits that had developed genetic resistance, this was much less evident in kittens than in adult animals. The related leporipoxvirus, fibroma virus, produces only a benign skin tumor in adult rabbits but a rapidly progressive fatal inflammatory type of disease in newborn rabbits.[106] There is no obvious explanation for these age-related effects except perhaps that the mechanisms of specific (immunological) and nonspecific resistance function less effectively at the extremes of life.

C. HORMONAL EFFECTS

Pregnancy has a very pronounced effect on the severity of poxvirus infections. In variola major, for example, pregnant women were much more likely than any other group of people to suffer from hemorrhagic-type smallpox, which was almost always lethal.[9] Pregnant women have elevated levels of 17-dihydroxycorticosteroids, which have an antiinflammatory effect, depress the immune response, and inhibit interferon production. Studies in rabbits infected with vaccinia virus showed that cortisone diminished the local inflammatory reaction and increased viral titers in the blood and internal organs,[107] and vaccination can produce severe effects in humans receiving corticosteroid therapy.

D. INTERFERONS

Interferon was discovered and named because of its antiviral effect in influenza virus infection of the chick embryo[108] and later in the same year it was shown to protect rabbits against intradermal infection with vaccinia virus.[109] The situation has turned out to be infinitely more complex than originally thought. Not only are three groups of interferons now recognized (α-interferon produced by fibroblasts, β-interferon by leukocytes, and γ-interferon by lymphocytes and NK cells), but antiviral activity is only one of many functions, their major activity being the control of cellular functions related to growth, regulation, and immunoregulation. Most studies of the effects of interferons on poxvirus infections have been carried out in cultured cells[30] and cannot be related directly to pathogenesis.

Looking for an effect of interferon in the recovery process from poxvirus infection in intact animals, Blanden[89,90] showed that passively administered interferon had no effect on recovery from mousepox. However, the biological effects of early production of interferon at the site of viral infection may differ considerably from those that occur after systemic infection, but these proved difficult to study until vaccinia virus recombinants were used to provide both the virus and an interleukin or interferon at the sites of viral replication. Nude mice could be used to exclude the effects of T-cell-dependent immune responses. Thus Kohonen-Corish et al.[110] showed that in athymic nude mice and sublethally irradiated mice expression of γ-interferon by vaccinia virus enabled the mice to recover from infection whereas mice infected with control virus died. The effect was apparently not due to NK cells. Subsequently Karapiah et al.[111] used a vaccinia virus recombinant expressing interleukin-2 and a monoclonal antibody to γ-interferon to demonstrate that the antiviral effect of interleukin-2 was due in part to the release of γ-interferon. This approach provides a powerful way of analysing the effects of a variety of interferons and cytokines on poxvirus infections.

REFERENCES

1. **Aragao, H. de B.,** Myxoma dos coelhos, *Mem. Inst. Oswaldo Cruz,* 20, 225, 1927.
2. **Goodpasture, E. W.,** Borreliotoses. Fowlpox, molluscum contagiosum, variola-vaccinia, *Science,* 77, 119, 1933.
3. **Fenner, F. and Burnet, F. M.,** A short description of the poxvirus group (vaccinia and related viruses), *Virology,* 4, 305, 1957.
4. **Fenner, F., Wittek, R., and Dumbell, K. R.,** *The Orthopoxviruses,* Academic Press, San Diego, 1989.
5. **Baxby, D.,** *Jenner's Smallpox Vaccine. The Riddle of the Origin of Vaccinia Virus,* Heinemann, London, 1981.
6. **Otten, L.,** Trockenlymphe, *Z. Hyg. Infektionskr.* 107, 677, 1927.
7. **Collier, L. H.,** The preservation of vaccinia virus, *Bacteriol. Rev.,* 18, 74, 1954.
8. **Collier, L. H.,** The development of a stable smallpox vaccine, *J. Hyg.,* 53, 76, 1955.
9. **Fenner, F., Henderson, D. A., Arita, I., Jezek, Z., and Ladnyi, I. D.,** *Smallpox and Its Eradication,* World Health Organization, Geneva, 1988.
10. **Leake, J. P.,** Questions and answers on smallpox and vaccination, *Public Health Rep.,* 60, 221, 1927.
11. **Wilson, G. S.,** *The Hazards of Immunization,* Athlone, London, 1967.
12. **Jenner, E.,** *An Inquiry into the Cause and Effects of the Variolae Vaccinae, a Disease Discovered in Some of the Western Counties of England, Particularly Gloucestershire, and Known by the Name of Cowpox,* London, 1798, as reprinted in **Camac, C. N. B., Ed.,** *Classics of Medicine and Surgery,* Dover, New York, 1959, 213.
13. **Downie, A. W.,** Study of the lesions produced experimentally by cowpox virus, *J. Pathol. Bacteriol.,* 48, 361, 1939.
14. **Guarnieri, G.,** Ricerche sulla patogenesi ed etiologia dell' infezione vaccinica e variolosa, *Arch. Sci. Med.,* 16, 403, 1892.
15. **Prowazek, S.,** Untersuchungen uber die Vaccine, *Arb. Reichgesundh.,* 22, 535, 1905.
16. **Baxby, D.,** The genesis of Edward Jenner's *Inquiry* of 1798: a comparison of the two unpublished manuscripts and the published version, *Med. Hist.,* 29, 193, 1985.
17. **Wokatsch, R.,** Vaccinia virus, in *Strains of Human Viruses,* Majer, M. and Plotkin, S. A., Eds., S. Karger, Basel, 1972, 241.
18. **Kelsch, Teissier, and Camus,** De la variole-vaccine. Recherche expérimentelle presentée à l'Académie de médicine, *Bull. Acad. Med.,* 62, 13, 1909.
19. **Herrlich, A., Mayr, A., Mahnel, H., and Munz, E.,** Experimental studies on the transformation of the variola virus into the vaccinia virus, *Arch. Gesamte Virusforsch.,* 12, 579, 1963.
20. **Razzell, P.,** *Edward Jenner's Cowpox Vaccine: the History of a Medical Myth,* Caliban, Firle, 1977.
21. **Bedson, H. S. and Dumbell, K. R.,** Hybrids derived from the viruses of variola major and cowpox, *J. Hyg.,* 62, 147, 1964.
22. **Amano, H. and Tagaya, I.,** Isolation of cowpox virus clones deficient in the production of type A inclusions: relationship to the production of diffusible LS antigen, *J. Gen. Virol.,* 54, 203, 1981.
23. **Patel, D. D., Pickup, D. J., and Joklik, W. K.,** Isolation of cowpox virus A-type inclusions and characterization of their major protein component, *Virology,* 149, 174, 1986.
24. **Levaditi, C., Lépine, P., and Schoen, R.,** Virulence et neurotropisme du virus vaccinal, *C. R. Soc. Biol.,* 107, 802, 1931.
25. **Levaditi, C., Fasquelle, R., Beguignon, R., and Reinié, L.,** Influence des sélecteurs sur le potential encéphalitogène du vaccin jennerian, *C. R. Acad. Sci. Ser. D,* 207, 688, 1938.
26. **Green, H. S. N.,** A pandemic of rabbit-pox, *Proc. Soc. Exp. Biol. Med.,* 30, 892, 1933.

27. **Wittek, R., Menna, A., Schumperli, D., Stoffel, S., Muller, H. K., and Wyler, R.,** *Hind* III and *Sst*1 restriction sites mapped on rabbit poxvirus and vaccinia virus DNA, *J. Virol.*, 23, 669, 1977.
28. **Fenner, F.,** The biological characteristics of several strains of vaccinia, cowpox and rabbitpox virus, *Virology,* 5, 502, 1958.
29. **Ghendon, Y. Z. and Chernos, V. I.,** Comparative study of genetic markers of some poxvirus strains, *Acta Virol. Engl. Ed.*, 8, 359, 1964.
30. **Buller, R. M. L. and Palumbo, G. J.,** Poxvirus pathogenesis, *Microbiol. Rev.*, 55, 80, 1991.
31. **Fenner, F.,** The pathogenesis of the acute exanthems, *Lancet,* 2, 915, 1948.
32. **Bedson, H. S. and Duckworth, M. J.,** Rabbit pox: an experimental study of the pathways of infection in rabbits, *J. Pathol. Bacteriol.*, 85, 1, 1963.
33. **Cho, C. T. and Wenner, H. A.,** Monkeypox virus, *Bacteriol. Rev.*, 37, 1, 1973.
34. **Roberts, J. A.,** Histopathogenesis of mousepox II. Cutaneous infection, *Br. J. Exp. Pathol.*, 43, 462, 1962.
35. **Roberts, J. A.,** Histopathogenesis of mousepox. I. Respiratory infection, *Br. J. Exp. Pathol.*, 43, 451, 1962.
36. **Wallace, G. D., Werner, R. M., Golway, P. L., Hernandez, D. M., Alling, D. W., and George, D. A.,** Epizootiology of an outbreak of mousepox at the National Institutes of Health, *Lab. Anim. Sci.*, 31, 609, 1981.
37. **Westwood, J. C. N., Boulter, E. A., Bowen, E. T. A., and Maber, H. B.,** Experimetnal respiratory infection with pox viruses. I. Clinical, virological and epidemiological studies, *Br. J. Exp. Pathol.*, 47, 453, 1966.
38. **Bennett, M.,** Cowpox in cats, *In Practice,* 11, 343, 1989.
39. **Marennikova, S. S., Maltseva, N. N., Korneeva, V. I., and Garanina, N. M.,** Pox outbreak among carnivora (Felidae) and Edentata, *J. Infect. Dis.*, 135, 358, 1977.
40. **Fenner, F.,** The clinical features of mouse-pox (infectious ectromelia of mice) and the pathogenesis of the disease, *J. Pathol. Bacteriol.*, 60, 529, 1948.
41. **Mims, C. A.,** Aspects of the pathogenesis of virus diseases, *Bacteriol. Rev.*, 28, 30, 1964.
42. **Bhatt, P. N. and Jacoby, R. O.,** Mousepox in inbred mice innately resistant or susceptible to lethal infection with ectromelia virus. III. Experimental transmission of infection and derivation of virus-free progeny from previously infected dams, *Lab. Anim. Sci.*, 37, 23, 1987.
43. **Christensen, L. R., Bond, E., and Matanic, B.,** "Pock-less" rabbit pox, *Lab. Anim. Care,* 17, 281, 1967.
44. **Boulter, E. A. and Appleyard, G.,** Differences between extracellular and intracellular forms of poxvirus and their implications, *Prog. Med. Virol.*, 16, 86, 1973.
45. **Payne, L. G.,** Significance of extracellular enveloped virus in the *in vitro* and *in vivo* dissemination of vaccinia, *J. Gen. Virol.*, 50, 89, 1980.
46. **Payne, L. G.,** The existence of an envelope on extracellular cowpox virus and its antigenic relationship to the vaccinia envelope, *Arch. Virol.*, 90, 125, 1986.
47. **Buller, R. M. L.,** Answer, in *Viral and Mycoplasmal Infections of Laboratory Rodents. Effects on Biomedical Research,* Bhatt, P. N., Jacoby, R. O., Morse, H. C., III, and New, A. E., Eds., Academic Press, Orlando, 1986, 551.
48. **Downie, A. W., McCarthy, K., McDonald, A., MacCallum, F. O., and Macrae, A. D.,** Virus and virus antigen in the blood of smallpox patients, *Lancet,* 2, 164, 1953.
49. **Fenner, F. and Woodroofe, G. M.,** The pathogenesis of infectious myxomatosis: The mechanism of infection and the immunological response in the European rabbit *(Oryctolagus cuniculus), Br. J. Exp. Pathol.*, 34, 400, 1953.
50. **Boulter, E. A., Maber, H. B., and Bowen, E. T. W.,** Studies on the physiological disturbances occurring in experimental rabbit pox: an approach to rational therapy, *Br. J. Exp. Pathol.*, 42, 433, 1961.

51. **Ueda, Y., Ito, M., and Tagaya, I.,** A specific surface antigen induced by poxvirus, *Virology,* 38, 180, 1969.
52. **Ueda, Y., Tagaya, I., Amano, H., and Ito, M.,** Studies on the early antigens induced by vaccinia virus, *Virology,* 49, 794, 1972.
53. **Kato, S., Takahashi, M., Kameyama, S., and Kamahora, J.,** A study on the morphological and cyto-immunological relationship between the inclusions of variola, cowpox, rabbitpox, vaccinia (variola origin) and vaccinia IHD and a consideration of the term "Guarnieri body", *Biken J.,* 2, 353, 1959.
54. **Cairns, J.,** The initiation of vaccinia infection, *Virology,* 11, 603, 1960.
55. **Lees, D. N. and Stephen, J.,** Ectromelia virus-induced changes in primary cultures of mouse hepatocytes, *J. Gen. Virol.,* 66, 2171, 1985.
56. **Amano, H., Ueda, Y., and Tagaya, I.,** Orthopoxvirus strains deficient in surface antigen induction, *J. Gen. Virol.,* 44, 265, 1979.
57. **Payne, L. G.,** Identification of the vaccinia hemagglutinin polypeptide from a cell system yielding large amounts of extracellular enveloped virus, *J. Virol.,* 31, 147, 1979.
58. **Marchal, J.,** Infectious ectromelia. A hitherto undescribed virus disease of mice, *J. Pathol. Bacteriol.,* 33, 713, 1930.
59. **Eaves, G. and Flewett, T. H.,** The structure of fowl-pox inclusions, *J. Hyg.,* 53, 102, 1954.
60. **Ichihashi, Y. and Matsumoto, S.,** The relationship between poxvirus and A-type inclusion body during double infection, *Virology,* 36, 262, 1968.
61. **Ichihashi, Y., Matsumoto, S., and Dales, S.,** Biogenesis of poxviruses: role of A-type inclusions and host cell membranes in virus dissemination. *Virology,* 46, 507, 1971.
62. **Baxby, D.,** Laboratory characteristics of British and Dutch strains of cowpox virus, *Zentralbl. Veterinaermed. Reihe B,* 22, 480, 1975.
63. **Shida, H., Tanabe, K., and Matsumoto, S.,** Mechanism of virus occlusion into A-type inclusion during poxvirus infection, *Virology,* 76, 217, 1977.
64. **Ichihashi, Y. and Dales, S.,** Biogenesis of poxviruses: interrelationship between hemagglutinin production and polykaryocytosis, *Virology,* 46, 533, 1971.
65. **Woodruff, A. M. and Goodpasture, E. W.,** The susceptibility of the chorioallantoic membrane of chick embryos to infection with the fowlpox virus, *Am. J. Pathol.,* 7, 209, 1931.
66. **Keogh, E. V.,** Titration of vaccinia virus on chorioallantoic membrane of chick embryo and its application to immunological studies of neuro-vaccinia, *J. Pathol. Bacteriol.,* 43, 441, 1936.
67. **Beveridge, W. I. B. and Burnet, F. M.,** The cultivation of viruses and rickettsiae in the chick embryo, *Spec. Rep. Med. Res. Counc. G. B. Ser.,* 256, 1, 1946.
68. **Buller, R. M. L., Chakrabarti, S., Moss, B., and Fredrickson, T.,** Cell proliferative response to vaccinia virus is mediated by VGF, *Virology,* 164, 182, 1988.
69. **Burnet, F. M. and Lush, D.,** The propagation of the virus of infectious ectromelia of mice in the developing egg, *J. Pathol. Bacteriol.,* 43, 105, 1936.
70. **Bedson, H. S. and Dumbell, K. R.,** The effect of temperature on the growth of pox viruses in the chick embryo, *J. Hyg.,* 59, 457, 1961.
71. **Porterfield, J. S. and Allison, A. C.,** Studies with poxviruses by an improved plaque technique, *Virology,* 10, 233, 1960.
72. **Pickup, D. J., Ink, B. S., Hu, W., Ray, C. A., and Joklik, W. K.,** Hemorrhage in lesions caused by cowpox virus is induced by a viral protein that is related to plasma protein inhibitors of serine proteases, *Proc. Natl. Acad. Sci. U.S.A.,* 83, 7698, 1986.
73. **Palumbo, G. J., Pickup, D. J., Frederickson, T. N., McIntyre, L. J., and Buller, R. M. L.,** Inhibition of an inflammatory response is mediated by a 38-kDa protein of cowpox virus, *Virology,* 172, 262, 1989.
74. **Howard, W. T. and Perkins, R. G.,** Studies on the etiology and pathology of vaccinia in the rabbit and in man, *J. Med. Res.,* 14, 51, 1905.

75. **Stagles, M. J., Watson, A. A., Boyd, J. F., More, I. A. R., and McSeveney, D.,** The histopathology and electron microscopy of a human monkeypox lesion, *Trans. R. Soc. Trop. Med. Hyg.*, 79, 192, 1985.
76. **Greene, H. S. N.,** Rabbit Pox. I. Clinical manifestations and cause of disease, *J. Exp. Med.*, 60, 427, 1934.
77. **Horgan, E. S. and Haseeb, M. A.,** Some observations on accidental vaccination on the hands of workers in a vaccine lymph institute, *J. Hyg.*, 43, 273, 1944.
78. **Downie, A. W. and McCarthy, K.,** The viruses of variola, vaccinia, cowpox and ectromelia — neutralization tests on the chorioallantois with unabsorbed and absorbed immune sera, *Br. J. Exp. Pathol.*, 31, 789, 1950.
79. **McNeill, T. A.,** The neutralization of pox viruses. II. Relations between vaccinia, rabbitpox, cowpox and ectromelia, *J. Hyg.*, 66, 549, 1968.
80. **Boulter, E. A., Zwartouw, H. T., Titmuss, D. H. I., and Maber, H. B.,** The nature of the immune state produced by inactivated vaccinia virus in rabbits, *Am. J. Epidemiol.*, 94, 612, 1971.
81. **Appleyard, G., Hapel, A. J., and Boulter, E. A.,** An antigenic difference between intracellular and extracellular rabbitpox virus, *J. Gen. Virol.*, 13, 9, 1971.
82. **Boulter, E. A. and Appleyard, G.,** Differences between extracellular and intracellular forms of poxvirus and their implications, *Prog. Med. Virol.*, 16, 86, 1973.
83. **Fenner, F.,** Studies in mousepox (infectious ectromelia of mice). IV. Quantitative investigations on the spread of virus through the host in actively and passively immunized animals, *Aust. J. Exp. Biol. Med. Sci.*, 27, 1, 1949.
84. **Boulter, E. A., Westwood, J. C. N., and Maber, H. B.,** Value of serotherapy in a virus disease (rabbitpox), *Lancet*, 2, 1012, 1961.
85. **Turner, G. S. and Squires, E. J.,** Inactivated smallpox vaccine: Immunogenicity of inactivated intracellular and extracellular virus, *J. Gen. Virol.*, 13, 19, 1971.
86. **Ada, G. L., Jackson, D. C., Blanden, R. V., Tha Hla, R., and Bowern, N. W.,** Changes in the surface of virus-infected cells recognized by T cells. I. Minimal requirements for lysis of ectromelia-infected P-815 cells, *Scand. J. Haematol.*, 5, 23, 1976.
87. **Jackson, D. C., Ada, G. L., and Tha Hla, R.,** Cytotoxic T cells recognize very early minor changes in ectromelia virus-infected target cells, *Aust. J. Exp. Biol. Med. Sci.*, 54, 349, 1976.
88. **Zinkernagel, R. M. and Althage, A.,** Anti-viral protection by virus-immune cytotoxic T cells: infected target cells are lysed before infectious virus progeny are assembled, *J. Exp. Med.*, 145, 644, 1977.
89. **Blanden, R. V.,** Mechanisms of recovery from a generalized viral infection: mousepox. I. The effects of antithymocyte serum, *J. Exp. Med.*, 132, 1035, 1970.
90. **Blanden, R. V.,** Mechanisms of recovery from a generalized viral infection: mousepox. II. Passive transfer of recovery mechanisms with immune lymphoid cells, *J. Exp. Med.*, 133, 1074, 1971.
91. **Blanden, R. V.,** Mechanisms of recovery from a generalized viral infection: mousepox. III. Regression of infectious foci, *J. Exp. Med.*, 133, 1090, 1971.
92. **Blanden, R. V.,** T cell response to viral and bacterial infection, *Transplant. Rev.*, 19, 56, 1974.
93. **Cole, G. A. and Blanden, R. V.,** Immunology of poxvirus infections, in *Comprehensive Immunology, Volume 9, Immunology of Human Infections, Part 2: Viruses and Parasites; Immunodiagnosis and Prevention of Infectious Diseases,* Nahmias, A. J. and O'Reilly, R. J., Eds., Plenum Press, New York, 1982, 1.
94. **Wagner, H. and Rollinghoff, M.,** T-T-cell interactions during in vitro cytotoxic allograft responses. I. Soluble products from activated Lyl^+ T cells trigger autonomously antigen-primed $Ly23^+$ T cells to proliferation and cytolytic activity, *J. Exp. Med.*, 148, 1523, 1978.

95. **Buller, R. M. L., Holmes, K. L., Hugin, A., Frederickson, T. N., and Morse, H. C., III,** Induction of cytotoxic T-cell responses *in vivo* in the absence of CD4 helper cells, *Nature,* 328, 77, 1987.
96. **O'Neill, H. C. and Brenan, M.,** A role for early cytotoxic T cells in resistance to ectromelia virus infection in mice, *J. Gen. Virol.,* 68, 2669, 1987.
97. **Fulginiti, V. A., Kempe, C. H., Hathaway, W. E., Pearlman, D. S., Serber, O. F., Jr., Ellen, J. J., Joyner, J. J., and Robertson, A.,** Progressive vaccinia in immunologically deficient individuals, in *Immunological Deficieincy Diseases in Man,* Bergsma, D. and Good, R. A., Eds., The National Foundation, New York, 1968, 129.
98. **Kempe, C. H.,** Acceptance of the Howland award, *Pediatr. Res.,* 14, 1155, 1980.
99. **Kumar, L., Abdul Salam, N. M., Datta, U., and Walia, B. N. S.,** Cell-mediated immuno-deficiency with normal immunoglobulins (Nezelof's syndrome) with progressive vaccinia, *Indian Pediatr.,* 14, 69, 1977.
100. **Freed, E. R., Richard, J. D., and Escobar, M. R.,** Vaccinia necrosum and its relationship to impaired immunologic responsiveness, *Am. J. Med.,* 52, 420, 1972.
101. **Hansson, O., Johansson, S. G., and Vahlquist, B.,** Vaccinia gangrenosa with normal humoral antibodies. A case possibly due to deficient cellular immunity treated with N-methylisatin beta-thiosemicarbazone (compound 33T57, Marboran), *Acta Paediatr. Scand.,* 55, 264, 1966.
102. **Marshall, I. D.,** The influence of ambient temperature on the course of myxomatosis in rabbits, *J. Hyg.,* 57, 484, 1959.
103. **Roberts, J. A.,** Enhancement of the virulence of attenuated ectromelia virus in mice maintained in a cold environment, *Aust. J. Exp. Biol. Med. Sci.,* 42, 657, 1964.
104. **Fenner, F.,** Studies in mousepox (infectious ectromelia of mice). VII. The effect of the age of the host on the response to infection, *Aust. J. Exp. Biol. Med. Sci.,* 27, 45, 1949.
105. **Fenner, F. and Marshall, I. D.,** Passive immunity in myxomatosis of the European rabbit *(Oryctolagus cuniculus):* the protection conferred on kittens born by immune does, *J. Hyg.,* 52, 321, 1954.
106. **Duran-Reynals, F.,** Production of degenerative inflammatory or neoplastic effects in the newborn rabbit by the Shope fibroma virus, *Yale J. Biol. Med.,* 13, 99, 1940.
107. **Bugbee, L. M., Like, A. A., and Stewart, R. B.,** The effects of cortisone on intradermally induced vaccinia infection in rabbits, *J. Infect. Dis.,* 106, 166, 1960.
108. **Isaacs, A. and Lindenmann, J.,** Virus interference. I. The interferon, *Proc. R. Soc. London Ser. B,* 147, 258, 1957.
109. **Isaacs, A. and Westwood, M. A.,** Inhibition by interferon of the growth of vaccinia virus in the rabbit skin, *Lancet,* 2, 324, 1959.
110. **Kohonen-Corish, M. R. J., King, N. J. C., Woodhams, C. E., and Ramshaw, I. A.,** Immunodeficient mice recover from infection with vaccinia virus expressing interferon-γ, *Eur. J. Immunol.,* 20, 157, 1990.
111. **Karupiah, G., Blanden, R. V., and Ramshaw, I. A.,** Interferon-γ is involved in the recovery of athymic nude mice from recombinant vaccinia virus/interleukin 2 infection, *J. Exp. Med.,* 172, 1495, 1990.
112. **Pickup, D. J.,** Personal communication, 1987.
113. **Esposito, J. J.,** Personal communication, 1987.

Chapter 2

MOLECULAR BIOLOGY OF POXVIRUSES

Bernard Moss

TABLE OF CONTENTS

I. INTRODUCTION

The rational development and proper use of poxviruses as expression vectors depend on an understanding of the molecular biology of these complex and fascinating viruses. The common and distinctive properties of poxviruses include possession of a large and complex enveloped virion containing enzymes for mRNA synthesis, a linear double-stranded DNA genome of 130 to 300 kilobase pairs with covalently linked termini, and a cytoplasmic site of replication. Vaccinia virus, a member of the orthopoxvirus genus, has been studied more intensively than other poxviruses and is the one most extensively used as a vector. However, the concepts obtained from vaccinia virus apply to all orthopoxviruses and probably apply to members of the other five poxvirus genera as well. In this chapter, emphasis has been placed on those aspects of the molecular biology of poxviruses that have particular relevance to their use as expression vectors. Interested readers may wish to consult other recent reviews of the biology of this family.[1–5]

II. THE VIRION

A. APPEARANCE

Vaccinia virions appear as oval- or brick-shaped bodies about 200 to 400 nm long.[6] A scanning electron micrograph of progeny virions at the surface of an infected cell is shown in Figure 1. On thin-section, a lipoprotein bilayer surrounding a core is seen.[7] In some views, the core appears biconcave with two structures called lateral bodies embedded in the concavities. Infectious vaccinia virus particles recovered from the cytoplasm of infected cells have been referred to as intracellular naked virions (INV), in contrast to the extracellular enveloped virions (EEV) that are harvested from the culture medium.[8,9] The EEV have an additional loosely appended lipoprotein envelope.

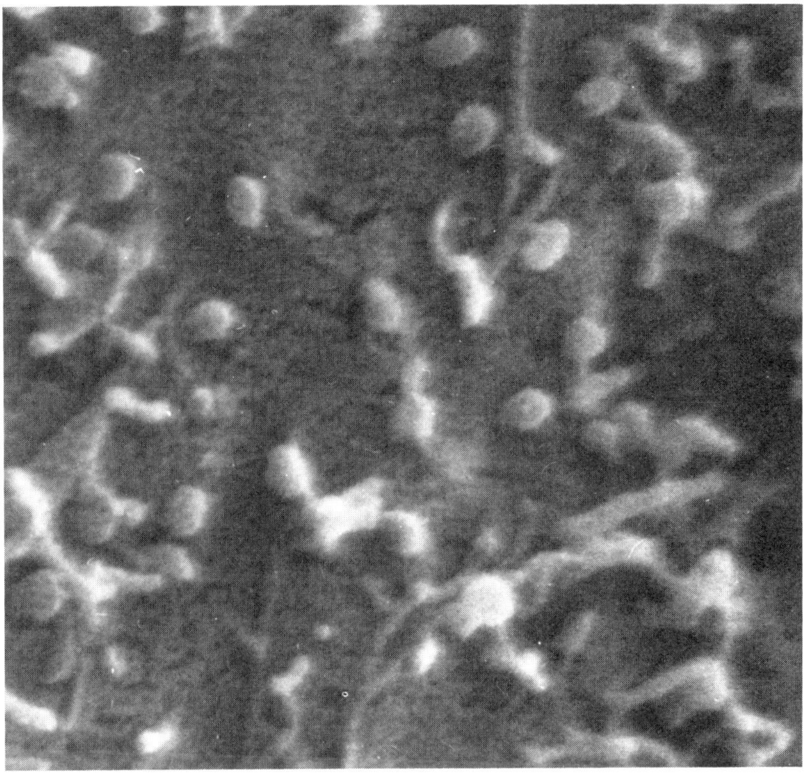

FIGURE 1. Scanning electron microscopy of the surface of a cell at 16 h after infection with vaccinia virus strain IHD-J. Numerous brick-shaped progeny virions are seen. (The micrograph was made at Advanced Biotechnologies, Inc. with material prepared by Rafael Blasco.)

B. PROTEIN COMPONENTS

The precise number of different polypeptides comprising the virion has not been determined. About 30 bands have been resolved from purified INV by one-dimensional polyacrylamide gel electrophoresis[10] and more than 100 bands have been resolved by two-dimensional procedures.[11,12] An additional eight polypeptides of 210 kiloDaltons (K), 110, 89, 42, 37, 21.5, 21, and 20K, have been found in the EEV envelope.[9] The latter are all glycosylated except for the major 37K polypeptide, which is acylated.[13] The 89K polypeptide has both N- and O-linked oligosaccharide chains[14] and has been identified as the hemagglutinin.[15] A polypeptide of 54 to 58K forms tubules on the surface of INV, is a target for proteolytic activation of infectivity, and is capable of eliciting neutralizing antibodies.[16,17] A 14K protein that forms trimers on the surface of INV appears to play a role in virus fusion with the cell membrane and also is capable of eliciting neutralizing antibody.[18,19] An INV transmembrane protein of approximately 35K has sequence similarities to carbonic anhydrase[20] and another has similarities to the respiratory syncytial virus G protein.[21] There are one or two glycosylated INV proteins[10,22] with one or more N-acetylglucosamine residues.[23] A 25K INV membrane protein is myristoylated as is a more internal 35K protein.[24,25] Four proteins of 74, 62, 25, and 11K together account for about 70% of the viral core by weight;[26–29] the three largest are derived from higher molecular weight precursors[28,30,31] and the fourth contains phosphorylated serines.[31,32] The 11 and 25K proteins have a strong affinity for DNA,[33,34] but no histone-like proteins have been identified. Also in the core are a large number of enzymes, some of which are obviously involved in transcription or modification of mRNA (Table 1).

C. DNA GENOME

Special features of the orthopoxvirus genome include terminal hairpin loops that link together the two DNA strands, short tandem repeats near the ends of the genome, a variable length inverted terminal repetition, and a highly conserved central region (Figure 2). Electron microscopic analysis of partially and fully denatured vaccinia virus DNA molecules indicated that the two DNA strands are cross-linked at or near their ends.[35] Evidence for terminal cross-links has been obtained with the DNA of every poxvirus examined. Direct sequence analysis of the ends of the vaccinia virus genome revealed hairpin loops that are A+T-rich, contain several unpaired bases, and exist in two isomeric forms that are inverted and complementary in sequence (Figure 2).[36] This type of structure is probably common to poxviruses since similar termini were found in the leporipoxvirus Shope fibroma virus[37] and the distantly related African swine fever virus.[38]

The orthopoxviruses, including vaccinia virus,[39] cowpox virus,[40] and raccoonpox virus,[41] have similarly arranged sets of short repeats separated by a unique region of 254 to 325 base pairs (Figure 2). By contrast, the leporipoxvirus Shope fibroma virus has a small number of imperfect repeats.[42] Unequal crossing over within or between the sets of tandem repeats accounts in part for length heterogeneity of the terminal fragments of orthopoxvirus DNA.[43]

The DNA present at the two ends of the genome, which is identical but inverted in sequence, comprise the inverted terminal repetition (Figure 2).[44,45] The length of this structure varies greatly in different poxviruses. In vaccinia virus, this 10 kilobase pairs segment contains several sets of tandem repeats as well as some coding regions.[39,46]

Restriction endonuclease maps indicate that the central regions of orthopoxvirus genomes are highly conserved, with greater variation occuring near the ends.[47,48] Except for the regions close to the termini, there is usually little and sometimes no DNA separating coding sequences.[49,50] The absence of introns, the short promoter sequences, and the small sizes of many open reading frames, account for the packing of about 200 genes into the approximately 200,000 base pairs vaccinia virus genome. Some clustering of genes,

TABLE 1
Enzymes and Factors Packaged in Vaccinia Virions

RNA polymerase

500K — Large subunits of 147 and 132K; small subunits of 35 to 17K; role in transcription; 2 large subunits homologous with prokaryotic and eukaryotic RNA polymerases; 35K subunit homologous with transcription elongation factor; purified enzyme requires Mn^{2+} and transcribes only single stranded DNA

Early transcription factor

150K — Subunits 82 and 70K; required for transcription of early genes; promoter-binding; associated DNA-dependent ATPase activity

Poly(A) polymerase

80K — Subunits of 55 and 30K; polyadenylates 3′ end of mRNA; specificity for ATP; primer dependent; biphasic reaction; no sequence specificity

Capping enzyme complex

127K — Subunits of 97 and 33K; forms cap at 5′ end of mRNA; three activities: RNA triphosphatase, RNA guanylyltransferase, RNA (guanine-7-) methyltransferase; 97K subunit forms covalent GMP complex; required for transcription termination

RNA (nucleoside-2′)methyltransferase

38K — Methylates ribose of second nucleotide in cap structure

Nucleoside triphosphate phosphohydrolase I

72K — Role unknown; hydrolyzes ATP or dATP to release ADP or dADP and Pi; DNA-dependent

Nucleoside triphosphate phosphohydrolase II

68K — Role unknown; hydrolyzes rNTPS or dNTPS to release NDPs and Pi; stimulated by DNA or RNA; antigenically distinct from nucleoside triphosphate phosphohydrolase I; presumed to be virus encoded

Nicking-joining enzyme

100–50K Subunits — Suggested role in formation of hairpin ends of genome; cleaves supercoiled DNA and covalently links ends; activated by protease digestion; single-strand deoxyribonuclease activity optimum at low pH; presumed to be virus encoded

DNA topoisomerase

37K — Possible roles in transcription and replication; type I activity; relaxes negatively and positively supercoiled DNA; virus encoded; homology with yeast type I topoisomerase

Protein kinase

62K — Role unknown; partially purified; phosphorylates serine and threonine residues of virus acceptor protein; activated by protamine

with regard to their time and direction of transcription, occurs in vaccinia virus.[50–53] Many nonessential genes occur near the ends of the genome, as judged by the spontaneous appearance[54–56] and construction[57,58] of mutants with large deletions. The size flexibility of the vaccinia virus genome is also illustrated by the ability to insert at least 25,000 base pairs of extra DNA.[59]

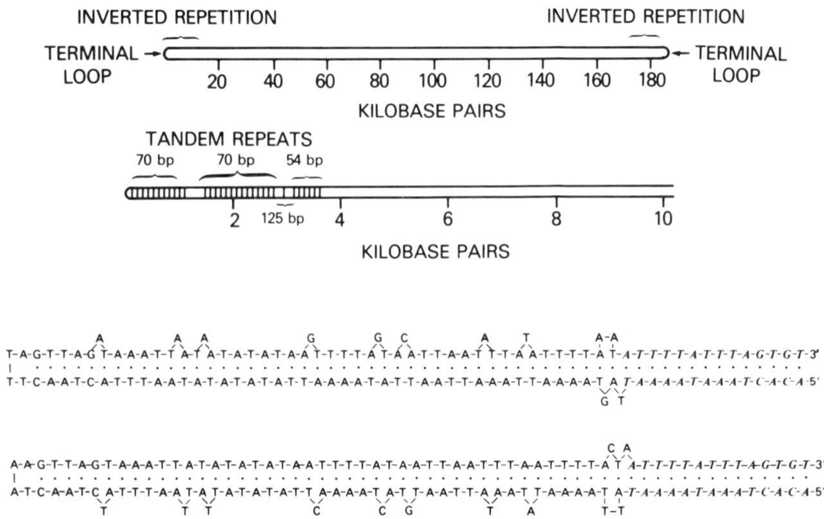

FIGURE 2. Structural features of vaccinia virus DNA. A representation of the entire linear double-stranded DNA genome and an expansion of the 10,000 base pair inverted terminal repetition are shown at the top. (From Moss, B. et al., Mechanisms of DNA Replication and Recombination, in *Proc. 1983 UCLA Symp.,* Alan R. Liss, New York, 1983, 449. With permission.) Nucleotide sequences of the two alternative inverted and complementary forms of the terminal hairpin loops of vaccinia virus DNA are shown at the bottom. (From Baroudy, B. M. et al., *Cell,* 28, 315, 1982. With permission. Copyright held by Cell Press.)

The latter feature demonstrates the large capacity of poxvirus genomes for foreign DNA.

III. VIRUS ENTRY

The wide host range of some orthopoxviruses indicates that they either use several alternative receptors for binding or one common to most cells. Studies of poxvirus entry are complicated by the existence of two infectious forms, INV and EEV. Entry by EEV appears to be more rapid and the latter form may be responsible for virus spread in infected animals.[60,61]

Electron microscopy has provided support for both endocytosis[62] and cell surface fusion[63] as entry mechanisms for INV. These modes of entry are not mutually exclusive, since biochemical evidence suggests that INV fuses in a pH-independent manner.[64] Direct support for neutral pH fusion of both INV and EEV has been obtained by a sensitive and quantitative fluorescence assay.[65] The latter study also demonstrated more rapid fusion of EEV than INV.

The mechanisms of poxvirus binding and entry are not well understood and may be quite complex since up to 30 vaccinia viral envelope proteins have been identified[12] and both INV and EEV forms exist. A 32K INV protein has been shown to play a role in virus binding, although variants that lack it entirely are still infectious.[66] The 54K surface tubule protein also has been implicated in the binding reaction.[16] A 14K envelope protein, that exists as a homotrimer, is important for membrane fusion.[18] A monoclonal antibody to the 14K protein[18] blocks fusion of INV but not EEV.[65] Brief treatment of vaccinia virions with proteolytic enzymes decreases adsorption but paradoxically increases infectivity, apparently by dissociating the hemagglutinin and increasing fusion.[67]

There have been few attempts to identify the cell surface receptors for binding poxviruses. Vaccinia virus encodes a growth factor called vaccinia growth factor (VGF) with homology to epidermal growth factor (EGF).[68–70] Evidence that occupancy of the EGF receptor partially inhibits vaccinia virus infectivity has led to suggestions that it might serve as virus receptor.[71,72] However, mutant forms of vaccinia virus lacking VGF have similar plaquing efficiencies to wild-type vaccinia virus in a variety of cell types, including those with high or low densities of EGF receptor.[73]

IV. REGULATION OF GENE EXPRESSION

A. GENERAL STRATEGY

Members of the poxvirus family are the only known DNA viruses (with the exception of African swine fever virus) that synthesize their mRNA in the cytoplasm of the cell. This capability depends on their encoding enzymes and transcription factors and incorporating those necessary for early mRNA synthesis into the infectious virus particle. The enzymes present in vaccinia virus cores comprise a complete system for the transcription of the prereplicative or early class of genes[74,75] and the polyadenylation[76] and capping[77] of the mRNAs. These virus-encoded proteins include a eukaryotic-like multisubunit DNA-dependent RNA polymerase, transcription and termination factors, RNA capping and methylating enzymes, and a poly(A) polymerase (Table 1). Thus the virus is prepared to express a subset of its genes immediately after penetration into the cytoplasm (Figure 3). Additional genes are expressed after DNA replication. Cognizance of the fact that poxviruses have a unique transcription system that is incompatible with that of other viruses and organisms is crucial for the design of expression vectors.

B. EARLY GENE EXPRESSION

1. Promoters

The regulatory sequences for initiation of transcription occur upstream of the RNA start sites of early genes and can be excised as cassettes and translocated for the expression of foreign genes.[78] Early promoters are about 30 base pairs long[79,80] and, on the basis of detailed mutagenesis studies,[81] have been divided into 3 regions: a 15 base pairs A-rich critical region in which most single nucleotide substitutions have a major effect on expression

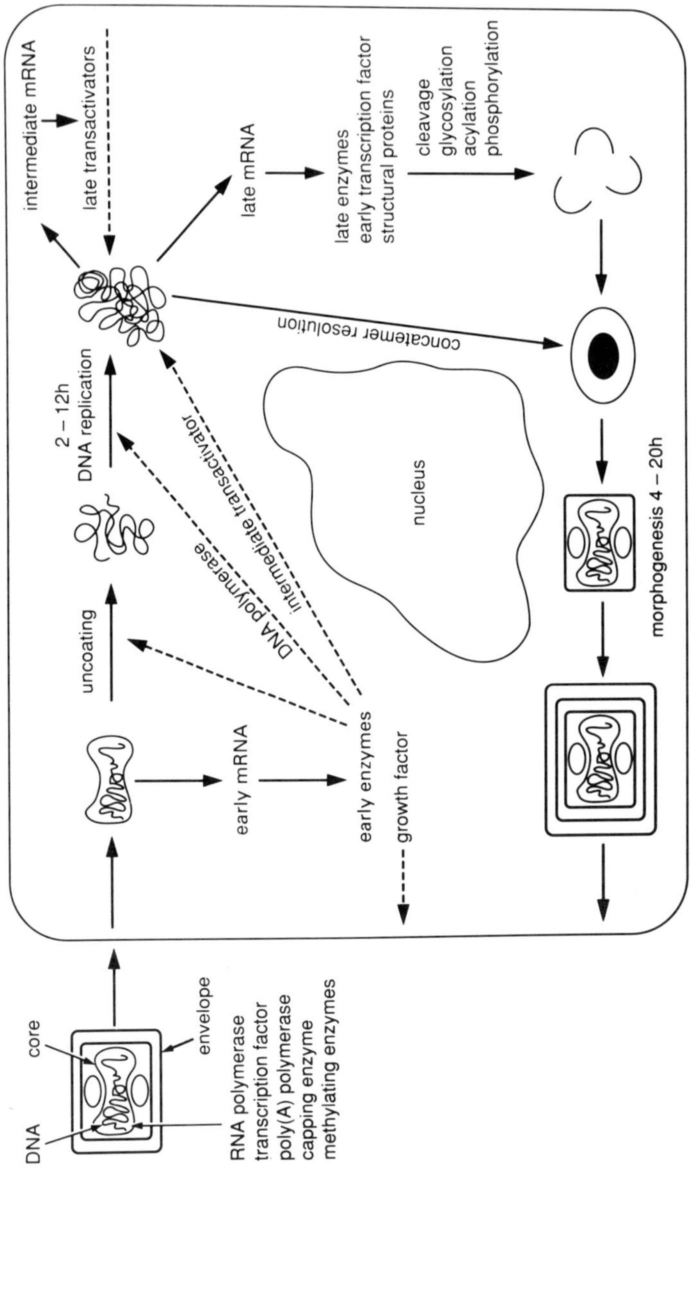

FIGURE 3. Outline of the replication cycle of vaccinia virus. (From Moss, B., Poxviridae and their replication, in *Virology*, Fields, B. N. et al., Eds., Raven Press, New York, 1990, 2079. With permission.)

separated by 11 base pairs of a less critical T-rich sequence from a 5 to 7 base pairs region within which initiation occurs. Some mutations increased promoter strength over that of the natural sequence.[81] An optimized promoter sequence found by mutagenesis corresponded closely to a consensus sequence determined by analysis of a large number of vaccinia virus genes.[81] Individual natural promoters, however, are generally suboptimal in strength, perhaps to provide a range of expression levels.

The sequence at the site of initiation of early RNA synthesis is not highly conserved. Instead, the poxvirus A-rich critical sequence determines the distance downstream at which initiation of RNA synthesis occurs. Analysis of total early RNAs indicated that the 5′ ends are exclusively or predominantly purines.[82] Because of this, substitution of pyrimidines for purines can cause the RNA initiation site to move several bases upstream or downstream.[81] Evidence that the 5′ ends of early mRNAs are actually start sites and not cleavage sites has come from incorporation of [β-^{32}P]GTP into RNA made *in vitro* by vaccinia virus cores.[83]

Detailed studies of promoters of poxviruses other than vaccinia virus have not been made yet. However, avipoxvirus promoters apparently can function in vaccinia virus and vice versa,[84–86] suggesting that the transcriptional regulatory sequences are conserved. If this is further documented, it will simplify the design of poxvirus vectors since the promoters optimized for vaccinia virus may be appropriate for other poxviruses.

2. Termination Motif

The 3′ ends of early vaccinia virus mRNAs are formed by a termination mechanism.[87] Using *in vitro* transcription systems, the termination signal was shown to be TTTTTNT, in which N can be any nucleotide including T.[88] This sequence is both necessary and sufficient for termination about 20 to 50 nucleotides downstream. Examination of DNA sequences revealed that the TTTTTNT motif is commonly present near the ends of early genes, but occasionally occurs within the body of the gene where it could cause premature termination. Since termination is not 100% efficient, internal termination signals might provide a regulatory role. There are some early genes that lack the TTTTTNT signal; in these cases termination occurs after the next gene downstream. The eukaryotic processing signal AAUAAA, which is present about 20 nucleotides before the polyadenylation site of eukaryotic mRNA, is not found near the 3′ ends of vaccinia virus mRNAs and evidently plays no role in poxvirus mRNA formation. The TTTTTNT sequence has been found in the appropriate locations of leporipoxvirus,[89] avipoxvirus,[90] and capripoxvirus[91] genes, suggesting that it is generally used as a transcription termination signal by poxviruses.

The termination signal has an important practical implication for the use of poxviruses as expression vectors, since on a purely random basis the sequence TTTTTNT should occur commonly in foreign genes inserted into expression vectors. Indeed, that is the case and can lead to premature termination of early (but not late) transcripts.[92] Nevertheless, the nucleotide sequence can be altered without changing the coding properties of the gene.

3. Enzymes and Factors

a. RNA Polymerase

RNA polymerase, purified from vaccinia virions, is incapable of using double-stranded DNA as a template, but can transcribe single-stranded DNA in the presence of Mn^{2+}.[93,94] The enzyme sediments in glycerol gradients with an M_r of approximately 500,000 and contains two large and several small polypeptides, all of which appear to be virus-encoded.[95] The identities of the viral genes for six subunits (rpo147, rpo132, rpo35, rpo30, rpo22, and rpo18) have been reported thus far.[96–101] The two large subunits, encoding 147 and 132K polypeptides, are homologous to the corresponding size subunits of prokaryotic and eukaryotic RNA polymerases, with greatest similarity to the former. In contrast, the small subunits identified thus far do not appear to be homologous to the bacterial or yeast small subunits, although the 30K subunit of the vaccinia virus RNA polymerase may be related to a eukaryotic transcription elongation factor.[99]

b. Transcription Factor

The vaccinia virus early transcription factor (VETF) has been purified from virions and is a heterodimer of 82 and 70K polypeptides.[102] The factor has a DNA-dependent ATPase activity[103] that may account at least in part for a previously described requirement for ATP hydrolysis in transcription initiation.[104] The transcription factor binds to early promoters and single nucleotide substitutions in the critical region of the promoter that decrease transcription and also diminish DNA binding.[105] The genes encoding the VETF subunits have been identified.[106,107] Except for the presence of a characteristic ATP-binding motif in the large subunit, VETF is unrelated in sequence to any known eukaryotic or prokaryotic protein.

c. Capping and Methylating Enzymes

Vaccinia virus mRNAs, like those of eukaryotes, have a 5′-terminal cap structure that is required for translation.[77,108] Early mRNA is capped by a complex heterodimeric enzyme that is packaged in the virion.[109,110] The enzyme is composed of two subunits of 97 and 33K with three activities: an RNA triphosphatase, an RNA guanylyltransferase, and RNA(guanine-7-)-methyltransferase that catalyze the following reactions:

$$\text{pppN-(N)n} \longrightarrow \text{ppN(N)n} + \text{Pi} \quad (1)$$

$$\text{GTP} + \text{ppN(N)n} \longrightarrow \text{G(5')pppN-(N)n} + \text{PPi} \quad (2)$$

$$\text{AdoMet} + \text{G(5')pppN-(N)n} \longrightarrow \text{m7G(5')pppN-(N)n} + \text{AdoHcy} \quad (3)$$

Step 2 involves the formation of a covalent GMP complex with the large subunit of capping enzyme.[111,112] The genes encoding both subunits have been identified[113,114] and the active heterodimeric enzyme has been expressed in *Escherichia coli*.[115–117] The later studies indicate that Steps 1 and 2 can be carried out by the large subunit alone. The small subunit presumably contains the methyltransferase activity although this has not yet been shown directly. The fourth step in cap formation,

$$\text{AdoMet} + \text{m7G(5')pppN-(N)n} \rightarrow \text{m7G(5')pppNm(N)n} + \text{AdoHcy} \quad (4)$$

is carried out by a separate methyltransferase that is presumably virus-encoded as well.[118]

d. Termination Factor

The capping enzyme appears to play an essential role in termination of early transcription. *In vitro* studies indicated that purified RNA polymerase and VETF catalyze initiation and elongation, but not termination, of RNA.[102] For the latter function, the capping enzyme itself or a factor that copurified with it was required.[119] *In vitro* studies with halogenated ribonucleoside triphosphates indicated that the termination signal is recognized as the UUUUUNU sequence in nascent RNA.[120] The mechanism whereby capping enzyme is involved is still an enigma, however, since formation of the cap structure is not obligatory for termination. The entire process of transcriptional initiation, elongation, capping, and termination might be carried out by a complex that includes RNA polymerase, VETF, and capping enzyme.[121]

e. Poly(A) Polymerase

A poly(A) polymerase with an M_r of approximately 80,000, that has associated polypeptides of 55 and 33K, was purified from vaccinia virus cores[122] and infected cells.[123] The enzyme catalyzes the addition of adenylate residues to the 3′ ends of primers with little or no sequence specificity. It is selective for ATP, but in its absence other ribonucleoside triphosphates can be polymerized inefficiently.[124] Detailed kinetic studies revealed that polyadenylation is a biphasic reaction.

f. Additional Enzymes

Other enzymes have been purified from vaccinia virus cores but their biological roles are not yet understood (Table 1). These include nucleoside triphosphate phosphohydrolase I (NPH I), a virus-encoded DNA-dependent ATPase of 72K.[125–127] The finding that a conditionally lethal temperature-sensitive NPH I mutant is defective in intermediate and late gene expression suggests that the enzyme may be involved in transcription.[128,129]

Nucleoside triphosphate phosphohydrolase II (NPH II), an M_r 68,000 protein, is immunologically distinct from NPH I, is stimulated by a wide range of nucleic acids, and hydrolyzes all four nucleoside triphosphates rather than just ATP.[130] The gene encoding this protein has not yet been identified and there is no information regarding its function.

A 32K protein with type I DNA topoisomerase activity, that can relax both positively and negatively supercoiled DNA in the absence of ATP, has been purified from vaccinia virions.[131–133] Relaxation involves the formation of a transient covalent intermediate between the DNA of one strand and the enzyme.[132,133] Site directed mutagenesis implicated Tyr-274 as the active-site residue that forms the covalent adduct.[134] The enzyme exhibits a remarkable degree of sequence specificity, reminiscent of a restriction endonuclease.[135] The vaccinia virus topoisomerase gene is only one-third the size of eukaryotic topoisomerases, yet has regions of significant homology with the latter.[136] The sequence of the Shope fibroma virus topoisomerase gene also has been determined and compared with that of vaccinia virus.[137] The two poxvirus genes predict proteins with precisely the same number of amino acids, of which 61% are identical. The inability to isolate a viable recombinant vaccinia virus with a deleted topoisomerase gene containing a selectable marker provides circumstantial evidence that it is essential.[138] Although the topoisomerase is not required for transcription of DNA templates *in vitro,*[102] it might be needed with more topologically constrained natural templates *in vivo.* Equally plausible, however, are roles for the topoisomerase in DNA replication or packaging.

Additional core-associated enzymes include a protein kinase[139,140] and a deoxyribonuclease[141–143] with nicking joining activity,[131,144] neither of which have been genetically mapped.

C. INTERMEDIATE AND LATE GENE EXPRESSION

1. Role of DNA Replication

DNA replication triggers a profound shift in poxvirus gene expression. At this time, host protein synthesis is severely reduced, making it possible to readily see the changes by polyacrylamide gel electrophoresis of extracts of cells pulse-labeled at intervals with radioactive amino acids.[145–147] This effect also permits the visualization of highly expressed foreign proteins made by recombinant vaccinia viruses. Kinetic studies suggested one prereplicative or early class of proteins and two postreplicative classes comprised of intermediate proteins that are detected almost immediately upon DNA replication and late proteins that have a more delayed onset of appearance. Early protein synthesis is not adversely affected by inhibitors of DNA replication, whereas synthesis of both intermediate and late proteins are blocked by such treatment. The requirement for DNA replication is maintained when plasmids containing late genes are transfected into vaccinia virus infected cells.[148] By contrast, DNA replication is not required for expression of intermediate genes from transfected plasmid templates, thus providing a convenient way of discriminating between the two classes of postreplicative genes.[149] The latter finding, as well as *in vitro* studies, suggested that the proteins needed for intermediate gene expression are present before DNA replication, but that the DNA entering the cell with the virion cannot serve as a template; hence the need for replicated or transfected DNA. The reason why the virion DNA cannot serve as template for intermediate gene expression has not yet been clarified, but it is likely due to either the exclusion of protein factors from the partially degraded virus core or to as yet unidentified specific repressor proteins.

2. mRNAs

In general, early mRNAs have typical eukaryotic characteristics, including a discrete size, capped 5′ ends, and polyadenylated 3′ ends. The late mRNAs, however, are unusual; typically they are long and heterogeneous, appearing as smears on polyacrylamide gel electrophoresis.[150,151] This heterogeneity is probably related to the lack of functional recognition of the early termination sequence TTTTTNT, which occurs randomly in late genes. As both DNA strands are transcribed late in infection, self-annealing occurs, at least during RNA extraction as revealed by RNase resistance[152] and electron microscopy.[153] Characteristically, the late mRNAs have a capped 5′ poly(A) leader that is variable in length, but is a predominant size of about 35 nucleotides.[154–157] The mRNAs of foreign genes regulated by vaccinia virus late promoters have the properties of authentic late mRNAs, e.g., poly(A) leaders and length heterogeneity; therefore Northern blotting and 3′-end nuclease S1 analysis are not useful ways of characterizing such transcripts.

Intermediate mRNAs have not yet been well characterized with regard to length heterogeneity; 5′ poly(A) was reported absent from two species examined.[149]

3. Intermediate and Late Promoter Sequences

Inspection of DNA sequences revealed a common TAAAT motif that typically occurs at or very close to the start of late open reading frames (ORFs).[52,158] Functional analysis of late promoters revealed a gradual decrease in expression of a reporter gene as 5′ to 3′ deletions were made, but considerable activity remained with only 15 to 20 base pairs upstream of the putative RNA initiation sites.[26,79,159,160] Mutations within TAAAT decreased or abolished transcription.[158,161] Further mutagenesis suggested that the late promoter may be considered in terms of 3 regions: an upstream sequence of about 20 base pairs with some consecutive T or A residues, separated by a spacer region of about 6 base pairs from the TAAAT element within which transcription initiates.[161] Synthetic promoters, with 18 to 20 consecutive T residues, were stronger than any natural promoters examined. Decreases in the length of the 5′ poly(A) leader were noted when mutations were made downstream of the TAAAT[162] and when T residues flanking the AAA were mutated.[161] The poly(A) was absent or limited to a few nucleotides when the three A residues were mutated, although these mutations also drastically reduced transcription.[161] The data are most consistent with formation of the poly(A) leader by a mechanism involving backwards slippage of the RNA polymerase while transcribing the complement of the AAA sequence.[163] Occasionally, the TAAAT sequence is present at the transcription initiation site of early genes and then a 5′ poly(A) leader is formed, perhaps by a similar mechanism.[98,164] In most late genes, the TAAAT sequence overlaps the initiation codon forming a TAAATG sequence and therefore the poly(A) segment provides the entire untranslated leader. By not having to encode leader sequences for all late genes, a saving of a few thousand base pairs of DNA might be accomplished. However, an additional purpose for the poly(A) leader seems likely. One possibility is related to the presence of read-through complementary antisense RNAs late in infection. The capped 5′ poly(A) leader might provide a binding site for the 40S ribosomal subunit, which would then move unimpeded by antisense RNA to the first translation initiation codon where ribosome assembly could occur. Since the most highly expressed viral proteins are made late in infection, either natural[165,166] or synthetic[167] late promoters are particularly valuable for expression vectors.

Only five intermediate genes have been identified thus far,[149,168] precluding the derivation of a reliable consensus promoter sequence. Mutagenesis studies indicate that intermediate promoters, like early and late ones, are about 30 base pairs long, but they appear unique in that they contain two regions that are sensitive to mutagenesis instead of one.[169]

4. Intermediate and Late Transcription Factors

Extracts prepared from cells infected in the presence of inhibitors of DNA replication are competent to transcribe exclusively intermediate genes,[149] whereas extracts prepared late in infection in the absence of inhibitors transcribe late genes efficiently.[170,171] There has been some progress on the isolation of late transcription factors; when combined, three separate fractions eluted from a phosphocellulose column were able to transcribe late genes.[172] One of the fractions contained RNA polymerase as well as another factor. The RNA polymerases used for early and late transcription are probably similar or at least share subunits since the phenotypes of conditionally lethal mutations in virion RNA polymerase subunits include decreased expression of late genes.[173,174]

A transfection strategy was used to identify intermediate genes needed for late transcription.[168] Initial experiments demonstrated that a transfected reporter gene, regulated by a late promoter, was expressed in cells infected with vaccinia virus provided that purified vaccinia virus DNA also was transfected into those cells. It was then found, by repeated subcloning and transfection, that three intermediate class genes encoding polypeptides of 17, 26, and 30K were necessary and sufficient to enable expression of a transfected late gene in the presence of an inhibitor of DNA replication. Further studies revealed that the 30K protein is a transcription factor.[175]

D. CASCADE REGULATION OF GENE EXPRESSION

While a precise understanding of poxvirus gene expression is not yet available, the work cited above provides evidence for a cascade regulatory mechanism (Figure 3). Thus, RNA polymerase and factors specific for transcription of early genes are present in the virion and used immediately after infection. Among the products of the early genes may be still-unidentified factors that transactivate expression of intermediate genes. When DNA replication occurs, these putative intermediate factors permit the transcription of the three identified late transactivator genes. The late transactivators then permit expression of the late genes including those that encode the subunits of the early transcription factor. The early transcription factor, and other enzymes, are packaged in newly formed virions to complete the cycle.

A variety of other processes may be superimposed on the virus-specific transcriptional regulatory mechanisms considered above. For example, the stability of cellular and viral mRNAs progressively decrease during the course of infection.[176–178] Such enhanced rates of mRNA degradation, in concert with transcriptional changes, could accelerate changes in mRNA populations. Translational regulatory mechanisms also have been proposed.[179–181] In addition, vaccinia virus profoundly inhibits host protein synthesis, probably by a variety of mechanisms involving specific viral gene products.[182–184] Three genes that determine the cell host range of cowpox or vaccinia virus have been discovered;[185–187] for at least two of them, loss of the host range gene results in failure to maintain viral and cellular protein synthesis in the nonpermissive cell lines.[188–191] Observations that are compatible with a role for host RNA polymerase II (or some component of it) in poxvirus replication have been summarized elsewhere.[4]

V. DNA REPLICATION

A. SITE AND TIME OF DNA SYNTHESIS

Cytoplasmic foci of vaccinia virus DNA replication were visualized by autoradiographic and microscopic procedures.[17,192–194] Claims for the presence of some viral DNA in the nuclei of poxvirus-infected cells[195–197] have been disputed.[198] The synthesis of vaccinia virus DNA in enucleated cells[199,200] is consistent with the predominant or exclusive use of viral proteins for replication.

The DNA genome present in the core of the virion serves as a template for early RNA synthesis and then for DNA replication. Degradation of the virion structure, known as uncoating, presumably is required for access by DNA polymerase and other newly synthesized proteins.[201–203] Replication of vaccinia virus DNA can be detected within the first 2 h of infection,[204,205] but the original evidence for shut off of DNA synthesis at the time of assembly of virions was probably an artifact of the [^{3}H]thymidine and BudR incorporation methods used. The period of DNA accumulation appears to be quite long when analyzed by DNA hybridization methods.[173,174,206,207]

B. PROTEINS ASSOCIATED WITH DNA REPLICATION

Poxviruses encode many of the enzymes needed for replication of their DNA (reviewed by Traktman).[208] Three virus-encoded enzymes involved in nucleotide metabolism, that are normally expressed early in infection, are not essential for replication in tissue culture cells. Thymidine kinase genes, homologous with eukaryotic enzymes, have been identified in vaccinia virus,[209,210] fowlpox virus,[211] and Shope fibroma virus.[212] The 19K protein exists in a tetrameric form that is subject to feedback inhibition by the distal products of its own metabolic pathway.[213] The selection methods for the absence and presence of thymidine kinase have been useful for selecting recombinant viruses. A gene with homology to thymidylate kinase, which catalyzes the next step in the biosynthetic pathway, has been found in vaccinia virus.[214] Both the large and small subunits of vaccinia virus ribonucleotide reductase are homologous to the eukaryotic enzymes[215–217] and together catalyze the conversion of ribonucleotides to deoxyribonucleotides. Under hydroxyurea selection, the gene encoding the small subunit is reiterated.

The vaccinia virus DNA polymerase is expressed early in infection, has a molecular weight of about 110K, and has an associated 3′ exonuclease activity.[218] The genes from vaccinia[219] and fowlpox[220] are closely related to each other in sequence and significantly related to α-DNA polymerases from disparate sources. The mapping of phosphonoacetic acid[219] and aphidicolin-resistant[221] and temperature sensitive[222] mutations to single nucleotide substitutions within the DNA polymerase gene of vaccinia virus provided compelling evidence for the essential role of this enzyme.

Early reports of increased DNA ligase activity in cells infected with vaccinia virus[223] have recently been confirmed by the finding of the virus-encoded gene.[224,225] Surprisingly, however, the gene can be deleted and is not required for virus replication.[226] Other enzymes that might be involved in DNA replication include the previously mentioned core-associated topoisomerase and nicking-joining enzyme (Table 1).

Three complementation groups of temperature sensitive (ts) vaccinia virus mutants with DNA synthesis negative phenotypes have been isolated (reviewed by Condit and Niles).[227] One of the complementation groups, referred to above, was mapped to the DNA polymerase. A second was mapped to a gene predicted to encode a protein of 90K.[228,229] As anticipated, the protein is expressed early in infection. Furthermore, in cells infected with such ts mutants, an upwards shift in temperature leads to a rapid stop in DNA synthesis, suggesting that the product acts directly at the replication fork.

The third complementation group of ts mutants, with defects in DNA replication, has been mapped to a gene predicted to encode a protein of 34K with homology to the family of serine-threonine protein kinases.[207,230] The role of this gene in DNA replication is not yet known.

C. CONCATEMERIC FORMS OF DNA

Although the DNA that is packaged in a virus particle is a unit length genome, rapidly sedimenting and concatemeric forms of intracellular poxvirus DNA have been detected particularly at early stages of replication.[206,231–233] The DNA molecules are either extraordinarily long (>12 genome lengths) or consist of complex multibranched structures since they did not penetrate agarose gels during pulse-field electrophoresis.[128,234] The junctions between the unit genomes have been recovered after restriction endonuclease digestion of the cytoplasmic DNA and have been found to consist of duplex copies of the hairpin loop present at the ends of mature virion DNA.[235] Concatemeric forms of viral DNA accumulate when late RNA or protein synthesis is prevented either through the use of ts mutants of vaccinia virus or specific inhibitors.[128,234] The block is reversible, however, providing evidence that the concatemers are replicative intermediates.

D. MECHANISMS OF REPLICATION

Since DNA polymerases are primer dependent, special mechanisms are needed for replication of the ends of linear genomes. Poxviruses have solved the problem with terminal hairpin structures (reviewed by DeLange and McFadden).[236] Models for poxvirus DNA replication, that take into account the concatemeric form of intracellular DNA and the generation of incompletely base-paired terminal loops that connect the two strands of the mature genome, have been derived from earlier eukaryotic palindromic telomere models.[237,238] The self-priming model involves the introduction of a near terminal nick to form a 3′ end that can be extended to form a palindrome that folds back on itself to replicate the remainder of the genome.[206,232,233] The alternative *de novo* start model postulates an RNA to prime DNA synthesis around the hairpin ends.[206,233,239] Both models can predict the formation of concatemeric molecules. Concatemers might also form nonreplicatively by isomerization following nicking of hairpin structures. Reported data suggest that replication begins at each end of the genome possibly by the introduction of nicks and involving both strand displacement mechanisms and small DNA fragments covalently linked to RNA primers,[239–243] but is insufficient to discriminate between the different models.

The DNA genomes of many families of viruses contain specific origins of DNA replication that have been identified by their ability to convey autonomous replication to plasmids. It was noted, however, that plasmids transfected into poxvirus-infected cells are replicated in a nonsequence specific fashion.[244] Moreover, plasmid replication is not enhanced by any vaccinia virus DNA sequence.[245] While the data obtained with plasmids have not been shown to relate to the virus genome, they do raise the possibility that specific replication origins are not used. Random nicking, followed by primer extension, strand displacement, and perhaps strand invasion, could result in the formation of complex networks of DNA that are eventually resolved into unit genomes.

There have been few published attempts to study poxvirus DNA replication *in vitro*. Crude cytoplasmic extracts can carry out limited DNA synthesis[246,247] and seal gaps or nicks in full-length vaccinia virus DNA molecules that are not yet cross-linked.[248] ATP, Mg^{2+}, and deoxyribonucleoside triphosphates were required for the latter.

E. RESOLUTION OF CONCATEMER JUNCTIONS

Circular plasmids, containing a copy of the concatemer junction of vaccinia virus or Shope fibroma virus DNA, are converted into linear molecules with hairpin termini by a process called resolution when transfected into poxvirus infected cells.[37,249] This *in vivo* assay provided a convenient way of determining the structural and sequence requirements for resolution. Evidence has been obtained that a highly conserved sequence, located adjacent to the hairpin loops of poxviruses, is required for resolution.[245,250–253] Two separated copies of this sequence, in inverted orientation, are needed. Detailed mutagenesis data indicated that the crucial part of the sequence is (A/T)TTT (A/G)N_{7-9}AAAAAAA.[253] By contrast, there are no precise sequence requirements other than A/T-richness for the hairpin loop itself (Figure 2), which is not highly conserved among poxviruses.[250] DNA replication is not required for resolution[254] and the hairpin is a heteroduplex derived from the DNA of each strand of the concatemer junction.[255] Conservative strand exchange models, involving cruciform or nicked DNA intermediates that are consistent with the available data and result in the formation of flip-flop hairpin termini, have been described.[251,255]

The proteins involved in resolution have not been identified. However, a correlation was found between six different complementation groups of vaccinia virus ts mutants and a decreased ability to resolve the concatemer junctions of transfected plasmids[245] or replicated genomic DNA.[128,234] A common feature of these mutants, with one possible exception, is a general inhibition of late gene expression. Indeed, some of the mutants are in genes encoding RNA polymerase subunits. Decreased resolution also was caused by the drug isatin-β-thiosemicarbazone, which selectively blocks late protein synthesis.[234] These results implied that one or more late proteins are needed for resolution. Two possible candidate late proteins are the DNA type I topoisomerase and the nicking-joining DNase mentioned above. The topoisomerase was found to promote illegitimate recombination when expressed in bacteria.[256] The DNase was shown to nick plasmids and seal the ends of some of the molecules to form hairpins.[144,235,257] Recently, it was pointed out that part of the seqeunce required for resolution resembles a late promoter and mediates late transcription when placed in front of a reporter gene in a plasmid.[258] On this basis, the genetic evidence for late gene expression has been reinterpreted to indicate a physical requirement for late transcription at the resolution site.

VI. RECOMBINATION

Recombination occurs at a high rate in poxvirus-infected cells. Exchanges of genetic information are observed when cells are infected with different members of the same genus[259] and can result in the formation of a virus with new biological properties.[260,261] Recombination between the terminal regions of orthopoxvirus DNA may explain variations in the number of tandem repeats,[43] deletions and translocations,[262] and mirror image deletions.[263] Recombination also occurs between virus-derived genomic DNA and either subgenomic DNA fragments[264,265] or plasmids containing poxvirus sequences[266] that have been transfected into the cell. The latter approach has been exploited to map mutants and insert foreign genes.

The molecular mechanisms involved in recombination are not well understood. Direct repeats may be rapidly eliminated from poxvirus genomes with the formation of intra- and intermolecular recombination products.[267] Single and double crossover products were detected after recombination between transfected plasmids and viral genomes.[268] In addition, inter- and intramolecular recombinants were detected in vaccinia virus infected cells cotransfected with plasmids containing polymorphic restriction sites.[269] Recombination does not require late gene products and there appears to be a connection between recombination and replication.[254]

VII. VIRUS ASSEMBLY, MATURATION, AND RELEASE

Poxvirus assembly is a complex process that has not yet been subjected to detailed molecular analysis. Morphological studies have revealed electron dense areas of the cytoplasm, known as factories, within which viral membranes appear to form *de novo* and then develop into immature virus particles.[270] In the presence of the drug rifampicin, there is an accumulation of irregular viral membranes that lack spicules and fail to progress in maturation.[271–273] Remarkably, within minutes after removal of the drug, the membranes become spicule-coated and assume a regular convex shape. The gene responsible for resistance of mutant viruses to rifampicin has been mapped to one that encodes a 95K protein.[274–276] In the next stage of development, the immature viral envelopes appear circular (or spherical in three dimensions) with a dense nucleoprotein mass in a granular matrix (Figure 4). Occasional electron micrographs suggest that the nucleoprotein enters the immature envelopes just before they are completely sealed.[277]Proteolytic cleavage of several of the major core proteins is blocked by rifampicin, suggesting that processing is coupled to virus assembly.[30,278] The more mature vaccinia virus particles are condensed and brick-like in appearance (Figure 4). They are transported, in association with actin filaments, to the cell periphery where they may become enveloped by additional membranes that appear to be

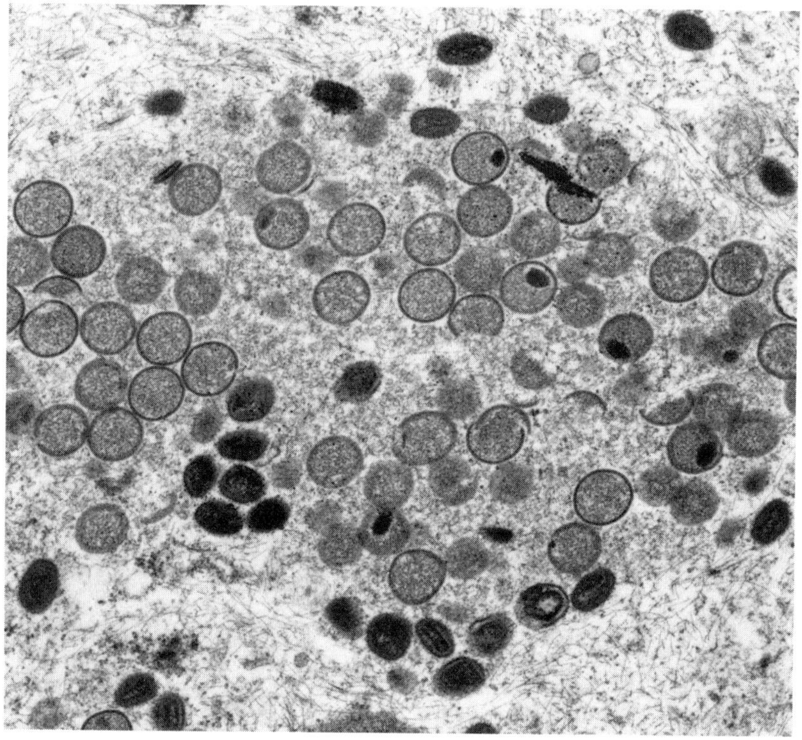

FIGURE 4. Transmission electron micrograph of a cell infected with vaccinia virus strain IHD-J for 16 h. There are numerous immature forms, which appear circular with a granular center; in some sections an electron dense nucleoid can be seen. The more mature particles have a condensed and brick-shaped appearance; in some the core structure is visible. (The micrograph was made at Advanced Biotechnologies, Inc. with material prepared by Rafael Blasco.)

derived from Golgi.[13,279] Some of the double-wrapped virions fuse with the plasma membrane, resulting in the externalization of the virion still enclosed by one layer of Golgi membrane. The latter has been referred to as the EEV earlier in this chapter.

Expression of a 14K protein is required for the egress of virions.[280] In addition, the release of virions can be prevented by a variety of drugs, including glycosylation inhibitors[281] monensin and cytochalasin D[282] and N1-isonicotinoyl-N2-3-methyl-4-chlorobenzoyl-hydrazine.[283,284]

VIII. STIMULATION OF CELL GROWTH

Many poxviruses induce hyperplastic cell responses and some even produce tumors in the skin of infected animals. The discovery that the deduced amino acid sequence of the 19K gene[285] of vaccinia virus, which is located within the inverted terminal repetition of strain WR, has a region of similarity to the active site of EGF and transforming growth factor α (TGFα),[286–288] provided a molecular mechanism to explain the growth promoting effects of poxviruses. This protein, called VGF, is made early in infection and secreted into the medium, where it binds to the EGF receptor of neighboring cells, and stimulates their autophosphorylation, and induces anchorage-independent cell growth.[289–291] VGF is glycosylated and processed at both ends.[290,292] A mutant vaccinia virus, with a deletion in the active site of VGF, no longer induced the proliferation of ectodermal and endodermal cells of the chorio-allantoic membrane of chicken embryos.[293] In addition, this mutation greatly attenuated the pathogenicity of vaccinia virus in mammals.[73] A gene related in structure to VGF has also been found in Shope fibroma virus.[294,295]

IX. MECHANISMS FOR PROTECTION AGAINST HOST DEFENSES

Poxviruses appear to have developed ways of impeding the host interferon, complement, and inflammatory responses which are designed to halt the spread of viruses and eliminate them. There is evidence that an early gene function of vaccinia virus produces an inhibitor that prevents the phosphorylation of the α-subunit of protein synthesis initiation factor eIF-2 by the interferon inducible double-stranded RNA-dependent protein kinase.[296–299] The level of the double-stranded RNA-dependent protein kinase also decreases during vaccinia virus infection.[300] In addition, there is evidence that vaccinia virus produces an inhibitor that diminishes the interferon inducible oligoadenylate synthetase.[301] The latter enzyme, upon activation by double-standed RNA, synthesizes a series of 2′ to 5′-linked adenylate oligomers that bind to and activate a latent cellular endoribonuclease which is capable of cleaving mRNA and rRNA and consequently decreasing protein synthesis. Other studies, however, indicated high levels of oligoadenylate under conditions in which vaccinia virus replication was unimpeded, but the cleavage of rRNA was delayed suggesting inhibition of the RNase.[302]

The complement proteins are thought to assist in the elimination of viruses by several different mechanisms. The predicted sequence of the 35K major secreted protein of vaccinia virus revealed that it most closely resembles a complement regulatory protein, C4B, which belongs to a family of eukaryotic proteins with 60 amino acid repeats.[303] *In vitro* data indicate that the purified protein binds to the fourth component of complement and inhibits hemolysis mediated by the classical complement pathway.[304] A second vaccinia virus gene with homology to a complement control protein was recently sequenced.[50]

Deletion of a cowpox virus gene encoding a 38K protein, belonging to the serine protease inhibitor family, caused an accelerated influx of heterophiles into the pock lesion in the chorioallantoic membrane of the chicken embryo suggesting an anti-inflammatory role for the protein.[305]

Vaccinia virus strain WR has three serine protease inhibitor genes[306–308] and deletion of either of two of these genes resulted in increased antibody responses to a foreign antigen after inoculation of animals.[309]

REFERENCES

1. **Fenner, F., Wittek, R., and Dumbell, K. R.,** *The Orthopoxviruses,* Academic Press, San Diego, 1988, 1.
2. **Moss, B.,** Regulation of vaccinia virus transcription, *Annu. Rev. Biochem.,* 59, 661, 1990.
3. **Moyer, R. W. and Turner, P. C.,** *Poxviruses,* Springer-Verlag, New York, 1990, 211.
4. **Dales, S.,** Reciprocity in the interactions between the poxviruses and their host cells, *Annu. Rev. Microbiol.,* 44, 173, 1990.
5. **Moss, B., Ahn, B.-Y., Amegadzie, B., Gershon, P. D., and Keck, J. G.,** Cytoplasmic transcription system encoded by vaccinia virus, *J. Biol. Chem.,* 266, 1355, 1991.
6. **Westwood, J. C. N., Harris, W. J., Zwartouw, H. T., Titmus, D. H. J., and Appleyard, G.,** Studies on the structure of vaccinia virus, *J. Gen. Microbiol.,* 34, 67, 1964.
7. **Dales, S. and Siminovitch, L.,** The development of vaccinia virus in Earle's L strain cells as examined by electron microscopy, *J. Biophys. Biochem. Cytol.,* 10, 475, 1961.
8. **Appleyard, G., Hapel, A. J., and Boulter, E. A.,** An antigenic difference between intracellular and extracellular rabbitpox virus, *J. Gen. Virol.,* 13, 9, 1971.
9. **Payne. K.,** Polypeptide composition of extracellular enveloped vaccinia virus, *J. Virol.,* 27, 28, 1978.
10. **Sarov, I. and Joklik, W. K.,** Studies on the nature and location of the capsid polypeptide of vaccinia virions, *Virology,* 50, 579, 1972.
11. **Essani, K. and Dales, S.,** Biogenesis of vaccinia: evidence for more than 100 polypeptides in the virion, *Virology,* 95, 385, 1979.
12. **Oie, M. and Ichihashi, Y.,** Characterization of vaccinia polypeptides, *Virology,* 113, 263, 1981.
13. **Hiller, G. and Weber, K.,** Golgi-derived membranes that contain an acylated viral polypeptide are used for vaccinia virus envelopment, *J. Virol.,* 55, 651, 1985.
14. **Shida, H. and Dales, S.,** Biogenesis of vaccinia: carbohydrate of the hemagglutinin molecule, *Virology,* 111, 56, 1981.
15. **Payne, L. G.,** Identification of the vaccinia hemagglutinin polypeptide from a cell system yielding large amounts of extracellular enveloped virus, *J. Virol.,* 31, 147, 1979.
16. **Stern, W. and Dales, S.,** Biogenesis of vaccinia. Isolation and characterization of a surface component that elicits antibody suppressing infectivity and cell-cell fusion, *Virology,* 75, 232, 1976.
17. **Oie, M. and Ichihashi, Y.,** Modification of vaccinia virus penetration proteins analyzed by monoclonal antibodies, *Virology,* 157, 449, 1987.
18. **Rodriguez, J. F. and Esteban, M.,** Mapping and nucleotide sequence of the vaccinia virus gene that encodes a 14-kilodalton fusion protein, *J. Virol.,* 61, 3550, 1987.
19. **Rodriguez, J. F., Paez, E., and Esteban, M.,** A 14,000-Mr envelope protein of vaccinia virus is involved in cell fusion and forms covalently linked trimers, *J. Virol.,* 61, 393, 1987.
20. **Niles, E. G. and Seto, J.,** Vaccinia virus gene D8 encodes a virion transmembrane protein, *J. Virol.,* 62, 3772, 1988.

21. **Gordon, J., Kovala, T., and Dales, S.,** Molecular characterization of a prominent antigen of the vaccinia virus envelope, *Virology,* 167, 361, 1990.
22. **Holowczak, J. A.,** Glycopeptides of vaccinia virus. I. Preliminary characterization and hexosamine content, *Virology,* 42, 87, 1970.
23. **Garon, C. F. and Moss, B.,** Glycoprotein synthesis in cells infected with vaccinia virus. II. A glycoprotein component of the virion, *Virology,* 46, 223, 1971.
24. **Franke, C. A., Reynolds, P. L., and Hruby, D. E.,** Fatty acid acylation of vaccinia virus proteins, *J. Virol.,* 63, 4285, 1989.
25. **Franke, C. A., Wilson, E. M., and Hruby, M. D.,** Use of a cell-free system to identify the vaccinia virus L1R gene product as the major late myristylated virion protein M25, *J. Virol.,* 64, 5988, 1990.
26. **Bertholet, C., Drillien, R., and Wittek, R.,** One hundred base pairs of 5′ flanking sequence of a vaccinia virus late gene are sufficient to temporally regulate late transcription, *Proc. Natl. Acad. Sci. U.S.A.,* 82, 2096, 1985.
27. **Rosel, J. and Moss, B.,** Transcriptional and translational mapping and nucleotide sequence analysis of a vaccinia virus gene encoding the precursor of the major core polypeptide 4b, *J. Virol.,* 56, 830, 1985.
28. **Weir, J. P. and Moss, B.,** Use of a bacterial expression vector to identify the gene encoding a major core protein of vaccinia virus, *J. Virol.,* 56, 534, 1985.
29. **Van Meir, E. and Wittek, R.,** Fine structure of the vaccinia virus gene encoding the precursor of the major core protein 4a, *Arch. Virol.,* 102, 19, 1988.
30. **Moss, B. and Rosenblum, E. N.,** Protein cleavage and poxvirus morphogenesis: tryptic peptide analysis of core precursors accumulated by blocking assembly with rifampicin, *J. Mol. Biol.,* 81, 267, 1973.
31. **Yang, W. P., Kao, S.-Y., and Bauer, W. R.,** Biosynthesis and posttranslational cleavage of vaccinia virus structural protein VP8, *Virology,* 167, 585, 1988.
32. **Rosemond, H. and Moss, B.,** Phosphoprotein component of vaccinia virions, *J. Virol.,* 11, 961, 1973.
33. **Kao, S.-Y., Ressner, E., Kates, J., and Bauer, W. R.,** Purification and characterization of a superhelix binding protein from vaccinia virus, *Virology,* 111, 500, 1981.
34. **Yang, W. P. and Bauer, W. R.,** Purification and characterization of vaccinia virus structural protein VP8, *Virology,* 167, 578, 1988.
35. **Geshelin, P. and Berns, K. I.,** Characterization and localization of the naturally occurring cross-links in vaccinia virus DNA, *J. Mol. Biol.,* 88, 785, 1974.
36. **Baroudy, B. M., Venkatesan, S., and Moss, B.,** Incompletely base-paired flip-flop terminal loops link the two DNA strands of the vaccinia virus genome into one uninterrupted polynucleotide chain, *Cell,* 28, 315, 1982.
37. **Delange, A. M., Reddy, M., Scraba, D., Upton, C., and McFadden, G.,** Replication and resolution of cloned poxvirus telomeres in vivo generates linear minichromosomes with intact viral hairpin termini, *J. Virol.,* 59, 249, 1986.
38. **Gonzalez, A., Talavera, A., Almendral, J. M., and Viñuela, E.,** Hairpin loop structure of African swine fever virus DNA, *Nucleic Acids Res.,* 14, 6835, 1986.
39. **Wittek, R. and Moss, B.,** Tandem repeats within the inverted terminal repetition of vaccinia virus DNA, *Cell,* 21, 277, 1980.
40. **Pickup, D. J., Bastia, D., Stone, H. O., and Joklik, W.,** Sequence of terminal regions of cowpox virus DNA: arrangement of repeated and unique sequence elements, *Proc. Natl. Acad. Sci. U.S.A.,* 79, 7112, 1982.
41. **Parsons, B. L. and Pickup, D. J.,** Tandemly repeated sequences are present at the end of the DNA of raccoonpox virus, *Virology,* 161, 45, 1987.
42. **Upton, C., DeLange, A. M., and McFadden, G.,** Tumorigenic poxviruses: genomic organization and DNA sequence of the telomeric region of the Shope fibroma virus genome, *Virology,* 160, 20, 1987.
43. **Moss, B., Winters, E., and Cooper, N.,** Instability and reiteration of DNA sequences within the vaccinia virus genome, *Proc. Natl. Acad. Sci. U.S.A.,* 78, 1614, 1981.

44. **Garon, C. F., Barbosa, E., and Moss, B.,** Visualization of an inverted terminal repetition in vaccinia virus DNA, *Proc. Natl. Acad. Sci. U.S.A.,* 75, 4863, 1978.
45. **Wittek, R., Menna, A., Muller, K., Schumperli, D., Bosley, P. G., and Wyler, R.,** Inverted terminal repeats in rabbit poxvirus and vaccinia virus DNA, *J. Virol.,* 28, 171, 1978.
46. **Wittek, R., Cooper, J., Barbosa, E., and Moss, B.,** Expression of the vaccinia virus genome — analysis and mapping of mRNAs encoded within the inverted terminal repetition, *Cell,* 21, 487, 1980.
47. **Mackett, M. and Archard, L. C.,** Conservation and variation in orthopoxvirus genome structure, *J. Gen. Virol.,* 45, 683, 1979.
48. **Esposito, J. J. and Knight, J. C.,** Orthopoxivrus DNA: a comparison of restriction profiles and maps, *Virology,* 143, 230, 1985.
49. **Earl, P. L. and Moss, B.,** Vaccinia virus, in *Genetic Maps 1989,* Vol. 5, O'Brien, S. J., Ed., Cold Spring Harbor Laboratory, Cold Spring Harbor, NY, 1989, 1.138.
50. **Goebel, S. J., Johnson, G. P., Perkus, M. E., Davis, S. W., Winslow, J. P., and Paoletti, E.,** The complete DNA sequence of vaccinia virus, *Virology,* 179, 247, 1990.
51. **Belle Isle, H., Venkatesan, S., and Moss, B.,** Cell-free translation of early and late mRNAs selected by hybridization to cloned DNA fragments derived from the left 14 million to 72 million daltons of the vaccinia virus genome, *Virology,* 112, 306, 1981.
52. **Rosel, J. L., Earl, P. L., Weir, J. P., and Moss, B.,** Conserved TAAATG sequence at the transcriptional and translational initiation sites of vaccinia virus late genes deduced by structural and functional analysis of the HindIII H genome fragment, *J. Virol.,* 60, 436, 1986.
53. **Lee-Chen, G. J. and Niles, E. G.,** Transcription and translation mapping of the 13 genes in the vaccinia virus HindIII D fragment, *Virology,* 163, 52, 1988.
54. **Moyer, R. W. and Rothe, C. T.,** The white pock mutants of rabbit poxvirus. I. Spontaneous host range mutants contain deletions, *Virology,* 102, 119, 1980.
55. **Panicali, D., Davis, S. W., Mercer, S. R., and Paoletti, E.,** Two major DNA variants present in serially propagated stocks of the WR strain of vaccinia virus, *J. Virol.,* 37, 1000, 1981.
56. **Kotwal, G. J. and Moss, B.,** Analysis of a large cluster of nonessential genes deleted from a vaccinia virus terminal transposition mutant, *Virology,* 167, 524, 1988.
57. **Perkus, M. E., Limbach, K., and Paoletti, E.,** Cloning and expression of foreign genes in vaccinia virus, using a host range selection system, *J. Virol.,* 63, 3829, 1989.
58. **Perkus, M. E., Goebel, S. J., Davis, S. W., Johnson, G. P., Norton, E. K., and Paoletti, E.,** Deletion of 55 open reading frames from the termini of vaccinia virus, *Virology,* 180, 406, 1991.
59. **Smith, G. L. and Moss, B.,** Infectious poxvirus vectors have capacity for at least 25,000 base pairs of foreign DNA, *Gene,* 25, 21, 1983.
60. **Boulter, E. A. and Appleyard, G.,** Differences between extracellular and intracellular forms of poxvirus and their implications, *Prog. Med. Virol.,* 16, 86, 1973.
61. **Payne, L. G.,** Significance of extracellular virus in the in vitro and in vivo dissemination of vaccinia virus, *J. Gen. Virol.,* 50, 89, 1980.
62. **Dales, S. and Kajioka, R.,** The cycle of multiplication of vaccinia virus in Earle's strain L cells. I. Uptake and penetration, *Virology,* 24, 278, 1964.
63. **Chang, A. and Metz, D. H.,** Further investigations on the mode of entry of vaccinia virus into cells, *J. Gen. Virol.,* 32, 275, 1976.
64. **Janeczko, R. A., Rodriguez, J. F., and Esteban, M.,** Studies on the mechanism of entry of vaccinia virus into animal cells, *Arch. Virol.,* 92, 135, 1987.
65. **Doms, R. W., Blumenthal, R., and Moss, B.,** Fusion of intra- and extracellular forms of vaccinia virus with the cell membrane, *J. Virol.,* 64, 4884, 1990.
66. **Maa, J.-S., Rodriguez, J. F., and Esteban, M.,** Structural and functional characterization of a cell surface binding protein of vaccinia virus, *J. Biol. Chem.,* 265, 1569, 1990.

67. **Oie, M., Shida, H., and Ichihashi, Y.,** The function of the vaccinia hemagglutinin in the proteolytic activation of infectivity, *Virology,* 176, 494, 1990.
68. **Blomquist, M. C. L., Hunt, L. T., and Baker, W. C.,** Vaccinia virus 19-kilodalton protein: relationship to several mammalian proteins, including two growth factors, *Proc. Natl. Acad. Sci. U.S.A.,* 81, 7363, 1984.
69. **Brown, J. P., Twardzik, D. R., Harquardt, H., and Todaro, G. J.,** Vaccinia virus encodes a polypeptide homologous to epidermal growth factor and transforming growth factor, *Nature,* 313, 491, 1985.
70. **Reisner, A. H.,** Similarity between the vaccinia virus 19K early protein and epidermal growth factor, *Nature,* 313, 801, 1985.
71. **Eppstein, D. A., Marsh, Y. V., Schreiber, A. B., Newman, S. R., Todaro, G. J., and Nestor, J. J.,** Epidermal growth factor receptor occupancy inhibits vaccinia virus infection, *Nature London,* 266, 550, 1985.
72. **Marsh, Y. V. and Eppstein, D. A.,** Vaccinia virus and the EGF receptor: a portal of entry for infectivity?, *J. Cell. Biochem.,* 34, 239, 1987.
73. **Buller, R. M. L., Chakrabarti, S., Cooper, J. A., Twardzik, D. R., and Moss, B.,** Deletion of the vaccinia virus growth factor gene reduces virus virulence, *J. Virol.,* 62, 866, 1988.
74. **Kates, J. R. and McAuslan, B. R.,** Poxvirus DNA-dependent RNA polymerase, *Proc. Natl. Acad. Sci. U.S.A.,* 58, 134, 1967.
75. **Munyon, W. E., Paoletti, E., and Grace, J. T., Jr.,** RNA polymerase activity in purified infectious vaccinia virus, *Proc. Natl. Acad. Sci. U.S.A.,* 58, 2280, 1967.
76. **Kates, J. R. and Beeson, J.,** Ribonucleic acid synthesis in vaccinia virus. II. Synthesis of polyriboadenylic acid, *J. Mol. Biol.,* 50, 19, 1970.
77. **Wei, C. M. and Moss, B.,** Methylated nucleotides block 5′-terminus of vaccinia virus mRNA, *Proc. Natl. Acad. Sci. U.S.A.,* 72, 318, 1975.
78. **Mackett, M., Smith, G. L., and Moss, B.,** General method for production and selection of infectious vaccinia virus recombinants expressing foreign genes, *J. Virol.,* 49, 857, 1984.
79. **Cochran, M. A., Puckett, C., and Moss, B.,** In vitro mutagenesis of the promoter region for a vaccinia virus gene: evidence for tandem early and late regulatory signals, *J. Virol.,* 54, 30, 1985.
80. **Weir, J. P. and Moss, B.,** Determination of the promoter region of an early vaccinia virus gene encoding thymidine kinase, *Virology,* 158, 206, 1987.
81. **Davison, A. J. and Moss, B.,** The structure of vaccinia virus early promoters, *J. Mol. Biol.,* 210, 749, 1989.
82. **Keith, J., Gershowitz, A., and Moss, B.,** Dinucleotide sequences at the 5′ ends of vaccinia virus mRNAs synthesized in vitro, *J. Virol.,* 36, 601, 1980.
83. **Venkatesan, S. and Moss, B.,** In vitro transcription of the inverted terminal repetition of the vaccinia virus genome: correspondence of initiation and cap sites, *J. Virol.,* 37, 738, 1981.
84. **Kumar, S. and Boyle, D. B.,** Activity of a fowlpox virus late gene promoter in vaccinia and fowlpox virus recombinants, *Arch. Virol.,* 112, 139, 1990.
85. **Boyle, D. B. and Coupar, B. E. H.,** Construction of recombinant fowlpox viruses as vectors for poultry vaccines, *Virus Res.,* 10, 343, 1988.
86. **Taylor, J., Weinberg, R., Languet, B., Desmettre, P. ,and Paoletti, E.,** Recombinant fowlpox virus inducing protective immunity in non-avian species, *Vaccine,* 6, 497, 1988.
87. **Rohrmann, G., Yuen, L., and Moss, B.,** Transcription of vaccinia virus early genes by enzymes isolated from vaccinia virions terminates downstream of a regulatory sequence, *Cell,* 46, 1029, 1986.
88. **Yuen, L. and Moss, B.,** Oligonucleotide sequence signaling transcriptional termination of vaccinia virus early genes, *Proc. Natl. Acad. Sci. U.S.A.,* 84, 6417, 1987.

89. **Upton, C., Macen, J. L., and McFadden, G.,** Mapping and sequencing of a gene from myxoma virus that is related to those encoding epidermal growth factor and transforming growth factor alpha, *J. Virol.*, 61, 1271, 1987.
90. **Tomley, F., Binns, M., Campbell, J., and Boursnell, M.,** Sequence analysis of an 11.2 kilobase, near-terminal, BamHI fragment of fowlpox virus, *J. Gen. Virol.*, 69, 1025, 1988.
91. **Gershon, P. D. and Black, D. N.,** The nucleotide sequence around the capripoxvirus thymidine kinase gene reveals a gene shared specifically with Leporipoxviruses, *J. Gen. Virol.*, 70, 525, 1989.
92. **Earl, P. L., Hügen, A. W., and Moss, B.,** Removal of cryptic poxvirus transcription termination signals from the human immunodeficiency virus type 1 envelope gene enhances expression and immunogenicity of a recombinant vaccinia virus, *J. Virol.*, 64, 2448, 1990.
93. **Baroudy, B. M. and Moss, B.,** Purification and characterization of a DNA-dependent RNA polymerase from vaccinia virions, *J. Biol. Chem.*, 255, 4372, 1980.
94. **Spencer, E., Shuman, S., and Hurwitz, J.,** Purification and properties of vaccinia virus DNA-dependent RNA polymerase, *J. Biol. Chem.*, 255, 5388, 1980.
95. **Jones, E. V., Puckett, C., and Moss, B.,** DNA-dependent RNA polymerase subunits encoded within the vaccinia virus genome, *J. Virol.*, 61, 1765, 1987.
96. **Broyles, S. S. and Moss, B.,** Homology between RNA polymerases of poxviruses, prokaryotes, and eukaryotes: nucleotide sequence and transcriptional analysis of vaccinia virus genes encoding 147-kDa and 22-kDa subunits, *Proc. Natl. Acad. Sci. U.S.A.*, 83, 3141, 1986.
97. **Patel, D. D. and Pickup, D. J.,** The second-largest subunit of the poxvirus RNA polymerase is similar to the corresponding subunits of procaryotic and eucaryotic RNA polymerases, *J. Virol.*, 63, 1076, 1989.
98. **Ahn, B.-Y., Jones, E. V., and Moss, B.,** Identification of the vaccinia virus gene encoding an 18-kilodalton subunit of RNA polymerase and demonstration of a 5′ poly(A) leader on its early transcript, *J. Virol.*, 64, 3019, 1990.
99. **Ahn, B.-Y., Gershon, P. D., Jones, E. V., and Moss, B.,** Identification of *rpo30*, a vaccinia virus RNA polymerase gene with structural similarity to a eukaryotic transcription factor, *Mol. Cell. Biol.*, 10, 5433, 1990.
100. **Amegadzie, B., Holmes, M., Cole, N. B., Jones, E. V., Earl, P. L., and Moss, B.,** Identification, sequence, and expression of the gene encoding the second-largest subunit of the vaccinia virus DNA-dependent RNA polymerase, *Virology*, 180, 88, 1991.
101. **Amegadzie, B., Cole, N., Ahn, B. Y., and Moss, B.,** Identification, sequence and expression of the gene encoding a Mr 35,000 subunit of the vaccinia virus DNA-dependent RNA polymerase, *J. Biol. Chem.*, 266, 13,712, 1991.
102. **Broyles, S. S., Yuen, L., Shuman, S., and Moss, B.,** Purification of a factor required for transcription of vaccinia virus early genes, *J. Biol. Chem.*, 263, 10,754, 1988.
103. **Broyles, S. S. and Moss, B.,** DNA-dependent ATPase activity associated with vaccinia virus early transcription factor, *J. Biol. Chem.*, 263, 10,761, 1988.
104. **Gershowitz, A., Boone, R. F., and Moss, B.,** Multiple roles for ATP in the synthesis and processing of mRNA by vaccinia virus: specific inhibitory effects of adenosine (β,γ-imido)triphosphate, *J. Virol.*, 27, 399, 1978.
105. **Yuen, L., Davison, A. J., and Moss, B.,** Early promoter-binding factor from vaccinia virions, *Proc. Natl. Acad. Sci. U.S.A.*, 84, 6069, 1987.
106. **Broyles, S. S. and Fesler, B. S.,** Vaccinia virus gene encoding a component of the viral early transcription factor, *J. Virol.*, 64, 1523, 1990.
107. **Gershon, P. D. and Moss, B.,** Early transcription factor subunits are encoded by vaccinia virus late genes, *Proc. Natl. Acad. Sci. U.S.A.*, 87, 4401, 1990.
108. **Muthukrishnan, S., Moss, B., Cooper, J. A., and Maxwell, E. S.,** Influence of 5′ terminal cap structure on the initiation of translation of vaccinia virus mRNA, *J. Biol. Chem.*, 253, 1710, 1978.

109. **Martin, S. S. and Moss, B.,** mRNA guanylyltransferase and mRNA (guanine-7)methyltransferase from vaccinia virions. Donor and acceptor substrate activities, *J. Biol. Chem.*, 251, 7313, 1976.
110. **Venkatesan, S., Gershowitz, A., and Moss, B.,** Modification of the 5′-end of mRNA: association of RNA triphosphatase with the RNA guanylyltransferase-RNA (guanine-7)methyltransferase complex from vaccinia virus, *J. Biol. Chem.*, 255, 903, 1980.
111. **Shuman, S. and Hurwitz, J.,** Mechanism of mRNA capping by vaccinia virus guanylyltransferase: characterization of an enzyme-guanylate intermediate, *Proc. Natl. Acad. Sci. U.S.A.*, 78, 187, 1981.
112. **Roth, M. J. and Hurwitz, J.,** RNA capping by the vaccinia virus guanylyltransferase. Structure of enzyme-guanylate intermediate, *J. Biol. Chem.*, 259, 13,488, 1984.
113. **Morgan, J. R., Cohen, L. K., and Roberts, B. E.,** Identification of the DNA sequences encoding the large subunit of the mRNA capping enzyme of vaccinia virus, *J. Virol.*, 52, 206, 1984.
114. **Niles, E. G., Lee-Chen, G.-J., Shuman, S., Moss, B., and Broyles, S. S.,** Vaccinia virus gene D12L encodes the small subunit of the viral mRNA capping enzyme, *Virology*, 172, 513, 1989.
115. **Guo, P. and Moss, B.,** Interaction and mutual stabilization of the two subunits of vaccinia virus mRNA capping enzyme coexpressed in *Escherichia coli*, *Proc. Natl. Acad. Sci. U.S.A.*, 87, 4023, 1990.
116. **Shuman, S.,** Catalytic activity of vaccinia messenger RNA capping enzyme subunits coexpressed in *Escherichia coli*, *J. Biol. Chem.*, 265, 11,960, 1990.
117. **Shuman, S. and Morham, S. G.,** Domain structure of vaccinia virus messenger RNA capping enzyme — activity of the MR 95,000 subunit expressed in *Escherichia coli*, *J. Biol. Chem.*, 265, 11,967, 1990.
118. **Barbosa, E. and Moss, B.,** mRNA(nucleoside-2′-)-methyltransferase from vaccinia virus. Purification and physical properties, *J. Biol. Chem.*, 253, 7698, 1978.
119. **Shuman, S., Broyles, S. S., and Moss, B.,** Purification and characterization of a transcription termination factor from vaccinia virions, *J. Biol. Chem.*, 262, 12372, 1987.
120. **Shuman, S. and Moss, B.,** Factor-dependent transcription termination by vaccinia virus RNA polymerase: evidence that the cis-acting termination signal is in nascent RNA, *J. Biol. Chem.*, 263, 6220, 1988.
121. **Broyles, S. S. and Moss, B.,** Sedimentation of an RNA polymerase complex from vaccinia virus that specifically initiates and terminates transcription, *Mol. Cell. Biol.*, 7, 7, 1987.
122. **Moss, B., Rosenblum, E. N., and Gershowitz, A.,** Characterization of a polyriboadenylate polymerase from vaccinia virions, *J. Biol. Chem.*, 250, 4722, 1975.
123. **Nevins, J. R. and Joklik, W. K.,** Isolation and partial characterization of the poly(A) polymerases from HeLa cells infected with vaccinia virus, *J. Biol. Chem.*, 252, 6939, 1977.
124. **Shuman, S. and Moss, B.,** Vaccinia virus poly(A) polymerase: specificity for nucleotides and nucleotide analogs, *J. Biol. Chem.*, 263, 8405, 1988.
125. **Paoletti, E., Rosemond-Hornbeak, H., and Moss, B.,** Two nucleic acid-dependent nucleoside triphosphate phosphohydrolases from vaccinia virus: purification and characterization, *J. Biol. Chem.*, 249, 3273, 1974.
126. **Rodriguez, J. F., Kahn, J. S., and Esteban, M.,** Molecular cloning, encoding sequence, and expression of vaccinia virus nucleic acid-dependent nucleoside triphosphatase gene, *Proc. Natl. Acad. Sci. U.S.A.*, 83, 9566, 1986.
127. **Broyles, S. S. and Moss, B.,** Identification of the vaccinia virus gene encoding nucleoside triphosphate phosphohydrolase I, a DNA-dependent ATPase, *J. Virol.*, 61, 1738, 1987.
128. **DeLange, A. M.,** Identification of temperature-sensitive mutants of vaccinia virus that are defective in conversion of concatemeric replicative intermediates to the mature linear DNA genome, *J. Virol.*, 63, 2437, 1989.

129. **Kunzi, M. S. and Traktman, P.,** Genetic evidence for involvement of vaccinia virus DNA-dependent ATPase I in intermediate and late gene expression, *J. Virol.*, 63, 3999, 1989.
130. **Paoletti, E. and Moss, B.,** Two nucleic acid-dependent nucleoside triphosphate phosphohydrolases from vaccinia virus. Nucleotide substrate and polynucleotide cofactor specificities, *J. Biol. Chem.*, 249, 3281, 1974.
131. **Bauer, W. R., Ressner, E. C., Kates, J., and Patzke, J. V.,** A DNA nicking-closing enzyme encapsidated in vaccinia virus: partial purification and properties, *Proc. Natl. Acad. Sci. U.S.A.*, 74, 1841, 1977.
132. **Shaffer, R. and Traktman, P.,** Vaccinia virus encapsidates a novel topoisomerase with the properties of a eucaryotic type I enzyme, *J. Biol. Chem.*, 262, 9309, 1987.
133. **Shuman, S., Golder, M., and Moss, B.,** Characterization of vaccinia virus DNA topoisomerase I expressed in *Escherichia coli, J. Biol. Chem.*, 263, 16401, 1988.
134. **Shuman, S., Kane, E. M., and Morham, S. G.,** Mapping the active-site tyrosine of vaccinia virus DNA topoisomerase gene, *Proc. Natl. Acad. Sci. U.S.A.*, 86, 9793, 1989.
135. **Shuman, S. and Prescott, J.,** Specific DNA cleavage and binding by vaccinia virus DNA topoisomerase-I, *J. Biol. Chem.*, 265, 17826, 1990.
136. **Shuman, S. and Moss, B.,** Identification of a vaccinia virus gene encoding a type I DNA topoisomerase, *Proc. Natl. Acad. Sci. U.S.A.*, 84, 7478, 1987.
137. **Upton, C., Opgenorth, A., Traktman, P., and McFadden, G.,** Identification and DNA sequence of the Shope fibroma virus DNA topoisomerase gene, *Virology*, 176, 439, 1990.
138. **Shuman, S., Golder, M., and Moss, B.,** Insertional mutagenesis of the vaccinia virus gene encoding a type I DNA topoisomerase: evidence that the gene is essential for virus growth, *Virology*, 170, 302, 1989.
139. **Kleiman, J. H. and Moss, B.,** Purification of a protein kinase and two phosphate acceptor proteins from vaccinia virions, *J. Biol. Chem.*, 250, 2420, 1975.
140. **Kleiman, J. H. and Moss, B.,** Characterization of a protein kinase and two phosphate acceptor proteins from vaccinia virions, *J. Biol. Chem.*, 250, 2430, 1975.
141. **Rosemond-Hornbeak, H. and Moss, B.,** Single-stranded deoxyribonucleic acid-specific nuclease from vaccinia virus. Endonucleolytic and exonucleolytic activities, *J. Biol. Chem.*, 249, 3292, 1974.
142. **Rosemond-Hornbeak, H., Paoletti, E., and Moss, B.,** Single-stranded deoxyribonucleic acid-specific nuclease from vaccinia virus. Purification and characterization, *J. Biol. Chem.*, 249, 3287, 1974.
143. **Pogo, B. G. T. and O'Shea, M. T.,** Further characterization of deoxyribonucleases from vaccinia virus, *Virology*, 77, 55, 1977.
144. **Reddy, M. K. and Bauer, W. R.,** Activation of the vaccinia virus nicking-joining enzyme by trypsinization, *J. Biol. Chem.*, 264, 443, 1989.
145. **Moss, B. and Salzman, N. P.,** Sequential protein synthesis following vaccinia virus infection, *J. Virol.*, 2, 1016, 1968.
146. **Pennington, T. H.,** Vaccinia virus polypeptide synthesis: sequential appearance and stability of pre- and post-replicative polypeptides, *J. Gen. Virol.*, 25, 433, 1974.
147. **Carrasco, L. and Bravo, R.,** Specific proteins synthesized during the viral lytic cycle in vaccinia virus-infected HeLa cells: analysis by high-resolution two-dimensional gel electrophoresis, *J. Virol.*, 58, 569, 1986.
148. **Cochran, M. A., Mackett, M., and Moss, B.,** Eukaryotic transient expression system dependent on transcription factors and regulatory DNA sequences of vaccinia virus, *Proc. Natl. Acad. Sci. U.S.A.*, 82, 19, 1985.
149. **Vos, J. C. and Stunnenberg, H. G.,** Derepression of a novel class of vaccinia virus genes upon DNA replication, *EMBO J.*, 7, 3487, 1988.
150. **Cooper, J. A., Wittek, R., and Moss, B.,** Extension of the transcriptional and translational map of the left end of the vaccinia virus genome to 21 kilobase pairs, *J. Virol.*, 39, 733, 1981.

151. **Mahr, A. and Roberts, B. E.,** Arrangement of the late RNAs transcribed from a 7.1 kilobase *Eco R1* vaccinia virus DNA fragment, *J. Virol.*, 49, 510, 1984.
152. **Colby, C., Jurale, C., and Kates, J. R.,** Mechanism of synthesis of vaccinia virus double-stranded ribonucleic acid *in vivo* and *in vitro*, *J. Virol.*, 7, 71, 1971.
153. **Boone, R. F., Parr, R. P., and Moss, B.,** Intermolecular duplexes formed from polyadenylated vaccinia virus RNA, *J. Virol.*, 30, 365, 1979.
154. **Bertholet, C., Van Meir, E., ten Heggeler-Bordier, B., and Wittek, R.,** Vaccinia virus produces late mRNAs by discontinuous synthesis, *Cell*, 50, 153, 1987.
155. **Schwer, B., Visca, P., Vos, J. C., and Stunnenberg, H. G.,** Discontinuous transcription or RNA processing of vaccinia virus late messengers results in a 5′ poly(A) leader, *Cell*, 50, 163, 1987.
156. **Patel, D. D. and Pickup, D. J.,** Messenger RNAs of a strongly-expressed late gene of cowpox virus contains a 5′-terminal poly(A) leader, *EMBO J.*, 6, 3787, 1987.
157. **Ahn, B.-Y. and Moss, B.,** Capped poly(A) leader of variable lengths at the 5′ ends of vaccinia virus late mRNAs, *J. Virol.*, 63, 226, 1989.
158. **Hänggi, M., Bannwarth, W., and Stunnenberg, H. G.,** Conserved TAAAT motif in vaccinia virus late promoters: overlapping TATA box and site of transcription initiation, *EMBO J.*, 5, 1071, 1986.
159. **Weir, J. P. and Moss, B.,** Determination of the transcriptional regulatory region of a vaccinia virus late gene, *J. Virol.*, 61, 75, 1987.
160. **Miner, J. N., Weinrich, S. L., and Hruby, D. E.,** Molecular dissection of cis-acting regulatory elements from 5′-proximal regions of a vaccinia virus late gene cluster, *J. Virol.*, 62, 297, 1988.
161. **Davison, A. J. and Moss, B.,** Structure of vaccinia virus late promoters, *J. Mol. Biol.*, 210, 771, 1989.
162. **de Magistris, L. and Stunnenberg, H. G.,** Cis-acting sequences affecting the length of the poly(A) head of vaccinia virus late transcripts, *Nucleic Acids Res.*, 16, 3141, 1988.
163. **Stunnenberg, H. G., de Magistris, L., and Schwer, B.,** The generation of poly(A) heads on vaccinia late mRNA: a proposal of a slippage mechanism, in *Molecular Biology of RNA*, Cech, T. R., Ed., Alan R. Liss, New York, 1989, 199.
164. **Ink, B. S. and Pickup, D. J.,** Vaccinia virus directs the synthesis of early mRNAs containing 5′ poly(A) sequences, *Proc. Natl. Acad. Sci. U.S.A.*, 87, 1536, 1990.
165. **Falkner, G. G., Chakrabarti, S., and Moss, B.,** pUV I: a new vaccinia virus insertion and expression vector, *Nucleic Acids Res.*, 15, 7192, 1987.
166. **Patel, D. D., Ray, C. A., Drucker, R. P., and Pickup, D. J.,** A poxvirus-derived vector that directs high levels of expression of cloned genes in mammalian cells, *Proc. Natl. Acad. Sci. U.S.A.*, 85, 9431, 1988.
167. **Davison, A. J. and Moss, B.,** New vaccinia virus recombination plasmids incorporating a synthetic late promoter for high level expression of foreign proteins, *Nucleic Acids Res.*, 18, 4285, 1990.
168. **Keck, J. G., Baldick, C. J., and Moss, B.,** Role of DNA replication in vaccinia virus gene expression: a naked template is required for transcription of three late transactivator genes, *Cell*, 61, 801, 1990.
169. **Hirschmann, P., Vos, J. C., and Stunnenberg, H. G.,** Mutational analysis of a vaccinia virus intermediate promoter in vivo and in vitro, *J. Virol.*, 64, 6063, 1990.
170. **Wright, C. F. and Moss, B.,** In vitro synthesis of vaccinia virus late mRNA containing a 5′ poly(A) leader sequence, *Proc. Natl. Acad. Sci. U.S.A.*, 84, 8883, 1987.
171. **Schwer, B. and Stunnenberg, H. G.,** Vaccinia virus late transcripts generated in vitro have a poly(A) head, *EMBO J.*, 7, 1183, 1988.
172. **Wright, C. F. and Moss, B.,** Identification of factors specific for transcription of the late class of vaccinia virus genes, *J. Virol.*, 63, 4224, 1989.
173. **Ensinger, M. J.,** Phenotypic characterization of temperature-sensitive mutants of vaccinia virus with mutations in a 135,000-Mr subunit of the virion-associated DNA-dependent RNA polymerase, *J. Virol.*, 61, 1842, 1987.

174. **Hooda-Dhingra, U., Thompson, C. L., and Condit, R. C.,** Detailed phenotypic characterization of five temperature-sensitive mutants in the 22- and 147-kilodalton subunits of vaccinia virus DNA-dependent RNA polymerase, *J. Virol.*, 63, 714, 1989.
175. **Wright, C. F., Keck, J. G., and Moss, B.,** A transcription factor for expression of vaccinia virus late genes is encoded by an intermediate gene, *J. Virol.*, 65, 3715, 1991.
176. **Sebring, E. D. and Salzman, N. P.,** Metabolic properties of early and late vaccinia messenger ribonucleic acid, *J. Virol.*, 1, 550, 1967.
177. **Oda, K. and Joklik, W. K.,** Hybridization and sedimentation studies on "early" and "late" vaccinia messenger RNA, *J. Mol. Biol.*, 27, 395, 1967.
178. **Rice, A. P. and Roberts, B. E.,** Vaccinia virus induces cellular mRNA degradation, *J. Virol.*, 47, 529, 1983.
179. **McAuslan, B. R.,** Control of induced thymidine kinase activity in the poxvirus infected cell, *Virology*, 20, 162, 1963.
180. **Hruby, D. E. and Ball, L. A.,** Control of expression of the vaccinia virus thymidine kinase gene, *J. Virol.*, 40, 456, 1981.
181. **Vassef, A., Ben-Hamida, F., Dru, A., and Beaud, G.,** Translational control of early protein synthesis at the late stage of vaccinia virus infection, *Virology*, 118, 45, 1982.
182. **Tas, P. W. and Martini, O. H.,** Inhibition of the elongation step of protein synthesis by vaccinia virus, *FEBS Lett.*, 153, 427, 1983.
183. **Person-Fernandez, A. and Beaud, G.,** Purification and characterization of a protein synthesis inhibitor associated with vaccinia virus, *J. Biol. Chem.*, 261, 8283, 1986.
184. **Su, M. J. and Bablanian, R.,** Polyadenylated RNA sequences from vaccinia virus infected cells selectively inhibit translation in a cell-free system: structural properties and mechanism of inhibition, *Virology*, 179, 679, 1990.
185. **Gillard, S., Spehner, D., and Drillien, R.,** Mapping of a vaccinia virus host range sequence by insertion into the viral thymidine kinase gene, *J. Virol.*, 53, 316, 1985.
186. **Spehner, D., Gillard, S., Drillien, R., and Kirn, A.,** A cowpox virus gene required for multiplication in Chinese hamster ovary cells, *J. Virol.*, 62, 1297, 1988.
187. **Perkus, M. E., Goebel, S. J., Davis, S. W., Johnson, G. P., Limbach, K., Norton, E. K., and Paoletti, E.,** Vaccinia virus host range genes, *Virology*, 179, 276, 1990.
188. **Drillien, R. D., Spehner, D., and Kirn, A.,** Host range restriction of vaccinia virus in Chinese hamster ovary cells: relationship to shutoff of protein synthesis, *J. Virol.*, 28, 843, 1978.
189. **Hruby, D. E., Lynn, D. L., Condit, R. C., and Kates, J. R.,** Cellular differences in the molecular mechanisms of vaccinia virus host range restriction, *J. Gen. Virol.*, 47, 485, 1980.
190. **Drillien, R., Koehren, F., and Kirn, A.,** Host range deletion mutant of vaccinia virus defective in human cells, *Virology*, 111, 488, 1981.
191. **Njayou, M., Drillien, R., and Kirn, A.,** Characteristics of the inhibition of protein synthesis by vaccinia virus in nonpermissive Chinese hamster ovary cells, *Ann. Virol.*, 133E, 393, 1982.
192. **Cairns, J.,** The initiation of vaccinia infection, *Virology*, 11, 603, 1960.
193. **Harford, C., Hamlin, A., and Riders, E.,** Electron microscopic autoradiography of DNA synthesis in cell infected with vaccinia virus, *Exp. Cell Res.*, 42, 50, 1966.
194. **Kato, S., Kameyama, S., and Kamahora, J.,** Autoradiography with tritium-labeled thymidine of pox virus and human amnion cell system in tissue culture, *Biken J.*, 3, 135, 1960.
195. **Archard, L. C.,** Synthesis of full-length, virus genomic DNA by nuclei of vaccinia-infected HeLa cells, *J. Gen. Virol.*, 64, 2561, 1983.
196. **Gafford, L. G., Mitchell, E. B., and Randall, C. C.,** Virus-specific RNA and DNA in nuclei of cells infected with fowlpox virus, *Virology*, 69, 1, 1976.
197. **LaColla, P. and Weissbach, A.,** Vaccinia virus infection of HeLa cells. I. Synthesis of vaccinia DNA in host cell nuclei, *J. Virol.*, 15, 305, 1975.

198. **Minnigan, H. and Moyer, R. W.,** Intracellular localization of rabbit poxvirus nucleic acid within infected cells as determined by in situ hybridization, *J. Virol.*, 55, 634, 1985.
199. **Pennington, T. H. and Follett, E. A.,** Vaccinia virus replication in enucleated BSC-1 cells: particle production and synthesis of viral DNA and proteins, *J. Virol.*, 13, 488, 1974.
200. **Prescott, D. M., Kates, J., and Kirkpatrick, J. B.,** Replication of vaccinia virus DNA in enucleated L-cells, *J. Mol. Biol.*, 59, 505, 1971.
201. **Joklik, W. K.,** The intracellular uncoating of poxvirus DNA. I. The fate of radioactively-labeled rabbitpox virus, *J. Mol. Biol.*, 8, 263, 1964.
202. **Joklik, W. K.,** The intracellular uncoating of poxvirus DNA. II. The molecular basis of the uncoating process, *J. Mol. Biol.*, 8, 277, 1964.
203. **Zaslavsky, V.,** Uncoating of vaccinia virus, *J. Virol.*, 55, 352, 1985.
204. **Salzman, N. P.,** The rate of formation of vaccinia deoxyribonucleic acid and vaccinia virus, *Virology,* 10, 150, 1960.
205. **Joklik, W. K. and Becker, Y.,** The replication and coating of vaccinia DNA, *J. Mol. Biol.*, 10, 452, 1964.
206. **Baroudy, B. M., Venkatesan, S., and Moss, B.,** Structure and replication of vaccinia virus telomeres, *Cold Spring Harbor Symp. Quant. Biol.*, 47, 723, 1982.
207. **Rempel, R. E., Anderson, M. K., Evans, E., and Traktman, P.,** Temperature-sensitive vaccinia virus mutants identify a gene with an essential role in viral replication, *J. Virol.*, 64, 574, 1990.
208. **Traktman, P.,** The enzymology of poxvirus DNA replication, *Curr. Top. Microbiol. Immunol.*, 163, 93, 1990.
209. **Weir, J. P. and Moss, B.,** Nucleotide sequence of the vaccinia virus thymidine kinase gene and the nature of spontaneous frameshift mutations, *J. Virol.*, 46, 530, 1983.
210. **Hruby, D. E., Maki, R. A., Miller, D. B., and Ball, L. A.,** Fine structure analysis and nucleotide sequence of the vaccinia virus thymidine kinase gene, *Proc. Natl. Acad. Sci. U.S.A.*, 80, 3411, 1983.
211. **Boyle, D. B., Coupar, B. E. H., Gibbs, A. J., Seigman, L. J., and Both, G. W.,** Fowlpox virus thymidine kinase: nucleotide sequence and relationship to other thymidine kinases, *Virology,* 156, 355, 1987.
212. **Upton, C. and McFadden, G.,** Identification and nucleotide sequence of the thymidine kinase gene of Shope fibroma virus, *J. Virol.*, 60, 920, 1986.
213. **Hruby, D. E.,** Inhibition of vaccinia virus thymidine kinase by the distal products of its own metabolic pathway, *Virus Res.*, 2, 151, 1985.
214. **Smith, G. L., de Carlos, A., and Chan, Y. S.,** Vaccinia virus encodes a thymidylate kinase gene: sequence and transcriptional mapping, *Nucleic Acids Res.*, 17, 7581, 1989.
215. **Slabaugh, M. B., Roseman, N., Davis, R., and Mathews, C.,** Vaccinia virus-encoded ribonucleotide reductase: sequence conservation of the gene for the small subunit and its amplification in hydroxyurea-resistant mutants, *J. Virol.*, 62, 519, 1988.
216. **Tengelsen, L. A., Slabaugh, M. B., Bibler, J. K., and Hruby, D. E.,** Nucleotide sequence and molecular genetic analysis of the large subunit of ribonucleotide reductase encoded by vaccinia virus, *Virology,* 164, 121, 1988.
217. **Schmitt, J. F. C. and Stunnenberg, H. G.,** Sequence and transcriptional analysis of the vaccinia virus HindIII I fragment, *J. Virol.*, 62, 1889, 1988.
218. **Challberg, M. D. and Englund, P. T.,** Purification and properties of the deoxyribonucleic acid polymerase induced by vaccinia virus, *J. Biol. Chem.*, 254, 7812, 1979.
219. **Earl, P. L., Jones, E. V., and Moss, B.,** Homology between DNA polymerases of poxviruses, herpesviruses, and adenoviruses: nucleotide sequence of the vaccinia virus DNA polymerase gene, *Proc. Natl. Acad. Sci. U.S.A.*, 83, 3659, 1986.
220. **Binns, M. M., Stenzler, L., Tomley, F. M., Campbell, J., and Boursnell, M. E. G.,** Identification by a random sequencing strategy of the fowlpoxvirus DNA polymerase gene, its nucleotide sequence and comparison with other viral DNA polymerases, *Nucleic Acids Res.*, 15, 6563, 1987.

221. **DeFilippes, F. M.,** Site of the base change in the vaccinia virus DNA polymerase gene which confers aphidicolin resistance, *J. Virol.*, 63, 4060, 1989.
222. **Traktman, P., Kelvin, M., and Pacheco, S.,** Molecular genetic analysis of vaccinia virus DNA polymerase mutants, *J. Virol.*, 63, 841, 1989.
223. **Sambrook, J. and Shatkin, A. J.,** Polynucleotide ligase activity in cells infected with simian virus 40, polyoma virus or vaccinia virus, *J. Virol.*, 4, 719, 1969.
224. **Smith, G. L., Chan, Y. S., and Kerr, S. M.,** Transcriptional mapping and nucleotide sequence of a vaccinia virus gene encoding a polypeptide with extensive homology to DNA ligases, *Nucleic Acids Res.*, 17, 9051, 1989.
225. **Kerr, S. M. and Smith, G. L.,** Vaccinia virus encodes a polypeptide with DNA ligase activity, *Nucleic Acids Res.*, 17, 9039, 1989.
226. **Colinas, R. J., Goebel, S. J., Davis, S. W., Johnson, G. P., Norton, E. K., and Paoletti, E.,** A DNA ligase gene in the Copenhagen strain of vaccinia virus is nonessential for viral replication and recombination, *Virology*, 179, 267, 1990.
227. **Condit, R. C. and Niles, E. G.,** Orthopoxivrus genetics, *Curr. Top. Microbiol. Immunol.*, 169, 1, 1990.
228. **Evans, E. and Traktman, P.,** Molecular genetic analysis of a vaccinia virus gene with an essential role in DNA replication, *J. Virol.*, 61, 3152, 1987.
229. **Roseman, N. A. and Hruby, D. E.,** Nucleotide sequence and transcript organization of a region of the vaccinia virus genome which encodes a constitutively expressed gene required for DNA replication, *J. Virol.*, 61, 1398, 1987.
230. **Traktman, P., Anderson, M. K., and Rempel, R. E.,** Vaccinia virus encodes an essential gene with strong homology to protein kinases, *J. Biol. Chem.*, 264, 21458, 1989.
231. **Archard, L. C.,** De novo synthesis of two classes of DNA induced by vaccinia virus infection of HeLa cells, *J. Gen. Virol.*, 42, 223, 1979.
232. **Moyer, R. W. and Graves, R. L.,** The mechanism of cytoplasmic orthopoxvirus DNA replication, *Cell*, 27, 391, 1981.
233. **Moss, B., Winter, E., and Jones, E. V.,** Replication of vaccinia virus, in *Mechanics of DNA Replication and Recombination*, Cozzarelli, N., Ed., Alan R. Liss, New York, 1983, 449.
234. **Merchlinsky, M. and Moss, B.,** Resolution of vaccinia virus DNA concatemer junctions requires late gene expression, *J. Virol.*, 63, 1595, 1989.
235. **Merchlinsky, M., Garon, C., and Moss, B.,** Molecular cloning and sequence of the concatemer junction from vaccinia virus replicative DNA: viral nuclease cleavage sites in cruciform structures, *J. Mol. Biol.*, 199, 399, 1988.
236. **DeLange, A. M. and McFadden, G.,** The role of telomeres in poxvirus DNA replication, *Curr. Top. Microbiol. Immunol.*, 163, 71, 1989.
237. **Cavalier-Smith, T.,** Palindromic base sequences and replication of eukaryotic chromosomes, *Nature*, 250, 467, 1974.
238. **Bateman, A. J.,** Simplification of palindromic telomere theory, *Nature*, 253, 379, 1975.
239. **Esteban, M., Flores, L., and Holowczak, J. A.,** Model for vaccinia virus DNA replication, *Virology*, 83, 467, 1977.
240. **Esteban, M. and Holowczak, J. A.,** Replication of vaccinia DNA in mouse L cells. I. In vivo DNA synthesis, *Virology*, 78, 57, 1977.
241. **Esteban, M., Flores, L., and Holowczak, J. A.,** Topography of vaccinia virus DNA, *Virology*, 82, 163, 1977.
242. **Pogo, B. G. T. and O'Shea, M. T.,** The mode of replication of vaccinia virus DNA, *Virology*, 86, 1, 1978.
243. **Pogo, B. G. T., O'Shea, M., and Friemuth, P.,** Initiation and termination of vaccinia virus DNA replication, *Virology*, 108, 241, 1981.
244. **DeLange, A. M. and McFadden, G.,** Sequence-nonspecific replication of transfected plasmid DNA in poxvirus-infected cells, *Proc. Natl. Acad. Sci. U.S.A.*, 83, 614, 1986.

245. **Merchlinsky, M. and Moss, B.,** Sequence-independent replication and sequence-specific resolution of plasmids containing the vaccinia virus concatemer junction: requirements for early and late trans-acting factors, in *Vol. 6, Eukaryotic DNA Replication,* Kelly, T. and Stillman, B., Eds., Cold Spring Harbor Laboratory, Cold Spring Harbor, NY, 1988, 87.
246. **Esteban, M. and Holowczak, J. A.,** Replication of vaccinia DNA in mouse L cells. II. In vitro DNA synthesis in cytoplasmic extracts, *Virology,* 78, 76, 1977.
247. **Lambert, D. M. and Mageen, W. E.,** Characterization of vaccinia virus DNA in vitro, *Virology,* 79, 342, 1977.
248. **Pogo, B. G. T.,** Terminal cross-linking of vaccinia DNA strands by an in vitro system, *Virology,* 100, 339, 1980.
249. **Merchlinsky, M. and Moss, B.,** Resolution of linear minichromosomes with hairpin ends from circular plasmids containing vaccinia virus concatemer junctions, *Cell,* 45, 879, 1986.
250. **DeLange, A. M. and McFadden, G.,** Efficient resolution of replicated poxvirus telomeres to native hairpin structures requires two inverted symmetrical copies of a core target DNA sequence, *J. Virol.,* 61, 1957, 1987.
251. **McFadden, G., Stuart, D., Upton, C., Dickie, P., and Morgan, A. R.,** Replication and resolution of poxvirus telomeres, in *Eukaryotic DNA Replication,* Vol. 6, Kelly, T. and Stillman, B., Eds., Cold Spring Harbor Laboratory, Cold Spring Harbor, New York, 1988, 77.
252. **Merchlinsky, M. and Moss, B.,** Nucleotide sequence required for resolution of the concatemer junction of vaccinia virus DNA, *J. Virol.,* 63, 4354, 1989.
253. **Merchlinsky, M.,** Mutational analysis of the resolution sequence of vaccinia virus DNA — essential sequence consists of 2 separate AT-rich regions highly conserved among poxviruses, *J. Virol.,* 64, 5029, 1990.
254. **Merchlinsky, M.,** Intramolecular homologous recombination in cells infected with temperature-sensitive mutants of vaccinia virus, *J. Virol.,* 63, 2030, 1989.
255. **Merchlinsky, M.,** Resolution of poxvirus telomeres: processing of vaccinia virus concatemer junctions by conservative strand exchange, *J. Virol.,* 64, 3437, 1990.
256. **Shuman, S.,** Vaccinia DNA topoisomerase I promotes illegitimate recombination in *Escherichia coli, Proc. Natl. Acad. Sci. U.S.A.,* 86, 3489, 1989.
257. **Lakritz, N., Fogelsong, P. D., Reddy, M., Baum, S., Hurwitz, J., and Bauer, W. R.,** A vaccinia virus DNase preparation which cross-links superhelical DNA, *J. Virol.,* 53, 935, 1985.
258. **Stuart, D., Graham, K., Schreiber, M., Macaulay, C., and McFadden, G.,** The target DNA sequence for resolution of poxvirus replicative intermediates is an active late promoter, *J. Virol.,* 65, 61, 1991.
259. **Fenner, F. and Comben, B. M.,** Genetic studies with mammalian poxviruses. I. Demonstration of recombination between two strains of poxviruses, *Virology,* 5, 530, 1958.
260. **Strayer, D. S., Cabirac, G., Sell, S., and Lebowitz, J. L.,** Malignant rabbit fibroma virus: observations on the culture and histopathologic characteristics of a new virus-induced rabbit tumor, *J. Natl. Cancer Inst.,* 71, 105, 1983.
261. **Block, W., Upton, C., and McFadden, G.,** Tumorigenic poxviruses: genomic organization of malignant rabbit virus, a recombinant between Shope fibroma virus and myxoma virus, *Virology,* 140, 113, 1985.
262. **Moyer, R. W., Graves, R. L., and Rothe, C. T.,** The white pock (μ) mutants of rabbit poxvirus. III. The terminal DNA sequence duplication and transposition in rabbit poxvirus, *Cell,* 22, 545, 1980.
263. **McFadden, G. and Dales, S.,** Biogenesis of poxviruses: mirror image deletions in vaccinia virus DNA, *Cell,* 18, 101, 1979.
264. **Sam, C. K. and Dumbell, K. R.,** Expression of poxvirus DNA in coinfected cells and marker rescue of thermosensitive mutants by subgenomic fragments of DNA, *Ann. Virol.,* 132E, 135, 1981.

265. **Nakano, E., Panicali, D., and Paoletti, E.,** Molecular genetics of vaccinia virus: demonstration of marker rescue, *Proc. Natl. Acad. Sci. U.S.A.,* 79, 1593, 1982.
266. **Weir, J. P., Bajszar, G., and Moss, B.,** Mapping of the vaccinia virus thymidine kinase gene by marker rescue and by cell-free translation of selected mRNA, *Proc. Natl. Acad. Sci. U.S.A.,* 79, 1210, 1982.
267. **Ball, L. A.,** High-frequency homologous recombination in vaccinia virus DNA, *J. Virol.,* 61, 1788, 1987.
268. **Spyropoulos, D. D., Roberts, B. E., Panicali, D. L., and Cohen, L. K.,** Delineation of the viral products of recombination in vaccinia virus-infected cells, *J. Virol.,* 62, 1046, 1988.
269. **Evans, D. H., Stuart, D., and McFadden, G.,** High levels of genetic recombination among cotransfected plasmid DNAs in poxvirus-infected mammalian cells, *J. Virol.,* 62, 367, 1988.
270. **Dales, S. and Pogo, B. G. T.,** *Biology of Poxviruses,* Springer-Verlag, New York, 1981.
271. **Moss, B., Rosenblum, E. N., Katz, E., and Grimley, P. M.,** Rifampicin: a specific inhibitor of vaccinia virus assembly, *Nature,* 224, 1280, 1969.
272. **Grimley, P. M., Rosenblum, E. N., Mims, S. J., and Moss, B.,** Interruption by rifampin of an early stage in vaccinia virus morphogenesis: accumulation of membranes which are precursors of virus envelopes, *J. Virol.,* 6, 519, 1970.
273. **Nagayama, A., Pogo, B. G. T., and Dales, S.,** Biogenesis of vaccinia: separation of early stages from maturation by means of rifampicin, *Virology,* 40, 1039, 1970.
274. **Tartaglia, J., Piccini, A., and Paoletti, E.,** Vaccinia virus rifampicin-resistance locus specifies a late 63,000 Da gene product, *Virology,* 150, 45, 1986.
275. **Baldick, C. J. and Moss, B.,** Resistance of vaccinia virus to rifampicin conferred by a single nucleotide substitution near the predicted NH_2 terminus of a gene encoding an M_r 62,000 polypeptide, *Virology,* 156, 138, 1987.
276. **Miner, J. N. and Hruby, D. E.,** Rifampicin prevents virosome localization of L65, an essential vaccinia virus polypeptide, *Virology,* 170, 227, 1989.
277. **Morgan, C.,** The insertion of DNA into vaccinia virus, *Science,* 193, 591, 1976.
278. **Katz, E. and Moss, B.,** Formation of a vaccinia virus structural polypeptide from a higher molecular weight precursor: inhibition by rifampicin, *Proc. Natl. Acad. Sci. U.S.A.,* 66, 677, 1970.
279. **Morgan, C.,** Vaccinia virus reexamined: development and release, *Virology,* 73, 43, 1976.
280. **Rodriguez, J. F. and Smith, G. L.,** IPTG-dependent vaccinia virus: identification of a virus protein enabling virion envelopment by Golgi membrane and egress, *Nucleic Acids Res.,* 18, 5347, 1990.
281. **Payne, L. G. and Kristensson, K.,** Effect of glycosylation inhibitors on the release of enveloped vaccinia virus, *J. Virol.,* 41, 367, 1982.
282. **Payne, L. G. and Kristenson, K.,** The effect of cytochalasin D and monensin on enveloped vaccinia virus release, *Arch. Virol.,* 74, 11, 1982.
283. **Payne, L. G. and Kristensson, K.,** Mechanism of vaccinia virus release and its specific inhibition by N_1-isonicotinoyl-N_2-3-methyl-4-chlorobenzoylhydrazine, *J. Virol.,* 32, 614, 1979.
284. **Kato, N., Eggers, H. J., and Rolly, H.,** Inhibition of release of vaccinia virus by N_1-isonicotinoyl-N_2-3-methyl-4-chlorobenzoylhydrazine, *J. Exp. Med.,* 129, 795, 1979.
285. **Venkatesan, S., Gershowitz, A., and Moss, B.,** Complete nucleotide sequences of two adjacent early vaccinia virus genes located within the inverted terminal repetition, *J. Virol.,* 44, 637, 1982.
286. **Blomquist, M. C., Hunt, L. T., and Barker, W. C.,** Vaccinia virus 19-kilodalton protein: relationship to several mammalian proteins, including two growth factors, *Proc. Natl. Acad. Sci. U.S.A.,* 81, 7363, 1984.

287. **Brown, J. P., Twardzik, D. R., Marquardt, H., and Todaro, G. J.,** Vaccinia virus encodes a polypeptide homologous to epidermal growth factor and transforming growth factor, *Nature,* 313, 491, 1985.
288. **Reisner, A. H.,** Similarity between the vaccinia virus 19K early protein and epidermal growth factor, *Nature,* 313, 801, 1985.
289. **Twardzik, D. R., Brown, J. P., Ranchalis, J. E., Todaro, G. J., and Moss, B.,** Vaccinia virus-infected cells release a novel polypeptide functionally related to transforming and epidermal growth factors, *Proc. Natl. Acad. Sci. U.S.A.,* 82, 5300, 1985.
290. **Stroobant, P., Rice, A. P., Gullick, W. J., Cheng, D. J., Kerr, I. M., and Waterfield, M. D.,** Purification and characterization of vaccinia virus growth factor, *Cell,* 42, 383, 1985.
291. **King, C. S., Cooper, J. A., Moss, B., and Twardzik, D. R.,** Vaccinia virus growth factor stimulates tyrosine protein kinase activity of A431 cell epidermal growth factor receptors, *Mol. Cell Biol.,* 6, 332, 1986.
292. **Chang, W., Lim, J. G., Hellstrom, I., and Gentry, L. E.,** Characterization of vaccinia virus growth factor biosynthetic pathway with an antipeptide antiserum, *J. Virol.,* 62, 1080, 1988.
293. **Buller, R. M. L., Chakrabarti, S., Moss, B., and Frederickson, T.,** Cell proliferative response to vaccinia virus is mediated by VGF, *Virology,* 164, 182, 1988.
294. **Chang, W., Upton, C., Hu, S.-L., Purchio, A. F., and McFadden, G.,** The genome of Shope fibroma virus, a tumorigenic poxvirus, contains a growth factor gene with sequence similarity to those encoding epidermal growth factor and transforming growth factor alpha, *Mol. Cell. Biol.,* 7, 535, 1987.
295. **Chang, W., Macauley, C., Hu, S.-L., Tam, J. P., and McFadden, G.,** Tumorigenic poxviruses: characterization of the expression of a epidermal growth factor related gene in Shope fibroma virus, *Virology,* 179, 926, 1990.
296. **Whitaker-Dowling, P. and Youngner, J. S.,** Characterization of a specific kinase inhibitor factor produced by vaccinia virus which inhibits the interferon-induced protein kinase, *Virology,* 137, 171, 1984.
297. **Rice, A. P. and Kerr, I. M.,** Interferon-mediated, double-stranded RNA-dependent protein kinase is inhibited in extracts from vaccinia virus-infected cells, *J. Virol.,* 50, 229, 1984.
298. **Paez, E. and Esteban, M.,** Resistance of vaccinia virus to interferon is related to an interference phenomenon between the virus and the interferon system, *Virology,* 134, 12, 1984.
299. **Akkaraju, G. R., Whitaker-Dowling, P., Youngner, J. S., and Jagus, R.,** Vaccinia specific kinase inhibitor factor prevents translational inhibition by double-stranded RNA in rabbit reticulocyte lysate, *J. Biol. Chem.,* 264, 10321, 1989.
300. **Hovanessian, A. G., Galabru, J., Meurs, E., Buffet-Janvresse, C., Svab, J., and Robert, N.,** Rapid decrease in the levels of the double-stranded RNA-dependent protein kinase during virus infections, *Virology,* 159, 126, 1987.
301. **Paez, E. and Esteban, M.,** Nature and mode of action of vaccinia virus products that block activation of the interferon-mediated ppp(A2′p)nA-synthetase, *Virology,* 134, 29, 1984.
302. **Rice, A. P., Roberts, W. K., and Kerr, I. M.,** 2-5A accumulates to high levels in interferon-treated, vaccinia virus-infected cells in the absence of any inhibition of virus replication, *J. Virol.,* 50, 220, 1984.
303. **Kotwal, G. J. and Moss, B.,** Vaccinia virus encodes a secretory polypeptide structurally related to complement control proteins, *Nature,* 335, 176, 1988.
304. **Kotwal, G. J., Isaacs, S. N., Mckenzie, R., Frank, M. M., and Moss, B.,** Inhibition of the complement cascade by the major secretory protein of vaccinia virus, *Science,* 250, 827, 1990.

305. **Palumbo, G. J., Pickup, D. J., Frederickson, T. N., McIntyre, L. J., and Buller, R. M. L.,** Inhibition of an inflammatory response is mediated by 38-kDa protein of cowpox virus, *Virology,* 172, 262, 1989.
306. **Boursnell, M. E. G., Foulds, I. J., Campbell, J. I., and Binns, M. M.,** Non-essential genes in the vaccinia virus HindIII K fragment: a gene related to serine protease inhibitors and a gene related to the 37K vaccinia virus major envelope antigen, *J. Gen. Virol.,* 69, 2995, 1988.
307. **Kotwal, G. J. and Moss, B.,** Vaccinia virus encodes two proteins that are structurally related to members of the plasma serine protease inhibitor superfamily, *J. Virol.,* 63, 600, 1989.
308. **Smith, G. L., Howard, S. T., and Chan, T. S.,** Vaccinia virus encodes a family of genes with homology to serine protease inhibitors, *J. Gen. Virol.,* 70, 2333, 1989.
309. **Zhou, J., Crawford, L., McLean, L., Sun, X. Y., Stanley, M., Almond, N., and Smith, G. L.,** Increased antibody responses to human papillomavirus type-16 L1 protein expressed by recombinant vaccinia virus lacking serine protease inhibitor genes, *J. Gen. Virol.,* 71, 2185, 1990.

Chapter 3

THE DESIGN, CONSTRUCTION, AND USE OF VACCINIA VIRUS RECOMBINANTS

Geoffrey L. Smith and Michael Mackett

TABLE OF CONTENTS

I. INTRODUCTION

Recombinant vaccinia viruses that express genes from other organisms were first described nearly 10 years ago.[1,2] The principal interest in these viruses stemmed initially from their potential use as new live vaccines against a broad range of infectious agents.[3–7] Subsequently, vaccinia has become a popular and widely used laboratory expresion system. This popularity derives from several properties of recombinant vaccinia viruses, such as the retention of infectivity and ability to grow high titer virus stocks (10^{10} plaque forming units, pfu, per milliliter), the virion thermostability, the large capacity for foreign DNA,[8] the rapid and high level gene expression, and the broad host range *in vivo* and *in vitro*. Vaccinia virus recombinants have pioneered the principle of constructing genetically engineered live recombinant vaccines, a concept that has subsequently been applied to other poxviruses (described elsewhere in this book) and to a broad range of other organisms, e.g., poliovirus,[9] varicella zoster virus,[10] Salmonella species,[11] and recently bacillus Calmette-Guérin (BCG).[12]

Introduction of foreign DNA into vaccinia virus is hindered by the large size of the vaccinia genome (190 kb) and the fact that, unlike other DNA viruses, the naked virus DNA is noninfectious. Consequently, simple restriction endonuclease cleavage of virus DNA, religation with a foreign gene, and transfection into susceptible cells to form recombinant virus is not practicable. Instead, recombinant virus genomes are formed by homologous recombination within virus infected cells by using techniques developed for marker rescue.[13–15] Expression of the foreign gene is achieved either by engineering its protein coding sequences downstream of a defined and translocated vaccinia virus promoter[1,3] or by inserting the foreign gene close to an endogenous promoter at its natural locus within the vaccinia genome.[2,6] The use of vaccinia promoters is necessary because the virus encodes its own DNA-dependent RNA polymerase,[16,17] which does not recognize promoters from sources other than poxviruses and the distantly related African swine fever virus.

II. BASIC PROCEDURE

The large size of the virus genome and the noninfectious nature of poxvirus DNA make it necessary to construct recombinant viruses by a two-stage procedure. This involves construction of a plasmid in bacteria followed by transfer of the foreign gene from the plasmid into the virus by homologous recombination in infected cells (Figure 1). Within the plasmid, the standard arrangement is to position the protein coding region of the foreign gene downstream of a vaccinia virus promoter so that the initiating methionine will become the first AUG codon of the RNA transcript. The promoter foreign gene hybrid is flanked within the plasmid by vaccinia DNA derived from a genomic locus that is nonessential for virus replication. This DNA directs insertion of the foreign gene into that nonessential site within the virus genome. The rapid construction of recombinant vaccinia viruses has been aided by the availability of plasmids, called insertion vectors, which possess a defined vaccinia virus promoter with unique downstream cloning sites and flanking, nonessential virus DNA[18–23] (Figures 2 to 5). A foreign gene of choice may be easily cloned into such vectors and the derivative plasmid transfected into vaccinia virus infected cells. Within these cells, homologous recombination occurs between the sequences flanking the promoter/foreign gene hybrid of the plasmid and the corresponding sequences of the infecting vaccinia genome so that a recombinant genome containing the foreign DNA is formed. This is replicated and packaged into infectious progeny virus. Overall, this process is inefficient and the vast majority of progeny virus is parental (>99%). Consequently, methods are needed to select or detect the recombinant viruses (see Section V below).

Technically, the usual procedure is to infect cells at low multiplicity (0.01 to 0.1 pfu/cell) and then to introduce the plasmid DNA (2 $\mu g/10^6$ cells) into these cells 2 h after infection by transfection as calcium phosphate precipitated DNA. Alternatively, lipofection or electroporation may be used to insert the DNA into infected cells. Recombinant viruses are isolated by harvesting the transfected cells 2 d later, releasing the virus by freeze/thawing, and plating this onto fresh cell monolayers with suitable drugs to select or identify the recombinant viruses. The frequency of recombinant viruses has been reported to be increased by using a low multiplicity of infection,[15] by including 1 μg of intact virus DNA with the plasmid, and by leaving the infected monolayers

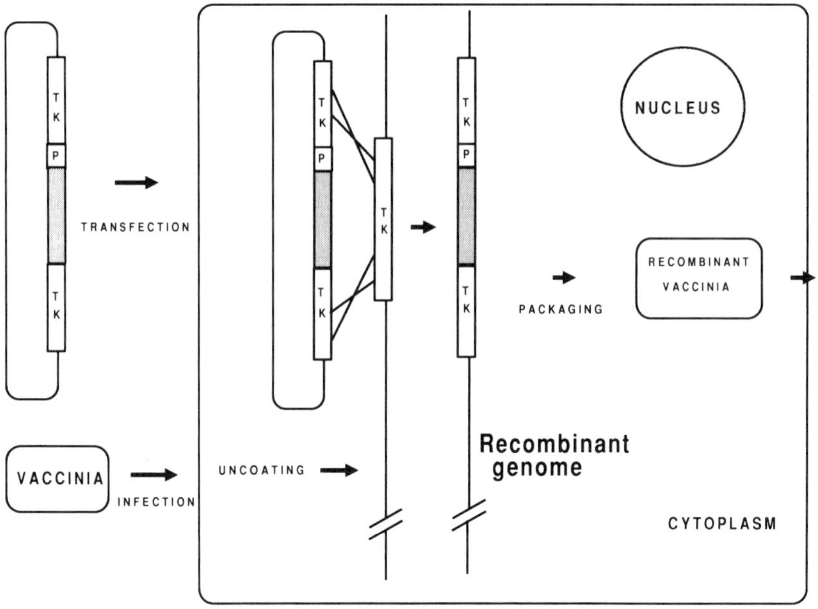

FIGURE 1. Schematic representation of the generation of vaccinia virus recombinants. A plasmid containing a foreign gene under control of a vaccinia virus promotor (P) and flanked by vaccinia virus DNA, in this case from the TK gene, is introduced by transfection into vaccinia virus infected cells. The foreign gene and promoter are inserted into the virus genome by homologous recombination and the resulting recombinant DNA genome is packaged to form live recombinant vaccinia virus.

for a full 48 h after infection.[24] If virus DNA is used it should be from the parental virus used for infection. Its inclusion is beneficial presumably because not all cells are transfectable, but those that are take up both virus DNA and plasmid enabling increased recombination frequencies. The use of single stranded insertion vectors based on bacteriophage M13 is reported to produce recombinant viruses at a higher frequency than those obtained with plasmids.[25,26] This method is limited by the size of inserts that are stable in M13.

III. DESIGN OF PLASMID INSERTION VECTORS

The plasmid insertion vector determines the site within the virus genome at which foreign DNA is integrated, by the choice of nonessential flanking DNA, and the temporal and quantitative regulation of that gene, by the type of vaccinia virus promoter selected. Multiple sites for insertion into the virus genome and many different promoters have been used (see Tables 1 and 2). Some of the options available are described.

A. CHOICE OF PROMOTER

Vaccinia virus transcriptional regulation is described in detail in Chapter

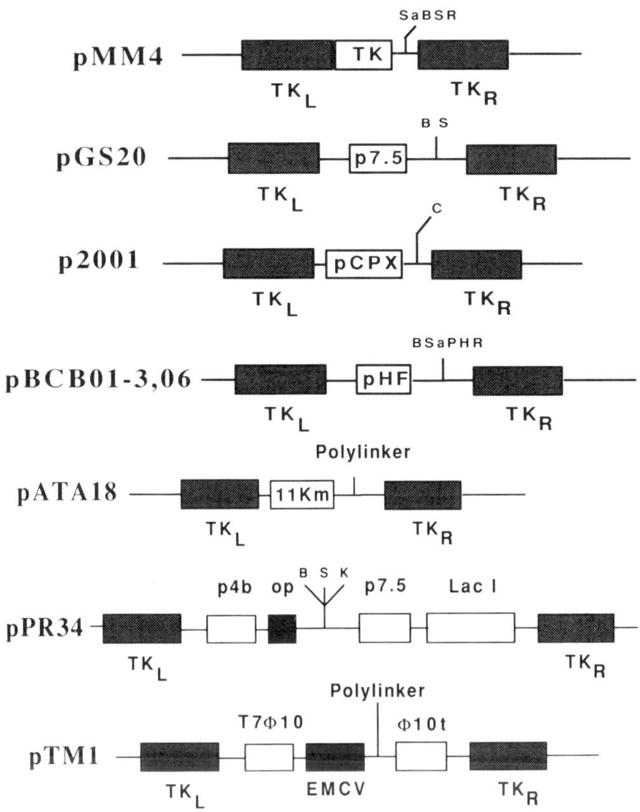

FIGURE 2. Plasmid vectors for the insertion and expression of foreign genes in the TK locus of vaccinia virus. Plasmid sequences are indicated by straight lines on either side of the TK gene. All vectors contain flanking vaccinia sequences derived from the left (TKL) and right (TKR) sides of the TK gene, an ampicillin resistance gene, and an *E. coli* origin of replication. pMM4 and pGS20 contain the TK or 7.5K promoters, respectivly.[18] p2001 contains the cowpox virus A type inclusion body gene promoter (pCPX).[45] pBCB01-03,06 contains a vaccinia virus promoter (pHF) derived from adjacent to the *Bam*HI site in *Hin*dIII F restriction fragment.[19] pATA18 contains a modified 11K promoter (11Km).[39] Unique restriction sites for the insertion of foreign genes are Sa, *Sal*I; B, *Bam*HI; S, *Sma*I; R, *Eco*RI; C, *Cla*I; P, *Pst*I; H, *Hin*dIII; K, *Kpn*I. The polylinker of pATA18 is derived from pUC18. pPR34 allows IPTG inducible gene expression; it contains the vaccinia 4b promoter upstream of a 21 bp *E. coli* lac operator (op) followed by unique restriction sites and the *E. coli* lac I repressor gene driven by the vaccinia virus 7.5K promoter.[41] pTM1 allows high level expression using the T7 RNA polymerase; it contains the T7 RNA polymerase promoter (T7ϕ10) and terminator (ϕ10t) flanking the 5′ noncoding region of the EMCV genome and a restriction site polylinker.[140] Transcription of all the promoters is from left to right.

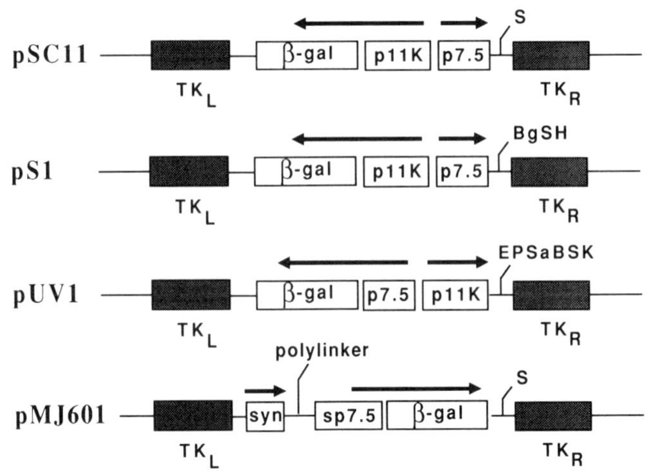

FIGURE 3. Plasmid vectors for the coinsertion and expression of β-gal and foreign genes in vaccinia virus. Plasmids pSC11,[20] pS1 (a derivative of pSC11),[198] pUV1,[21] and pMJ601[23] all contain flanking sequences from the left (TKL) and right (TKR) sides of the TK gene, an ampicillin resistance gene and *E. coli* origin of replication. Arrows indicate the direction of the transcription of the vaccinia promoters p11K and p7.5, and a synthetic vaccinia promoter, syn.[23] Plasmid sequences are indicated by the straight lines on either side of the vaccinia TK gene. Restriction sites are indicated by: S, *Sma*I; Bg, *Bgl*II; H, *Hin*dIII; E, *Eco*RI; P, *Pst*I; Sa, *Sal*I; and K, *Kpn*I.

2 of this book.[27] For construction of vaccinia virus recombinants, the most important concept is that the virus DNA-dependent RNA polymerase does not recognize promoters from other species and therefore foreign genes must be linked to vaccinia virus promoters to ensure expression. This was first shown with insertion of the herpes simplex virus thymidine kinase (HSV TK) gene into vaccinia virus and selection of recombinant viruses in the presence of methotrexate.[1] Recombinants could not be isolated under selective pressure unless the HSV gene was correctly orientated with respect to the 7.5K vaccinia virus promoter.

Four types of vaccinia virus promoters have been described: early, intermediate, late, and constitutive. Early promoters are defined as those active immediately after infection and are transcribed by the virion-associated RNA polymerase. Intermediate promoters are transcribed from virus DNA only after the onset of DNA replication,[28] but may be expressed from a plasmid transfected into infected cells in the absence of DNA replication.[28,29] The proteins necessary for their expression are, therefore, present before DNA replication, but cannot transcribe intermediate promoters from input virus DNA until after it has been replicated. Interestingly, three intermediate genes encode transcription factors required for late gene expression.[29] Late promoters are expressed after DNA replication has commenced and require prior expression of early and intermediate genes. Constitutive promoters are active

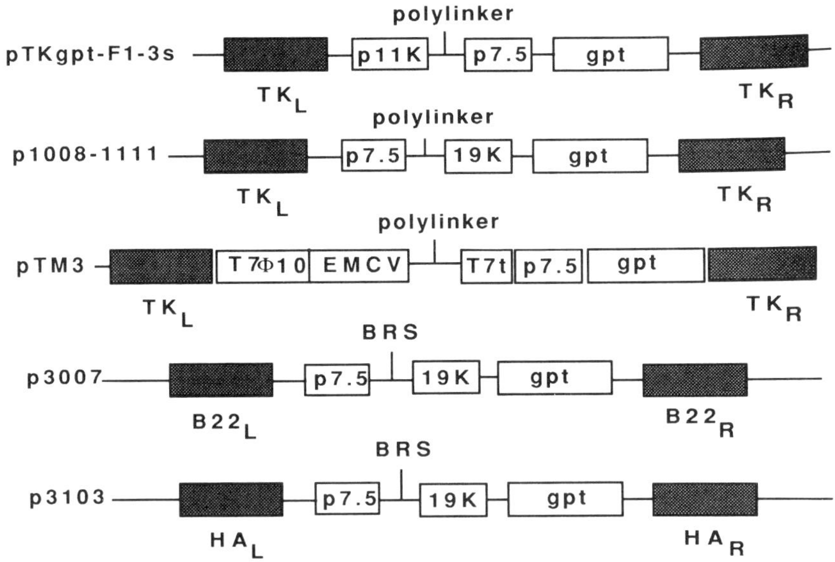

FIGURE 4. Plasmid vectors for the coinsertion and expression of *E. coli* guanine phosphoribosyltransferase (gpt) and foreign genes in vaccinia virus; pTKgpt-F1-3s;[22] p1008-1111;[199] p1008 contains the 7.5K promoter lacking a translational initiation codon adjacent to the polylinker restriction sites *Bam*HI, *Sma*I, *Hin*dIII, and *Eco*RI. p1109-1111 contain the same polylinker as p1108 and an ATG translational initiation codon. They differ from each other in that in each case the ATG is displaced by one base pair relative to the downstream restriction sites. pTM3;[140] p3007 and p3103.[199] All plasmids contain the ampicillin resistance gene and *E. coli* origin of replication. Plasmid sequences are indicated by the straight lines on either side of the vaccinia sequences from the TK gene (TK L-left or R-right), the hemagglutinin gene (HA L-left or R-right) or the serpin gene B22R (formerly called B24R,[200] but renamed 22R[89]) (B22 L-left or R-right). S, *Sma*I; R, *Eco*RI; B, *Bam*HI. All transcription from the vaccinia promoters 7.5K, 11K, and 19K and the bacteriophage T7ϕ10 promoter is from left to right.

before and after DNA replication and contain distinct early and late RNA start sites. Detailed mutagenesis studies for early,[30,31] late,[32–34] constitutive,[35] and intermediate promoters[36] have been described. A common feature of these promoters is the small size (about 30 bp) of the minimal region required for transcription at the normal level and from the correct initiation site. Consequently only short promoter sequences are needed for expression of foreign genes from recombinant vaccinia viruses.

The choice of promoter naturally depends upon the purpose for which the recombinant vaccinia virus is required. For high level expression the best natural vaccinia promoters derive from late genes encoding the major structural proteins of the virus particle, for instance the 11 kDa[20,37–39] and the 4b proteins.[34,40,41] Promoters from other late genes such as those encoding the 28-kDa precursor of the 25-kDa virus core polypeptide have given lower levels of expression.[32,42] Alternatively, increased expression levels have been

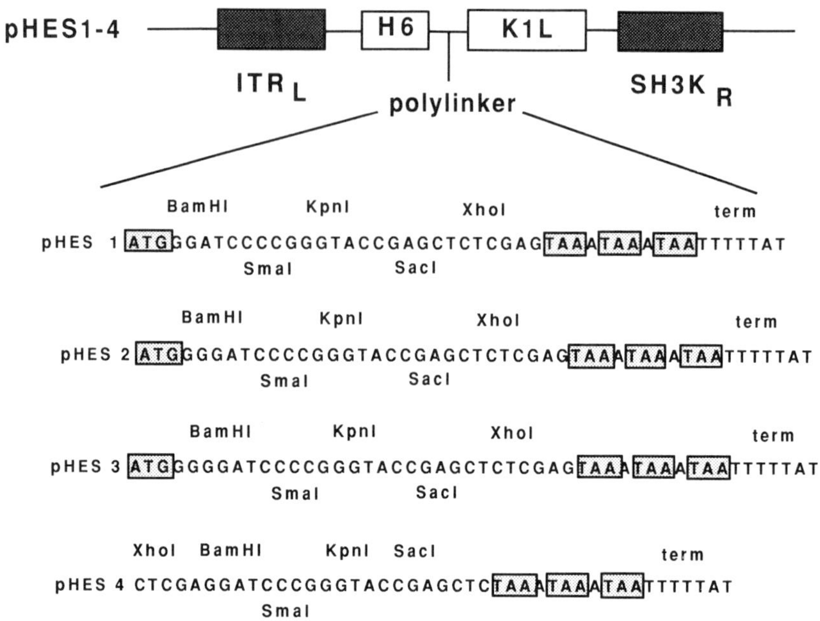

FIGURE 5. Plasmid vector enabling host range selection of foreign genes in vaccinia virus. Plasmids pHES1-4[96] allow expression from a synthetic early/late promoter (H6) of genes inserted into the polylinker region. pHES1-3 contain a single ATG translational initiation signal (boxed) displaced by a single base pair from each other, whereas pHES4 lacks an ATG. All plasmids contain a translational termination codon in all three frames (boxed). The host range gene K1L is indicated and recombination takes place between the inverted terminal repeat sequences (ITR L) on the left side of the plasmid or sequences from the *Hin*dIII K fragment (SH3K R) on the right. The direction of transcription of the H6 promoter is from left to right.

obtained with promoters derived from other poxviruses and by construction of synthetic promoters based on exhaustive mutagenesis data. Many poxviruses produce cytoplasmic A type inclusion bodies which are composed predominantly of a single virus polypeptide that surrounds new virus particles. These are functionally similar to the baculovirus polyhedrin protein. The gene encoding this protein has been identified and sequenced in cowpox[43–45] and an entomopoxvirus *Choristoneura biennis*.[46] Use of the promoter from either gene has given high level expression in vaccinia virus infected cells and demonstrates the functional compatibility of promoters from vaccinia virus, cowpox, and the distantly related insect poxviruses. A detailed mutagenesis study of a late vaccinia virus promoter[33] identified mutations which increased as well as decreased expression. This information was used to construct synthetic promoters with enhanced activity.[23] For optimal expression, the distance between the 5′ end of the mRNA and the AUG codon initiating the open reading frame should be minimized, and all other AUG codons should be removed from this region. Additionally, if the gene is to be expressed

TABLE 1
Promoters Used to Drive Foreign Gene Expression in Vaccinia Virus

Promoter	Ref.[a]
Early	
42K	49
19K	52
25K	51
7.5K derivatives	31
Early/Late	
7.5K	1
Synthetic H6	96
Late	
11K	20
11K derivatives	37–39
Synthetic (11K derived)	23, 33
28K	42
4b	34, 41
Cowpox Virus A type inclusion body	45
Other	
T7ϕ10	125
T7ϕ10-EMCV hyrbrid	128
4b-lac operator hybrid	41, 131
11K-lac operator hybrid	130

[a] References are not intended to be exhaustive. The example given is either representative or the initial description of the use of that particular promoter.

early during infection, the early transcriptional termination signal TTTTTNT[47,48] should be eliminated from the protein coding region. Another way of obtaining high levels of gene expression has been to use a hybrid system based on the bacteriophage T7 RNA polymerase (see Section VI).

Whether late promoters are intrinsically stronger than early promoters or whether the higher expression derives solely from increased copies of virus DNA template is unclear. Some early promoters that have been used are those of the TK gene:[3,18] the 42 kDa protein encoded within the inverted terminal repeat,[49] a serine protease inhibitor,[50] a 25 kDa protein,[51] the 19 kDa precursor of the vaccinia virus growth factor,[52] and those from within the *Hin*dIII F fragment.[6] Figure 2 shows a selection of plasmid vectors using different promoters designed for expression of foreign genes from within the vaccinia virus TK gene locus.

The first vaccinia virus promoter (p7.5K) to be functionally defined[1,35,53] and extensively used for expression of foreign genes derives from a gene which maps within the inverted terminal repeat of the vaccinia virus genome.[54]

TABLE 2
Sites for Insertion of Foreign DNA Within the Vaccinia Virus Genome

Gene name[a]	Function	Ref.
C23L-K1L		68, 96
C13L	7.5K gene[b]	67
C11R	Vaccinia growth factor[c]	65
C3L	Complement control protein	196
N1L	13.8K virokine	66
M1L		62
M2L		62
K1L	Human host range protein	62
K2L	Serine protease inhibitor	62, 76
K4L		62
K7R		62
F1L		62
F4L	Ribonucleotide reductase (small subunit)	62
F9L		62
E1L		62
I4L	Ribonucleotide reductase (large subunit)	70
J2R	Thymidine kinase[d]	1
D8L	35 kDa membrane protein	74
A26L/A27L[e]	A type inclusion body protein	45
A40R	Lectin-like glycoprotein	80
A43R	Glycoprotein	80a
A44L	3-β-hydroxy steroid dehydrogenase	73
A48R	Thymidylate kinase	69
A50R	DNA ligase	71, 72
56R	Hemagglutinin	77, 79
B5R	Complement control protein	81
	Host range	120
B12R	Protein kinase related	82
B13R[e]	Serine protease inhibitor	75
B22R[f]	Serine protease inhibitor	75
B15R-B29R		68

[a] Gene names according to the complete sequence of vaccinia virus strain Copenhagen.[197] The names indicate the *Hin*dIII restriction fragment to which the gene maps (fragments are labeled alphabetically in order of decreasing size) and the direction of transcription (R or L).

[b] In some virus strains (e.g. WR) this gene is located within the inverted terminal repeat.

[c] In some virus strains this is a 35K secreted protein and maps within the inverted terminal repeat.

[d] The majority of recombinants have been generated by insertion into this gene.

[e] In the Copenhagen strain this gene is interrupted.

[f] In the Copenhagen strain this gene is present at the left end of the genome and is named C12L.

The promoter is constitutively expressed and contains distinct early and late RNA start sites,[35] the latter being located approximately 50 bp upstream. Translation of either early or late transcripts from this promoter yields the same polypeptide due to the absence of translation initiation codons between the different transcriptional start sites. Several plasmid expression vectors which utilize this promoter have been described[18–20] and widely used. The constitutive expression from this promoter was fortunate because this provided an enhanced level of gene expression and also because in some cases expression of a gene either early or late during infection has been essential.

Early gene expression has also been important for some applications. For instance, recognition of specific foreign peptide epitopes on vaccinia virus infected cells by cytotoxic T lymphocytes (CTL) may occur when expression is early but not late during infection.[55,56] Thus, to identify which foreign proteins are recognized by CTL, or for vaccine purposes where a T-cell response is important, early gene expression should be included.

Late transcription has been useful for expression of genes which contain the early transcriptional termination sequence TTTTTNT[47,48] within their protein coding sequences. For instance, the human immunodeficiency virus (HIV) gp 160 gene and the human papillomavirus (HPV) type 16 L1 gene both contain this cryptic sequence. Expression of the HPV L1 gene was greatly increased by either using a late promoter[40] or by removing the TTTTTNT motif by oligonucleotide-directed mutagenesis while retaining the correct translational reading frame.[49] Likewise, expression of HIV gp160 from an early promoter was improved considerably by mutagenesis to remove the cryptic motif. This also enhanced the immune responses to HIV gp160 in vaccinated animals.[57]

B. SITES FOR INSERTION OF FOREIGN DNA

Plasmids designed to transfer foreign genetic information into vaccinia contain an appropriate nonessential region of the virus genome. A wide variety of insertion sites for foreign genes have been identified, many of which are located near the genomic termini (Table 2). These regions were selected because poxvirus genetics[58] and comparative genome analyses[59] established that these regions are most variable and may be deleted or transposed to the opposite end of the genome without loss of virus growth in tissue culture. Therefore, the genes encoded here are nonessential. Large deletions from the left end of the vaccinia virus genome have been described[60–63] and many sites within this region have been used for foreign DNA insertion.[1,2,6,62,64–67] Similarly, large deletions occur near the right end of the genome.[68] By combining deletions from both termini, a total of 55 open reading frames have been deleted from the Copenhagen strain of virus without loss of infectivity.[68]

The central region of poxvirus genomes are more highly conserved and many of the encoded genes are essential for virus replication. Nonetheless, several nonessential genes have been identified and used as insertion sites for foreign genes. The site most frequently used is the TK gene,[1,3,18] since this enables genetic selection of recombinant viruses (see below). Several other genes that encode enzymes are nonessential *in vitro,* e.g., thymidylate kinase,[69] ribonucleotide reductase,[70] DNA ligase,[71,72] and 3β-hydroxy steroid dehydrogenase.[73] Other genes nonessential for replication are located throughout the genome and include gene D8L (which encodes a 35 kDa membrane protein of the virus particle),[74] three serine protease inhibitor (serpin) genes,[75,76] the virus hemagglutinin (HA),[77–79] the A type inclusion body gene,[45] genes SalF2R and SalF5R (which potentially encode glycoproteins),[80,80a] gene B5R (which encodes a complement control protein homologue),[81] and gene B12R (which encodes a protein related to protein kinases).[82] These sites are summarized in Table 2.

Although several of the above genes are nonessential for virus growth *in vitro,* they are important for virus virulence *in vivo,* since their insertional inactivation causes attenuation in animals. This was first noticed with TK,[83–85] but is true for other enzymes, too, such as ribonucleotide reductase,[70] DNA ligase,[86] and thymidylate kinase.[87] Deletion of the vaccinia growth factor (VGF) genes resulted in a virus that required 10^4-fold more virus than wild type (WT) to give the same sized lesion on rabbit skin.[65] Similarly, insertion into the "virokine" gene N1L[66] reduces by 10^5 the amount of virus that is lethal by intracranial inoculation in mice. Despite this marked attenuation, loss of this gene has little effect on the antigenicity of the virus. Another site that causes some attenuation is the HA gene.[78] However, recombinant viruses with the foreign gene inserted into this site are still able to induce antibody and T-cell responses to the foreign gene product.[88] Alternative sites for immunization studies are either of the serpin genes B13R or B22R (renamed from B24R[89]). Insertion of the HPV type 16 L1 gene into these sites led to an increase in antibody response to this protein.[75]

Because of the large number of sites available it is possible to insert and express several foreign genes in the same virus and to obtain simultaneous immune responses to the expressed antigens in inoculated animals.[64] WT vaccinia virus can accept at least 25 kb of foreign DNA without compensating deletions[8] and at least 32.7 and 14.9 kb of DNA can be deleted from the left and right hand ends of the virus genome, respectively.[68] Therefore, in theory, it should be possible to insert over 70 kb of foreign genetic material into a viable vaccinia virus.

IV. VIRUS STRAIN

Many vaccinia strains have been used as vectors for foreign genes. The most commonly used strains are Western Reserve (WR), Copenhagen, Wyeth, Lister, and attenuated Lister derivatives. WR is a laboratory strain chosen for its vigorous growth *in vivo* and *in vitro,* which was derived from the New York City Board of Health vaccine strain by passage in mouse brain. Copenhagen also grows well in cultured cells and was for some time used as a vaccine strain, although the higher frequency of vaccine-related complications led to its replacement with the milder Lister or Wyeth strains in some European countries. The Wyeth vaccine strain, like WR, derives from the New York City Board of Health strain and was one of the safest smallpox vaccines used for widespread immunization (see Chapters 1 and 7 in this volume). The Lister strain was widely used as a vaccine in Europe and has been used to express foreign antigens.[67] Attenuated Lister derivatives isolated in Japan (LC16m0 and LC16m8) are also being evaluated as carriers for vaccine antigens.[90–93] Recombinant viruses based upon Tian Tan (the Chinese temple of heaven vaccine strain),[51,94] various deletion mutants of the WR strain, e.g., S-variant,[2,60] the highly attenuated derivative 48-7,[95] and vP293 (a mutant lacking 21.7 kb at the left hand end of the WR genome)[96] have been reported. Similarly, the LIVP strain (possibly derived from Lister),[97] IHD-J,[98] and temperature sensitive (ts) mutants have been used.[99] It should be noted that the strain of virus used can have a significant effect on the outcome of an experiment. For instance, a recombinant vaccinia virus based on the WR strain and expressing the Epstein-Barr Virus (EBV) glycoprotein gp340 protected cotton top tamarins against an EBV-induced lymphoma,[100] whereas a recombinant generated in an identical manner using the Wyeth vaccine strain as parent failed to elicit any protective effect.

V. SELECTION PROTOCOLS

Methods to distinguish or select the recombinant from parental virus are needed because greater than 99% of progeny virus from a transfection experiment is of parental type. The different methods available are described below. Detailed protocols for some of these have been described.[101,102]

A. *IN SITU* HYBRIDIZATION

Recombinant virus plaques containing foreign DNA can be identified by *in situ* hybridization using a radiolabeled DNA probe.[2,6,8] Virus harvested from transfected cells is plated onto new monolayers to produce thousands of plaques. These are transferred to a nitrocellulose membrane and after making a replica filter the original is hybridized with the foreign gene probe. After autoradiography, virus corresponding to plaques showing positive hybridization is recovered from the replica filter and further plaque purified. This is a useful procedure which allows the identification of recombinants with DNA inserted into any nonessential locus, but could not be used to identify viruses containing modifications to endogenous genes. Its utility has been rather superceded by the development of dominant selectable markers (see below).

B. CONDITIONAL LETHAL PARENT VIRUSES

One approach designed to reduce the background of parental virus from transfected cells is the use of a conditional lethal parent vaccinia strain. Either ts[103–1-05] or drug-dependent[105] viruses have been used. Cells are infected with the defective virus under nonpermissive conditions and then transfected with the WT vaccinia DNA and the plasmid insertion vector. Virus replication can only occur in the small proportion of infected cells that take up transfected WT DNA which is able to complement the defect of the infecting virus. Since these transfection-competent cells will also take up plasmid DNA, recombinants will be formed. This procedure enables isolation without genetic selection and reduces the amount of background virus by up to 3000-fold.[105] It may also be used in conjunction with other selective methods (see below) but offers little advantage if a dominant selection procedure can be used (see below).

C. TK NEGATIVE VIRUS

This method for generating recombinants is based on insertion of the foreign gene in the virus TK locus[1,3,4,18] and has been widely used because insertion into this site enables genetic selection of TK^- recombinant virus. After transfection of the plasmid insertion vector into cells infected with WT (TK^+) virus, TK^- recombinants are selected from the progeny virus by plaque assay on TK^- cells, using 5-bromodeoxyuridine (BUdR). A complicating factor in this procedure is the production of spontaneous TK^- mutants which can represent from 10 to 90% of TK^- virus and which need to be distinguished from true recombinant viruses. The screen most frequently employed has been based on DNA hybridization in order to detect the foreign gene.[1,4] Individual TK^- virus plaques are grown in small monolayers of TK^- cells with BUdR selection until a marked cytopathic effect is seen, usually 2 d later. These cells are harvested, loaded onto nitrocellulose filters, and probed with ^{32}P-labeled foreign DNA to identify virus containing the gene of interest. Two further rounds of plaque purification are sufficient to give homogeneous stocks of recombinant virus. Many recombinants have been generated in this fashion, but the screening of TK^- viruses for true recombinants has been laborious without additional methods of selection.[106,107] Other ways of distinguishing recombinant viruses from TK^- mutants include the use of antibodies to identify the foreign antigen within virus infected cell extracts, or more recently, polymerase chain reaction (PCR) to detect genomic alterations at the site of insertion of the foreign DNA. Figure 2 shows a number of plasmid vectors designed to allow insertion into the TK locus of vaccinia and expression of the foreign gene of interest.

D. COLOR DISTINCTION BY EXPRESSION OF β-GALACTOSIDASE

The expression of *Escherichia coli* β-galactosidase (β-gal) in vaccinia virus has provided a convenient and rapid method for visually distinguishing recombinant viruses expressing this enzyme. The usefulness of this approach relies on the enzyme stability in eukaryotic cells, the low endogenous levels of expression in these cells, and highly sensitive assays for detection of β-gal expression. The chromogenic substrate 5-bromo-4-chloro-3-indoyl-β-D-galactosidase (X-gal) is converted by β-gal into a deep blue color. Therefore, plaques formed by viruses expressing β-gal may be easily visualized by including X-gal in the culture medium. Initially β-gal was inserted into vaccinia virus to allow distinction of recombinant TK^- plaques from spontaneous mutants.[20,108] A plasmid, pSC11, was constructed in which β-gal was expressed from the late 11K promoter while another foreign gene could be expressed from the 7.5K promoter.[20] Both genes were inserted into the TK locus and recombinants were selected visually by plaque assay on TK^- cells in the presence of BUdR and X-gal. This greatly aided the selection and plaque purification of recombinant viruses. Subsequent replacement of β-gal with a foreign gene may also be used to isolate recombinant viruses since the recombinant can be selected as a white plaque against a lawn of blue plaques.[20]

E. TK POSITIVE VIRUS

HSV TK was the first foreign gene to be expressed in vaccinia virus.[1,2] Selection depended on the insertion of the HSV TK gene into a TK^- parent vaccinia virus to confer a TK^+ phenotype and was achieved by plaque assay on TK^- cells using methotrexate. Unlike the occurrence of spontaneous TK^- mutants during selection in BUdR, mutations causing reversion to a TK^+ phenotype have not been reported. Consequently the isolation of TK^+ viruses is straightforward. The HSV TK gene has also been used to simultaneously coinsert other foreign genes simultaneously into TK^- virus.[109]

F. G-418 RESISTANCE

The antibiotic G-418 is an effective inhibitor of vaccinia virus replication. Addition of 2 mg/ml G-418 48 h before infection reduces WT vaccinia growth by greater than 95% and abolishes plaque formation.[110] However, this toxicity can be overcome by expression of the neomycin resistance gene (derived from transposon Tn5). Thus vectors that coinsert this gene together with another gene of interest into vaccinia virus have been described.[110] In principle this method offers advantages over selection of TK^- virus since the genes can be inserted into any nonessential site in the virus genome and selection is not restricted to the use of TK^- cell lines. In practice this technology has been used only infrequently and has been largely superceded by selection with mycophenolic acid (MPA) and the *E. coli* guanine phosphoribosyltransferase (*Eco*gpt) gene (see below).

G. EXPRESSION OF *Eco*gpt

MPA, an inhibitor of purine metabolism, blocks the replication of vaccinia virus.[22,111] Constitutive expression of *Eco*gpt from vaccinia virus promoters can overcome this blockage if the substrates xanthine and hypoxanthine are provided. Thus if the *Eco*gpt gene is expressed in tandem with another gene of interest, recombinant viruses expressing both genes can be positively selected using MPA. Several coinsertion vectors have been constructed which contain both the *Eco*gpt gene under control of an early vaccinia promoter and a second promoter adjacent to multiple restriction sites (see Figure 4). In some vectors the second promoter is functional late during infection and contains translational initiation and termination codons, while in other vectors it is early and the gene of interest supplies its own translational initiation and termination codons.

This method of selection has several advantages over other genetic selection procedures. For example, it works in many different cell lines and with different poxviruses, it can be used to insert into any site within the virus genome, and, unlike BUdR, MPA is nonmutagenic. A further advantage is that in the presence of MPA, plaques are only formed by viruses which have acquired the *Eco*gpt gene and spontaneous MPA-resistant mutants do not occur. In this respect the *Eco*gpt gene is a truly dominant selectable marker.

H. REVERSE *Eco*gpt SELECTION

An additional utility of the *Eco*gpt gene is that drugs which select against its expression have been described. 6-Thioguanine (6-TG), which is toxic to cells expressing hypoxanthine guanine phosphoribosyltransferase (HGPRT) or *Eco*gpt, can be used in HGPRT negative cells to select for recombinant viruses which have lost *Eco*gpt.[112] It is necessary to use HGPRT negative cells since growth of WT or *Eco*gpt-expressing vaccinia viruses are inhibited in the presence of 6-TG if the cellular HGPRT enzyme is active. A mouse HGPRT negative embryonic fibroblast cell line (STO cells) is suitable for selection of *Eco*gpt negative vaccinia virus in the presence of 6-TG.[112] Similarly, an HGPRT negative derivative of HeLa cells (D980R) can be used.[86] The availability of techniques to select for or against expression of *Eco*gpt has made it possible to introduce a mutation into the virus genome without leaving the dominant selectable marker gene in the genome. Thus the *Eco*gpt gene may be used to introduce multiple mutations in a sequential manner.

I. PLAQUE MORPHOLOGY

Two methods of selection based upon an altered plaque phenotype have been used. One method relies upon vaccinia virus mutants which have a small plaque size phenotype attributable to alterations in the gene encoding the 14K structural protein.[113] By reinserting the WT 14K gene into the small plaque mutant, recombinant viruses can be visually selected that have a normal large plaque phenotype.[114] In principle the gene could be inserted at any nonessential site in the vaccinia genome although only the TK locus has been used.

A second method is based upon virus-induced polykaryocytosis. Some strains of vaccinia virus induce extensive cell fusion. One gene shown to influence this phenotype is the serpin K2L, which is an antifusion factor. Deletion of K2L induces fusion and reintroduction of the gene abrogates the fusogenic phenotype.[76] This change in plaque morphology, in association with other selective methods, enables selection of recombinant viruses by examining the virus plaques under low power microscopy.

J. HEMADSORPTION

Vaccinia virus encodes a hemagglutinin, an 85K glycoprotein that is present on the surface of virus infected cells and in the envelope of the extracellular virus (EEV).[115,116] The gene is nonessential for virus replication in cell culture and loss of the gene renders the plaques formed by the deletion mutant unable to agglutinate chicken erythrocytes. This provides a means of distinguishing recombinant viruses which have the foreign gene inserted into the HA locus.[77,79,117] Monolayers formed by the recombinant viruses are incubated with a 0.5% suspension of red blood cells and plaques which fail to agglutinate are selected as possible recombinants. HA negative plaques need to be screened to ensure they are true recombinants and not spontaneous HA^- mutants.

K. HOST RANGE

Vaccinia virus encodes at least three genes that affect host range.[118–120] Two of these, K1L[121] and C7L,[119] affect growth on human cells. The presence of these genes can be used to select recombinant viruses. For instance, a recombinant virus from which 21.7 kb of DNA from near the left genomic terminus has been deleted, including C7L and K1L, was isolated using the neomycin resistance gene.[96] This recombinant, vP293, grew in primary chick fibroblasts and several monkey cell lines but failed to grow in MRC-5 cells. Reintroduction of vaccinia gene K1L into vP293 allowed selection of recombinant virus on MRC-5 cells,[96] and a series of plasmid vectors designed to reinsert the K1L gene into vaccinia together with any chosen foreign gene linked to a late vaccinia virus promoter was constructed. Unique restriction sites, an early transcriptional termination signal, and translational start and stop signals are included in the vectors to allow rapid single step cloning and expression of any open reading frame.

L. TRANSIENT SELECTION

Two methods for generating poxvirus recombinants that are based on the transient expression of a marker gene have been described. One approach used the *E. coli* LacZ (β-gal) gene.[122] The second used the *Eco*gpt gene as a dominant selectable marker.[123] The first method took advantage of the instability of direct repeats in poxvirus genomes.[24,124] A vector was assembled that contained the β-gal gene joined to the 7.5K promoter, followed by another copy of this promoter driving the measles virus fusion protein gene. Thus the β-gal gene was flanked by a direct repeat of the 297 base pair 7.5K promoter.

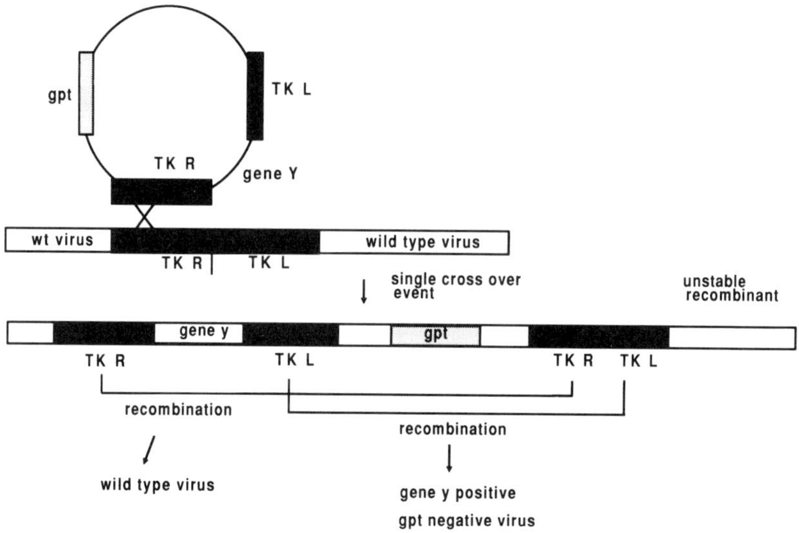

FIGURE 6. Selection of recombinant viruses by transient dominant selection.[123] The plasmid contains the vaccinia virus TK gene interrupted by a hypothetical gene (Y) with the *E. coli* gpt gene located distal to the vaccinia sequences. When recombination between the plasmid and virus genome occurs by a single crossover the whole plasmid is inserted into the virus genome. This virus is unstable due to the presence of direct repeats and resolves to either WT virus or a recombinant with gene Y inserted within the TK locus. All three types of virus genome are viable and can be packaged separately into particles, but only virus containing the gpt gene can form plaques under the selective conditions. Note that plasmids are formed during the resolution of the unstable intermediate virus.

First, recombinant fowlpoxviruses were selected that expressed β-gal. After further propagation of the virus, the β-gal gene was efficiently deleted by intramolecular recombination between the direct repeats, leaving a virus containing just the measles gene and one copy of the promoter at the insertion site.[122]

The second method has been termed transient dominant selection.[123] A plasmid is constructed which contains the gene to be inserted into the virus genome flanked by DNA from a nonessential locus in the conventional manner, and distal to the nonessential vaccinia DNA the *Eco*gpt gene linked to a vaccinia promoter (Figure 6). After transfection of this plasmid into virus infected cells, MPA-resistant recombinant viruses are selected. Resistance to MPA can only be acquired by a single recombination event that integrates the whole plasmid, including the *Eco*gpt gene, into the virus genome (Figure 6). Because this virus contains direct repeats of vaccinia DNA at the insertion site, it is unstable. If selection for MPA resistance is removed, the direct repeats recombine so that the intermediate virus resolves either to reform the parental virus, or to leave the foreign gene inserted within the nonessential site of the virus genome.[123] These viruses may be distinguished by either

DNA hybridization for the foreign DNA or by PCR to analyze the virus genome structure at the site of insertion. The frequency with which these two alternatives are formed depends upon the relative length of flanking vaccinia virus sequence on either side of the gene of interest. The major advantage of this system is that *Eco*gpt does not remain within the final virus genome and therefore it may be used sequentially to make successive genomic alterations at different loci. Even if multiple mutations are not needed, under some circumstances it is desirable not to have the selective marker present in the final virus if the virus is to be used as a vaccine, or if the virulence of the recombinant virus is to be tested without potential complication from the additional marker gene.

VI. TRANSFER OF PROKARYOTIC EXPRESSION SYSTEMS TO VACCINIA VIRUS

A. T7 RNA POLYMERASE

An elegant approach to obtain higher levels of foreign gene expression by vaccinia virus recombinants used the bacteriophage T7 RNA polymerase gene.[125] A transient expression system was described in which the T7 RNA polymerase gene was expressed from the 7.5K promoter within vaccinia virus. Cells infected with this virus were then transfected with a plasmid containing the target gene under control of a T7 promoter and terminator. Transient expression of the target protein was achieved at levels higher than those obtained by using vaccinia virus promoters.[53] This system provides a quick method for analyzing the function of a gene of interest without the labor-intensive construction of recombinant virus. Higher levels of expression were obtained by transferring the target gene regulated by a T7 promoter and terminator into another vaccinia virus and allowing expression to occur in coinfected cells[126] (see Figure 7).

Analysis of the T7 RNA polymerase transcripts within coinfected cells showed these RNAs to be very abundant, representing 30% of total cytoplasmic RNA at 24 h postinfection.[127] In view of this, the level of protein expression was less than anticipated. This was explained by the discovery that only 5 to 10% of the T7 transcripts contained a 5′ methylated cap structure and therefore the majority were unlikely to be translated. A hairpin loop at the 5′ end of the T7 transcripts was removed in an attempt to make the 5′ end more accessible to vaccinia virus capping enzymes, but the resultant RNA was unstable and yielded greatly reduced levels of expression.[127] The lack of translation of uncapped T7 transcripts was overcome by the inclusion of a sequence from the 5′ end of the encephalomyocarditis virus (EMCV) genome. EMCV mRNAs are uncapped and are translated by a mechanism allowing internal entry of ribosomes onto the mRNA. The presence of the EMCV ribosome entry sequence between the 5′ end of the T7 transcript and the AUG codon initiating translation of the downstream open reading frame allowed

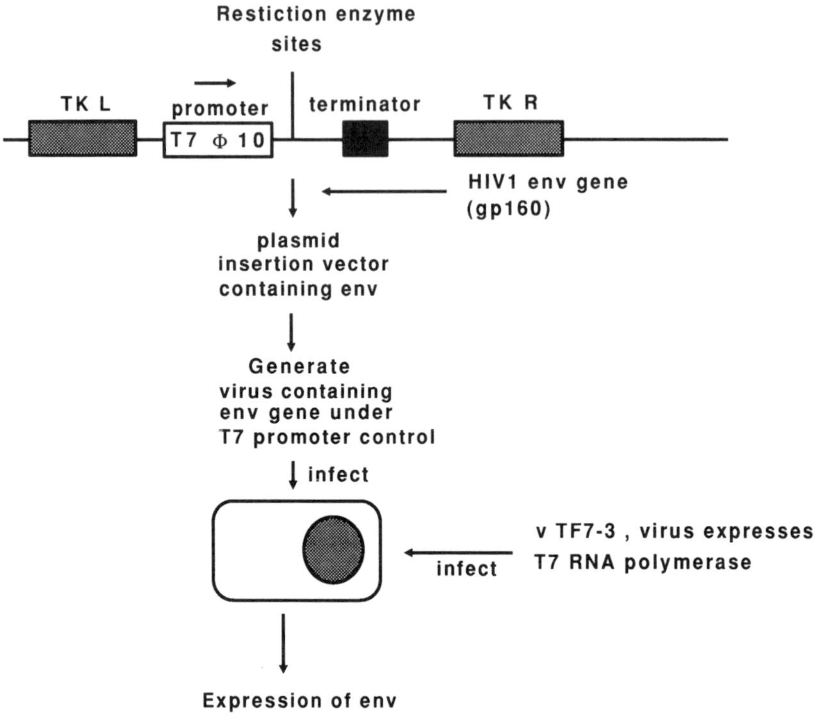

FIGURE 7. Use of T7 RNA polymerase to direct expression in recombinant vaccinia virus infected cells. A plasmid insertion vector which contains the T7ϕ10 promoter and terminator (T7t) transposed within the vaccinia virus TK gene is shown. Arrows indicate the direction of transcription of the T7 promoter and the vaccinia 7.5K promoter. The figure illustrates the use of the system for expression of HIV gp160.[167] A recombinant virus which contains the HIV gp160 gene is selected in BUdR. Expression of gp160 occurs only in cells coinfected with this virus and vTF7-3, a virus that expresses the T7 RNA polymerase.[125]

cap-independent translation of these transcripts and a 5- to 7-fold enhanced level of expression.[128] Hypertonic salt concentration further increased the level of translation.

Apart from high level expression, another advantage of this system is that gene products toxic to vaccinia virus infection may be expressed. Obviously, it is impossible to isolate a viable recombinant virus expressing such a gene, but when under the control of the T7 promoter the gene cannot be expressed unless cells are coinfected with the recombinant virus expressing T7 RNA polymerase. For example, attempts to express the vesicular stomatitis virus M gene using the 7.5K early/late promoter were unsuccessful, possibly due to M protein toxicity, but satisfactory levels of the protein were made using the T7 system.[129] One unexpected disappointment with the hybrid vaccinia/T7 RNA polymerase system was its failure to insert both T7 RNA polymerase and the target gene into the same recombinant virus,[126] thus limiting the utility of the expression system for vaccines. However, this has been overcome, at least *in vitro,* by the construction of inducible promoters within vaccinia virus (see below).

B. INDUCIBLE EXPRESSION WITH *ESCHERICHIA COLI* Lac OPERATOR AND REPRESSOR

Conditional expression of a foreign gene product in vaccinia has been obtained by use of a ts mutant defective in virus DNA synthesis.[99] If the foreign gene is controlled by a late vaccinia virus promoter, expression of the target gene is obtained only at the permissive temperature. However, no regulation was obtained by expression from an early promoter, and the system offers little advantage over using drugs to block DNA synthesis.

An alternative method to achieve regulated expression of exogenous or endogenous genes from vaccinia virus was described by two groups who transferred the inducible *E. coli* lac repressor/operator system into vaccinia.[41,130–132] The system is outlined in Figure 8. A lac operator sequence was placed downstream of the transcriptional initiation site of a late vaccinia virus promoter (either 11K[130] or 4b[41]). Either a synthetic, fully palindromic operator[130] or the naturally occurring, partially palindromic operator was used.[41] A reporter gene (β-gal or firefly luciferase) was then cloned downstream of the regulated promoter and the whole cassette introduced (by transfection or infection) into cells infected with a recombinant virus that expresses the lac I repressor protein from the constitutively active 7.5K promoter. The repressor protein forms an active tetramer and binds to the operator sequence, thus preventing transcription into the reporter gene. Addition of isopropylthiogalactopyranoside (IPTG) causes the repressor protein to dissociate from the operator and hence allows expression of the reporter gene.

With the 11K promoter, placement of the synthetic, palindromic operator 1 bp downstream of the conserved TAAAT promoter motif led to 99% repression of β-gal expression. Induction with IPTG resulted in expression to 20% of WT 11K promoter levels.[130] With the 4b promoter, positioning of the natural 21 bp operator 6 bp downstream of the promoter led to >97% repression of luciferase activity but >90% induction with IPTG.[41] Inclusion of a second operator in tandem repressed levels by >99.9%, but induction was reduced to 52% of WT levels. Vectors which allow the expression of both components of the inducible system and which have unique cloning sites for insertion of the regulated gene have been described[41] (see Figure 3). pPR34 contains the lac I gene linked to the 7.5K promoter and the inducible 4b promoter with a single operator, whereas pPR35 contains a double operator. Both vectors have unique cloning sites downstream and direct the whole cassette into the TK gene locus.[41]

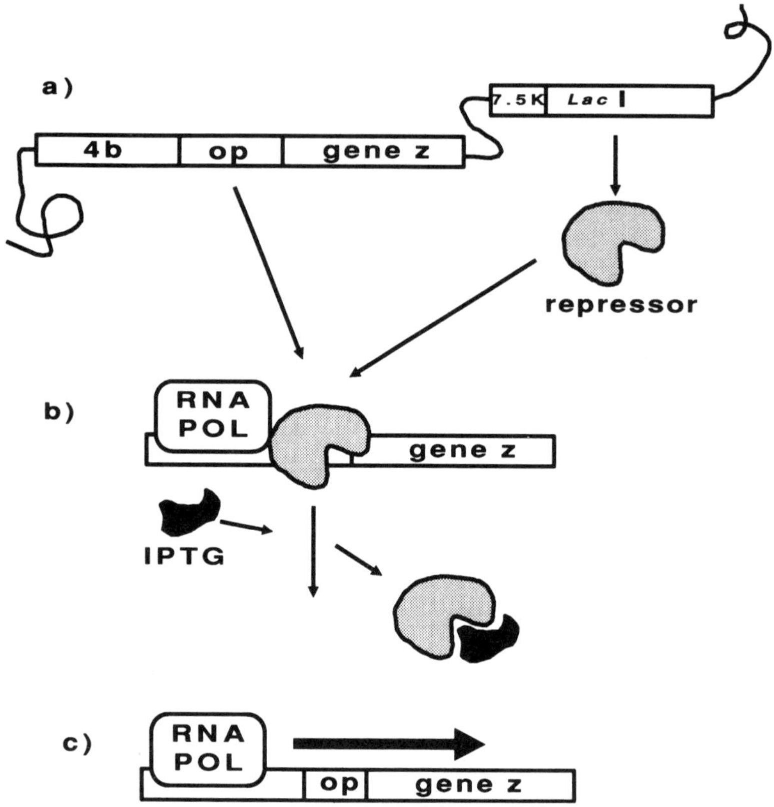

FIGURE 8. Inducible gene expression in vaccinia virus. A vaccinia virus recombinant expressing the *E. coli* Lac I repressor protein from the 7.5K promoter and with gene Z under control of an inducible promoter comprising the vaccinia virus 4b promoter and lac operator, is shown in (a). Upon infection the repressor protein is synthesized, oligomerizes into the active tetramer, and binds to the operator sequence downstream of the vaccinia 4b promoter (b). This blocks transcription of gene Z by vaccinia RNA polymerase (RNA POL). Addition of IPTG causes dissociation of the repressor from the operator and expression of gene Z (c).

VII. APPLICATIONS OF RECOMBINANT VACCINIA VIRUSES

Vaccinia virus is a popular expression system. Its use can be divided into three broad areas: for the study of poxvirus molecular biology; as an expression system for production or functional characterization of proteins; and as candidate live vaccines or tools for vaccine research. While the use of these viruses as new live vaccines remains controversial, they are already widely used experimental tools in the research laboratory.

A. STUDY OF POXVIRUS MOLECULAR BIOLOGY

1. Gene Mapping

The conservation of transcriptional control signals between poxviruses has enabled the expression of genes from other poxviruses without the need to engineer the protein coding sequences adjacent to a vaccinia promoter. This allows the identification of specific genes from other poxviruses by insertion of genomic restriction fragments into vaccinia. For example, the fowlpoxvirus TK gene was mapped by selection of a TK^+ phenotype after insertion of a specific fowlpoxvirus restriction fragment into a TK^- vaccinia strain.[133] Similarly, a vaccinia host range gene that permits growth on human cells was identified by insertion of restriction fragments into a vaccinia mutant nonpermissive for those cells and selection of progeny on human cells.[118,121]

2. Determining Gene Essentiality

Approximately 30% of the 200 vaccinia genes are nonessential for virus replication *in vitro,* but may affect virus virulence *in vivo.* Some of this information has been obtained from analyses of naturally occurring virus deletion mutants, while some has been determined experimentally by the construction of virus mutants with specific genes deleted or insertionally inactivated. The essentiality of a specific gene may simply be determined by insertion of a dominant selective marker gene into that locus, isolation of the recombinant virus, and analysis of its genome structure. This has been done using the HSV TK gene,[2,6,7,62] but more recently, the *Eco*gpt gene has been preferred,[50,66,69,72,134] since this does not require use of TK^- parent vaccinia nor TK^- cell lines for virus isolation. Isolation of a recombinant virus containing the selectable marker gene and with the target gene deleted or inactivated proves the gene is nonessential. However, care must be taken to check that no intact copy of the target gene remains in the virus genome, as would occur, for instance, during a single crossover event (Figure 6), and to check for the absence of plasmids.

Two types of plasmid are predicted to form, in addition to progeny virus genomes, during the resolution of the intermediate virus genome that contains the integrated plasmid sequences (Figure 6). One plasmid is identical to the original vector used for transfection, while the other contains a WT allele of the insertion site of the virus genome. Such plasmid structures were found in studies to insert the *Eco*gpt gene into the vaccinia virus DNA ligase gene.[72] After up to five rounds of plaque purification, some virus isolates contained the whole plasmid integrated into the virus genome and retained a WT copy of the ligase gene. These structures were evidently unstable since in addition to both possible forms of the intermediate virus, structures corresponding to WT and mutated DNA ligase alleles and other low molecular weight DNAs were present. The low molecular weight DNAs were shown to be plasmids by transformation of Hirt extracts of infected cells into *E. coli* and the recovery of ampicillin resistant colonies. The plasmids within these colonies contained either the WT or mutated allele of the DNA ligase locus, as predicted. Formation of intermediate recombinant viruses by single crossover structures has also been shown to occur at other sites within the virus genome, e.g., TK[24,72] and topoisomerase.[134] Plasmids can be formed by resolution of these genomes at the TK site, but not if BUdR is used as the selective agent.[72] This is explained by the toxicity of BUdR in virus-infected cells containing plasmids which express the WT TK gene.

The significance of the presence of these plasmids for construction of recombinant viruses is that they can induce heterogeneity into the recombinant virus stock and compromise the interpretation of the essentiality of a target gene. Any plasmid is replicated within vaccinia virus infected cells irrespective of whether it contains vaccinia virus sequences.[135,136] Therefore, plasmids produced during the resolution of intermediate genomes (formed by single crossover events) will replicate and then recombine with the virus genome to give rise to heterogeneity. Careful screening of virus isolates for the absence of plasmid sequences is necessary to avoid this problem.[69,72] The failure to isolate a virus deletion mutant does not prove essentiality, but merely makes it probable. If the transient dominant method of selection is used, careful analysis of which way the intermediate virus resolves in multiple plaque isolates (>20) helps to determine essentiality.

3. Analysis of Transcriptional Control Signals

Before large regions of the virus genome had been sequenced, a method was described to identify vaccinia virus promoters.[137] Randomly sheared virus DNA was cloned upstream of the HSV TK gene, and the resultant plasmids collectively transfected into cells infected with a TK^- vaccinia virus mutant. Progeny virus was then plated onto TK^- cells with methotrexate. Viable virus was only isolated in those cases where an active promoter had been cloned upstream. The promoter sequences from these clones were then isolated and sequenced.

Vaccinia virus promoters have been analyzed in detail by constructing recombinant viruses which possess a reporter gene linked to the promoter being tested. β-gal,[31,33,34,108] chloramphenicol acetyl transferase (CAT),[34–36,38,138] HSV TK,[133] dihydrofolate reductase,[139] or luciferase[41] have been used. A correlation between specific mutations and level of reporter gene activity is then made. Similarly, the early transcriptional termination signal was analyzed by specific mutation and correlation with functionality *in vitro* and *in vivo*.[47,48]

4. Gene Function

One way of determining the function of a gene is to delete it from the virus and then to assess the consequence of this deletion on virus replication *in vitro* and on virus virulence *in vivo*. This is straightforward for nonessential genes, but if the gene is essential, a conditional lethal mutant is needed. Several collections of vaccinia virus ts mutants have been isolated and char-

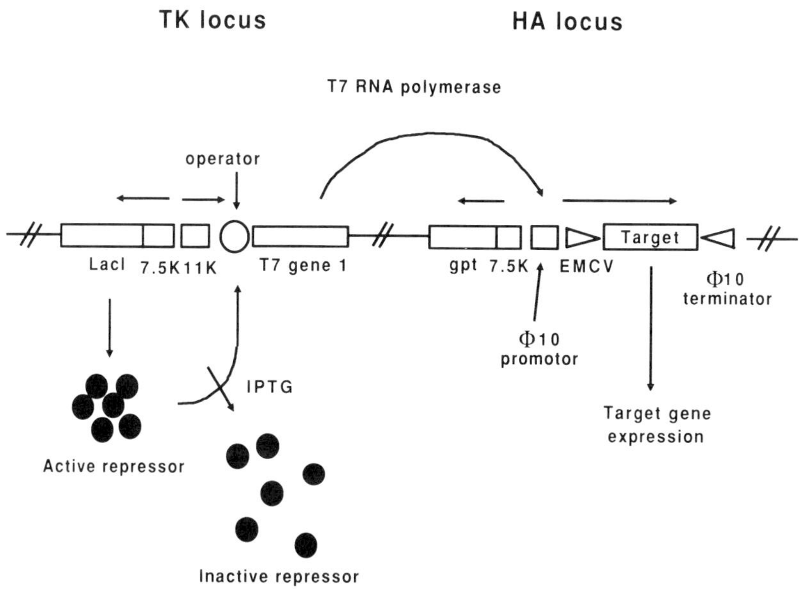

FIGURE 9. Schematic representation of vaccinia recombinant VAC/Op/T7[140] which contains the T7 RNA polymerase gene (gene 1) under control of the *E. coli* lac repressor and operator. Induction of T7 gene 1 expression is achieved by displacement of the Lac I repressor from the vaccinia 11K promoter hybrid by IPTG addition. This allows expression of T7 RNA polymerase which then transcribes the target gene from the T7ϕ10 promoter. The *Eco*gpt gene was used as a dominant selectable marker to construct the recombinant vaccinia virus. Arrows indicate the direction of transcription from the promoters used. HA and TK refer to the nonessential hemagglutinin and thymidine kinase genes used as sites to insert the foreign DNA.

acterized, but these only represent a fraction of all essential genes.[58] For essential genes without a conditional lethal mutant, another approach was needed. The IPTG inducible vectors described in Section VI.B have provided a means to study these genes.

Although there are other applications of these vectors (e.g., the regulation of genes which would otherwise be toxic to virus replication or to the cells and the inclusion of both components of the T7 RNA polymerase expression system into the same recombinant virus by regulating expression of T7 RNA polymerase, Figure 9[140]), the most useful application is the study of vaccinia virus gene function by the construction of recombinant viruses which express only a single inducible copy of the gene being studied.[131,132] Two approaches have been used to construct such a virus. One is outlined in Figure 10. A copy of the coding region of the gene to be regulated is obtained by PCR, cloned downstream of the inducible promoter, and inserted into the TK gene locus together with the lac I repressor protein. The intermediate virus contains two copies of the target gene, one under its natural promoter and one inducible within the TK locus. The natural gene is deleted from the virus by use of a

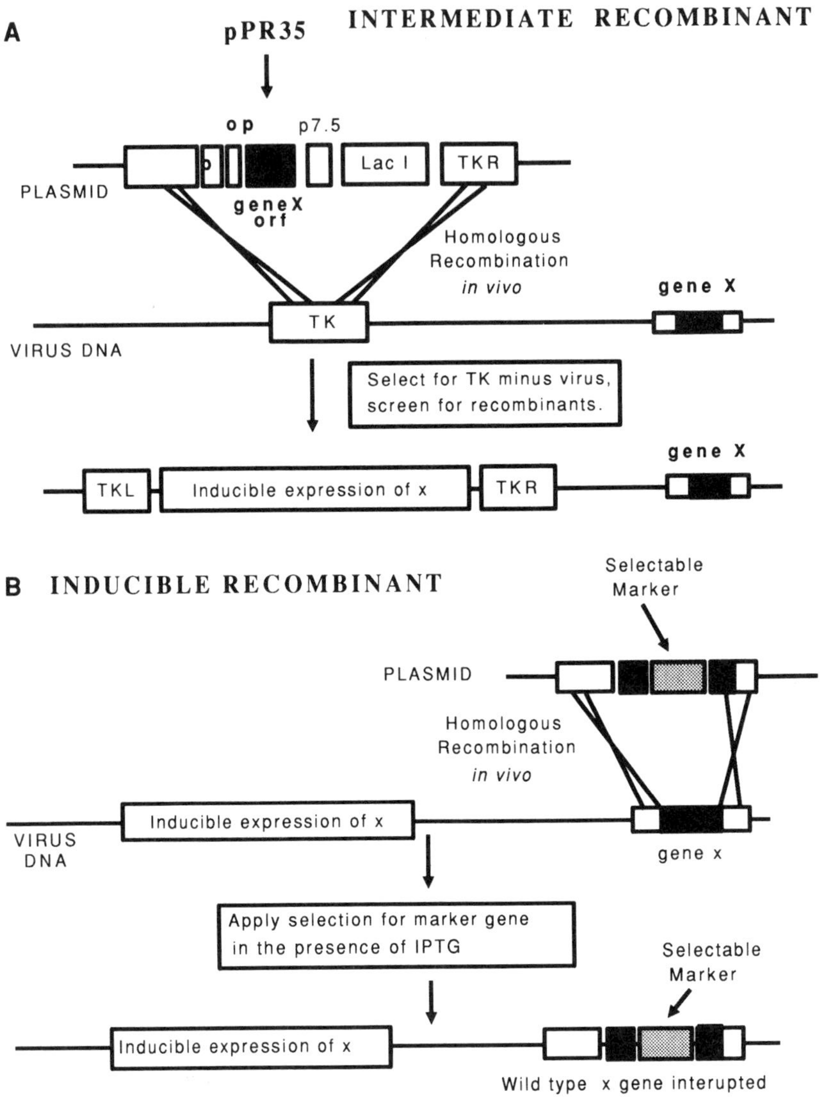

FIGURE 10. Generation of IPTG-dependent vaccinia virus. (A) A PCR copy of the coding region of the gene under study (gene X) is cloned under an inducible promoter in an insertion vector such as pPR35.[131] The resultant plasmid is transfected into vaccinia infected cells and an intermediate recombinant virus selected on TK^- cells with BUdR. This virus contains the original copy of the target gene and an inducible copy of the gene X within the TK locus. (B) The original copy of gene X is replaced by a dominant selectable marker such as *Eco*gpt. The IPTG-dependent virus is selected in the presence of IPTG and MPA. (Adapted from Rodriguez, J. F. and Smith, G. L., *Nucleic Acids Res.*, 18, 5347, 1990.)

dominant selectable marker, such as *Eco*gpt, and the recombinant viruses selected in the presence of MPA and IPTG.[131] The second method introduces the lac operator between the target gene promoter and the open reading frame by oligonucleotide directed mutagenesis *in vitro*. This altered gene is then inserted into a recombinant virus expressing lac I protein by transient dominant selection in the presence of IPTG to replace the natural copy of the gene.[123,132] If the gene being tested is essential, virus replication will be dependent upon IPTG.

The ability of an IPTG-inducible promoter to regulate vaccinia virus biology was first demonstrated by the inducible expression of 14K.[41] An inducible copy of the 14K protein was introduced into a mutant virus with a small plaque phenotype attributable to a mutation in 14K. In the presence of IPTG, the WT 14K protein was expressed and the virus produced WT plaques. Subsequently, a virus with only an inducible form of WT 14K was produced.[131] In the absence of IPTG no plaques were formed. Interestingly, normal amounts of intracellular naked virus (INV) were produced, but these were not wrapped by Golgi membrane and the virus did not egress from the cell, demonstrating that the 14K protein is critical for this process. A virus with an inducible copy of the 11K structural protein of the virus core has also been described.[132] This is also IPTG-dependent, but unlike the 14K inducible virus it does not produce any infectious virus in the absence of IPTG.

This system enables the functional characterization of any late vaccinia virus gene to be determined irrespective of whether a conditional lethal mutant for the gene exists. It offers advantages over the use of ts or drug dependent mutants, in that the function of the WT and not mutant amino acid sequence is studied.

For vaccine development, IPTG-dependent viruses offer a potential way to achieve virus attenuation. For instance, with the 14K-inducible virus, the synthesis of all virus proteins except 14K is normal. New INV particles are produced, but these cannot egress from the cell. Since EEV is required for virus virulence *in vivo,*[141] a virus defective in its synthesis will be attenuated. Nonetheless, the limited live infection should present all other antigens normally to the immune system and should therefore be immunogenic.

B. PROTEIN EXPRESSION AND FUNCTIONAL CHARACTERIZATION

1. Posttranslational Modifications

Vaccinia virus proteins undergo a wide variety of posttranslational modifications during infection[142] and it was therefore predicted that modifications of proteins expressed from vaccinia would occur as dictated by the primary amino acid sequence of the expressed protein and the natural capacity of the infected cell. This has been shown to be the case. Specific modifications include proteolysis,[143–145] oligomerization,[146,147] glycosylation,[4,42,148,149] γ-carboxylation,[104] phosphorylation,[150,151] and myristylation.[145] Transport of exogenous antigens within vaccinia virus infected cells has also been shown to be normal. Transport to the nucleus has been demonstrated with many proteins, including the influenza virus nucleoprotein,[152] the human papillomavirus L1 protein,[40] polyoma virus capsid protein VP1 and VP2,[153] and the human T-cell leukemia type 1 pX protein.[154] In the latter case, the amino acid sequence required for the localization of this protein to the nucleolus was determined. Several herpes virus glycoproteins have localized to the nuclear membrane.[155,156] Other proteins are transported to the cell surface, e.g., the influenza hemagglutinin, neuraminidase and matrix 2 protein,[152] the rabies glycoprotein,[157] and the Epstein-Barr virus gp340,[149] or are secreted from the cell, e.g., the hepatitis B virus surface antigen[3,7] and nerve growth factor.[144]

2. Transient Gene Expression

For the rapid study of protein function without the need to construct recombinant viruses, the gene of interest may be cloned downstream of a T7 RNA polymerase promoter and transfected into cells infected with a virus expressing the T7 RNA polymerase. This rapid transient expression has been widely used. Some applications include the study of HIV gp160 interactions with CD4,[158,159] the intracellular transport of normal or mutant glycoproteins,[160,161] complementation of ts or defective viruses,[162–164] virus morphogenesis,[163,165] and cell fusion.[166] Alternatively, if larger amounts of the protein are needed, or if expression of a gene is required repeatedly, the gene may be expressed from the T7 RNA polymerase promoter within vaccinia virus and the protein synthesized in cells coinfected with the virus and the recombinant expressing T7 RNA polymerase. With this system, greater than 50 mg of HIV gp160 was expressed and purified from 10^{11} infected cells.[167]

3. Generation of Libraries of Chimeric Genes

The high level of recombination that occurs in infected cells between directly repeated nucleotide sequences in poxvirus genomes has been utilized to generate a library of chimeric HIV enveloped genes derived from two related parent strains.[168] A vaccinia recombinant was constructed which contained two tandemly arrayed HIV envelope genes derived from different parent viruses. Due to the instability of the direct repeats, recombination occurs between the conserved regions of the two genes to form intertypic chimeras. After resolution, each progeny plaque will contain a single type of chimera which may be screened for properties such as cell fusion, CD4 binding, and antigenicity. This protocol is applicable to the mapping of epitopes and receptor binding sites, but is limited to those genes with sufficient nucleotide homology to enable frequent recombination.

4. Raising Monoclonal Antibodies to Native Gene Products

Recombinant vaccinia viruses expressing a specific protein are useful for raising monoclonal antibodies.[169,170] Animals are immunized with the recombinant virus, and the resultant hybridomas directed against the foreign gene product are easily selected using the pathogen from which the gene was derived as the target antigen for screening. This method has advantages over the use of bacterial-expressed products as immunogens, which may have been processed abnormally, producing epitopes not present in the natural protein. Alternatively, cells infected with a vaccinia virus recombinant expressing the foreign protein may be used as targets against which hybridomas raised with bacterially expressed antigens may be screened.[171] There is also a report of the isolation of human monoclonal antibody to HIV gp 160 from a volunteer immunized with antigen expressed from vaccinia virus,[172] although this is not a general method.

C. VACCINES

The potential application of recombinant vaccinia viruses and other poxviruses in immunoprophylaxis are considered in other chapters of this volume. In addition to the direct use of these viruses as immunizing vehicles, they are powerful laboratory reagents for the identification of antigens that can induce protective immune responses against pathogens and which may subsequently be incorporated into new vaccines. The principle is simple. A recombinant virus expressing the test antigen is constructed and used to immunize animals. No protein purification is needed since the protein is synthesized and simultaneously presented to the immune system as the virus replicates within the host. If an appropriate animal model exists, then these immunized animals may be challenged with the pathogen and the protective efficacy of the antigen assessed, as was done, for instance, with influenza virus,[4] HSV,[7,42] hepatitis B virus,[173] rabies virus,[157] and Epstein-Barr virus.[100] Alternatively, sera may be taken from the immunized animals and tested *in vitro* for neutralization of the infectivity of the pathogen, e.g., with human cytomegalovirus[156,174] or HIV.[175]

Vaccinia virus recombinants can be used to identify both B- and T-cell epitopes. In particular, recombinants are useful for the identification of peptide epitopes recognized in association with class I antigens of the major histocompatibility complex (MHC) by cytotoxic T cells (CTL).[176] CTL taken from immunized animals may be tested against syngeneic target cells infected with the corresponding pathogen and labeled with ^{51}chromium. Specific recognition of the cell causes cell lysis and release of radioactivity. Alternatively, CTL taken from an animal or human previously infected with the pathogen may be tested for recognition of the antigen expressed within syngeneic, vaccinia virus infected cells. For influenza virus, a library of vaccinia virus recombinants, each expressing one influenza virus antigen,[152] was used to systematically assess recognition by murine and human CTL.[55,177–183] Similar approaches have identified T-cell antigens for several other viruses[176] and the specific peptide epitopes from within whole proteins that are produced by proteolytic degradation.[184–187]

D. IMMUNOTHERAPY

Many tumors are induced following virus infection and present one or more virus-specified tumor antigen(s) on or within the transformed cell. Other tumors are not induced by an infectious agent, but are associated with the expression of an oncoprotein that is either mutated or abnormally expressed. These tumor antigens are potential targets for immunotherapy. Immune responses against these tumor antigens have been successfully induced by inoculation with vaccinia virus recombinants expressing the tumor antigen and have led to tumor prevention and/or regression.[188–195] There is no hazard of tumor induction by such treatment because each infected cell is killed by the cytocidal vaccinia virus infection.

Four types of virus-induced tumors have been prevented by such immunizations. The first report described a vaccinia virus recombinant that expressed the Friend murine leukemia virus (F-MuLV) *env* gene from the p7.5 promoter.[188] Infected cells synthesized an 85 kDa glycoprotein (gp85) that was processed to gp70 and p15E in a manner similar to that occurring during F-MuLV infection. Mice immunized with the recombinant virus developed envelope-specific CTL and T-cell proliferative responses and also antibodies that neutralized F-MuLV infectivity. After challenge with F-MuLV, the animals were protected against leukemia whereas control animals immunized with a recombinant vaccinia virus were not.

In another study, each of the polyoma virus tumor (T) antigens were expressed from a vaccinia virus recombinant, again from the p7.5K promoter.[189] These nuclear antigens are processed sufficiently within the cell to enable peptides derived from the mature proteins to be presented with class I MHC antigens on the cell surface. Rats immunized with a vaccinia virus recombinant expressing the large or middle T antigen were protected from polyoma virus-induced tumors, but immunization with a virus expressing the small T antigen conferred no protection. Importantly, immunization of rats with established polyoma-induced tumors caused regression and clearance of these malignancies. This demonstrated the potential application of vaccinia virus recombinants for immunotherapy of existing tumors.

Two other studies described vaccinia virus recombinants which expressed the early antigens of either bovine[194] or human[195] papillomaviruses from the 7.5K promoter. Recombinants which express the bovine papillomavirus (BPV) type 1 early genes E1, E2, E5, E6, or E7 were tested for their ability to induce antitumor immunity in Fischer rats challenged with BPV1-transformed syngeneic rat cells. Tumor growth was slowed or prevented in animals vaccinated with recombinants expressing BPV E5, E6, or E7, but not in those vaccinated with the other early genes.[194] Many of the more than 50 strains of human papillomavirus (HPV) induce benign skin warts which regress spontaneously, presumably due to immune recognition. However, other strains, notably types 16 and 18, are associated with development of cervical and penile malignant carcinoma. Like BPV, HPV cannot be grown in culture, and therefore tumors are induced in the experimental animal by transplantation with syngeneic cells transformed by the presence of the early antigens of HPV 16 in cooperation with the *ras* oncogene. Immunization of rats with the vaccinia recombinants expressing the E6 or E7 antigens protected a proportion of animals (18 to 44%) from tumors after challenge with various transformed cells. Immunization with both antigens simultaneously produced no greater protection, and E5 was ineffective in this system.

Other tumors derive from abnormal expression of cellular oncogenes. The rat *neu* oncogene encodes a 185 kDa cell surface protein and transfection of this gene into NIH 3T3 cells induces neoplastic transformation. The external domain, transmembrane anchor, and short cytoplasmic tail of this protein were expressed from an early promoter within a vaccinia virus recombinant.[190] Immunization of mice with the recombinant virus induced antibodies against the *neu* oncoprotein and prevented tumor development following challenge with *neu*-transformed cells. However, there was no protection in rats immunized with the recombinant virus and challenged with syngeneic rat cells expressing a transforming version of p185. Possibly the whole p185 antigen and not just the external domain are required for protection.

Malignant melanoma is an increasing problem and is associated with the abnormal expression of a cell surface 97 kDa glycosylated antigen p97. A recombinant virus containing the p97 gene under control of the 7.5K promoter synthesized a membrane glycoprotein recognized by a p97 specific monoclonal antibody.[192] Mice immunized with the recombinant virus produced antibodies against p97, and spleen cells taken from these animals proliferated *in vitro* in response to p97 protein or to syngeneic cells expressing the p97 antigen.[192] Immune responses were also produced in immunized monkeys (*Maca fascicularis*). Despite the trace amounts of p97 expressed in these monkey cells and the known cross reactivity of monkey and human p97 expressed in vaccinia, no untoward immune responses were observed during a follow-up period in these monkeys. Mice immunized with the recombinant virus rejected transplants of syngeneic mouse cells expressing p97.[191]

Human breast tumors are associated with the abnormal expression of an endogenous antigen called epithelial tumor antigen (ETA). Of these tumors, 91% express this antigen which contains multiple tandem repeats of a 20 amino acid sequence. Vaccinia virus recombinants which synthesize a membrane-bound or secretory form of the protein were constructed and tested for their ability to protect rats from tumors induced by rat cells transformed with ETA in cooperation with *ras*. Of the animals tested, 82% were protected from tumors with a recombinant expressing the membrane anchored form of ETA, whereas only 61% of the animals were protected with the soluble form.[193]

Collectively, these data clearly show the potential of vaccinia virus recombinants in tumor immunotherapy, in addition to immunoprophylaxis.

ACKNOWLEDGMENTS

We thank Kath Law for critical reading of the manuscript and the Medical Research Council, Wellcome Trust, Lister Institute of Preventive Medicine, and the Cancer Research Campaign for support.

REFERENCES

1. **Mackett, M., Smith, G. L., and Moss, B.,** Vaccinia virus: a selectable eukaryotic cloning and expression vector, *Proc. Natl. Acad. Sci. U.S.A.,* 79, 7415, 1982.
2. **Panicali, D. and Paoletti, E.,** Construction of poxviruses as cloning vectors: insertion of the thymidine kinase gene from herpes simplex virus into the DNA of infectious vaccinia virus, *Proc. Natl. Acad. Sci. U.S.A.,* 79, 4927, 1982.
3. **Smith, G. L., Mackett, M., and Moss, B.,** Infectious vaccinia virus recombinants that express hepatitis B virus surface antigen, *Nature,* 302, 490, 1983.
4. **Smith, G. L., Murphy, B. R., and Moss, B.,** Construction and characterization of an infectious vaccinia virus recombinant that expresses the influenza hemagglutinin gene and induces resistance to influenza virus infection in hamsters, *Proc. Natl. Acad. Sci. U.S.A.,* 80, 7155, 1983.
5. **Smith, G. L., Godson, N. G., Nussenzweig, V., Nussenzweig, R. S., Barnwell, J., and Moss, B.,** *Plasmodium knowlesi* sporozoite antigen: expression by infectious recombinant vaccinia virus, *Science,* 224, 397, 1984.
6. **Panicali, D., Davis, S. W., Weinberg, R. L., and Paoletti, E.,** Constructive of live vaccines by using genetically engineered poxviruses: biological activity of recombinant vaccinia virus expressing influenza virus hemagglutinin, *Proc. Natl. Acad. Sci. U.S.A.,* 80, 5364, 1983.
7. **Paoletti, E., Lipinskas, B. R., Samsanoff, C., Mercer, S., and Panicali, D.,** Construction of live vaccinia viruses using genetically engineered poxviruses: biological activity of vaccinia virus recombinants expressing the hepatitis B virus surface antigen and the herpes simplex virus glycoprotein D, *Proc. Natl. Acad. Sci. U.S.A.,* 81, 193, 1984.
8. **Smith, G. L. and Moss, B.,** Infectious poxvirus vectors have capacity for at least 25,000 base pairs of foreign DNA, *Gene,* 25, 21, 1983.
9. **Burke, K. L., Dunn, G., Ferguson, M., Minor, P. D., and Almond, J. W.,** Antigen chimaeras of poliovirus as potential new vaccines, *Nature,* 322, 81, 1988.
10. **Lowe, R. S., Keller, P. M., Keech, B. J., Davison, A. J., Whang, Y., Morgan, A. J., Kieff, E., and Ellis, R. W.,** Varicella-zoster virus as a live vector for the expression of foreign genes, *Proc. Natl. Acad. Sci. U.S.A.,* 84, 3896, 1987.
11. **Dougan, G., Hormaeche, C. E., and Maskell, D.,** Live oral Salmonella vaccines: potential use of attenuated strains as carriers of heterologous antigens to the immune system, *Parasite Immunol.,* 9, 151, 1987.
12. **Aldovini, A. and Young, R. A.,** Humoral and cell-mediated immune responses to live recombinant BCG-HIV vaccines, *Nature,* 351, 479, 1991.
13. **Sam, C. K. and Dumbell, K. R.,** Expression of poxvirus DNA in coinfected cells and marker rescue of thermosensitive mutants by subgenomic fragments of DNA, *Ann. Virol. Inst. Pasteur,* 132E, 135, 1981.
14. **Nakano, E., Panicali, D., and Paoletti, E.,** Molecular genetics of vaccinia virus: demonstration of marker rescue, *Proc. Natl. Acad. Sci. U.S.A.,* 79, 1593, 1982.

15. **Weir, J. P., Bajszar, G., and Moss, B.,** Mapping of the vaccinia virus thymidine kinase gene by marker rescue and cell-free translation of selected mRNA, *Proc. Natl. Acad. Sci. U.S.A.*, 79, 1210, 1982.
16. **Kates, J. and McAuslan, B. R.,** Poxvirus DNA-dependent RNA polymerase, *Proc. Natl. Acad. Sci. U.S.A.*, 58, 134, 1967.
17. **Munyon, W. E., Paoletti, E., and Grace, J. T.,** RNA polymerase activity in purified infectious vaccinia virus, *Proc. Natl. Acad. Sci. U.S.A.*, 58, 2280, 1967.
18. **Mackett, M., Smith, G. L., and Moss, B.,** General method for production and selection of infectious vaccinia virus recombinants expressing foreign genes, *J. Virol.*, 49, 857, 1984.
19. **Boyle, D. B., Coupar, B. E., and Both, G. W.,** Multiple-cloning-site plasmids for the rapid construction of recombinant poxviruses, *Gene,* 35, 169, 1985.
20. **Chakrabarti, S., Brechling, K., and Moss, B.,** Vaccinia virus expression vector: coexpression of β-galactosidase provides visual screening of recombinant virus plaques, *Mol. Cell. Biol.*, 5, 3403, 1985.
21. **Falkner, F. G., Chakrabarti, S., and Moss, B.,** pUV I: a new vaccinia virus insertion and expression vector, *Nucleic Acids Res.*, 15, 7192, 1987.
22. **Falkner, F. G. and Moss, B.,** *Escherichia coli gpt* gene provides dominant selection for vaccinia virus open reading frame expression vectors, *J. Virol.*, 62, 1849, 1988.
23. **Davison, A. J. and Moss, B.,** New vaccinia virus recombination plasmids incorporating a synthetic late promoter for high level expression of foreign proteins, *Nucleic Acids Res.*, 18, 4285, 1990.
24. **Spyropoulos, D. D., Roberts, B. E., Panicali, D. L., and Cohen, L. K.,** Delineation of the viral products of recombination in vaccinia virus-infected cells, *J. Virol.*, 62, 1046, 1988.
25. **Wilson, E. M., Hodges, W. M., and Hruby, D. E.,** Construction of recombinant vaccinia virus strains using single-stranded DNA insertion vectors, *Gene,* 49, 207, 1986.
26. **Wilson, E. M. and Hruby, D. E.,** A single-stranded insertion vector (SSIV) for introducing foreign genes into the vaccinia virus genome, *Nucleic Acids Res.*, 15, 4690, 1987.
27. **Moss, B.,** Molecular biology of poxviruses, in *Recombinant Poxviruses,* Binns, M. M. and Smith, G. L., Eds., CRC Press, Boca Raton, Florida, 1992.
28. **Vos, J. C. and Stunnenberg, H. G.,** Derepression of a novel class of vaccinia virus genes upon DNA replication, *EMBO J.*, 7, 3487, 1988.
29. **Keck, J. G., Baldick, C. J., and Moss, B.,** Role of DNA replication in vaccinia virus gene expression: a naked template is required for transcription of three late trans-activator genes, *Cell,* 61, 801, 1990.
30. **Weir, J. P. and Moss, B.,** Determination of the promoter region of an early vaccinia virus gene encoding thymidine kinase, *Virology,* 158, 206, 1987.
31. **Davison, A. J. and Moss, B.,** Structure of vaccinia virus early promoters, *J. Mol. Biol.*, 210, 749, 1989.
32. **Weir, J. P. and Moss, B.,** Determination of the transcriptional regulatory region of a vaccinia virus late gene, *J. Virol.*, 61, 75, 1987.
33. **Davison, A. J. and Moss, B.,** Structure of vaccinia virus late promoters, *J. Mol. Biol.*, 210, 771, 1989.
34. **Kent, R. K.,** Isolation and Analysis of the Vaccinia Virus p4b Gene Promoter, Ph.D. thesis, University of Cambridge, England, 1988.
35. **Cochran, M. A., Puckett, C., and Moss, B.,** *In vitro* mutagenesis of the promoter region for a vaccinia virus gene: evidence for tandem early and late regulatory signals, *J. Virol.*, 54, 30, 1985.
36. **Hirschmann, P., Vos, J. C., and Stunnenberg, H. G.,** Mutational analysis of a vaccinia virus intermediate promoter *in vivo* and *in vitro.*, *J. Virol.*, 64, 6063, 1990.
37. **Bertholet, C., Drillien, R., and Wittek, R.,** One hundred base pairs of 5′ flanking sequence of a vaccinia virus late gene are sufficient to temporally regulate late transcription, *Proc. Natl. Acad. Sci. U.S.A.*, 82, 2096, 1985.

38. **Bertholet, C., Stocco, P., Van Meir, E., and Wittek, R.,** Functional analysis of the 5′ flanking sequence of a vaccinia virus late gene, *EMBO J.*, 5, 1951, 1986.
39. **Stunnenberg, H., Lange, H., Philipson, L., van-Miltenburg, R. T., and van der Vliet, P. C.,** High expression of functional adenovirus DNA polymerase and precursor terminal protein using recombinant vaccinia virus, *Nucleic Acids Res.*, 16, 2431, 1988.
40. **Browne, H., Churcher, M., Stanley, M., Smith, G. L., and Minson, A. C.,** Analysis of the L1 gene product of HPV 16 by expression in a vaccinia virus recombinant, *J. Gen. Virol.*, 69, 1263, 1988.
41. **Rodriguez, J. F. and Smith, G. L.,** Inducible gene expression from vaccinia virus vectors, *Virology,* 177, 239, 1990.
42. **Cremer, K., Mackett, M., Wohlenberg, G., Notkins, A. L., and Moss, B.,** Vaccinia virus recombinants expressing herpes simplex virus type 1 glycoprotein D prevent latent herpes in mice, *Science,* 228, 737, 1985.
43. **Patel, D. D. and Pickup, D. J.,** Messenger RNAs of a strongly-expressed late gene of cowpox virus contain 5′-terminal poly(A) sequences, *EMBO J.*, 6, 3787, 1987.
44. **Funahashi, S., Sato, T., and Shida, H.,** Cloning and characterization of the gene encoding the major protein of the A-type inclusion body of cowpox virus, *J. Gen. Virol.*, 69, 35, 1988.
45. **Patel, D. D., Ray, C. A., Drucker, R. P., and Pickup, D. J.,** A poxvirus-derived vector that directs high levels of expression of cloned genes in mammalian cells, *Proc. Natl. Acad. Sci. U.S.A.*, 85, 9431, 1988.
46. **Pearson, A., Richardson, C., and Yuen, L.,** The 5′ noncoding region sequence of the *Choristoneura biennis* entomopoxvirus spheroidin gene functions as an efficient late promoter in the mammalian vaccinia expression system, *Virology,* 180, 561, 1991.
47. **Rohrmann, G., Yuen, L., and Moss, B.,** Transcription of vaccinia virus early genes by enzymes isolated from vaccinia virions terminates downstream of a regulatory sequence, *Cell,* 46, 1029, 1986.
48. **Yuen, L. and Moss, B.,** Oligonucleotide sequence signalling transcriptional termination of vaccinia virus early genes, *Proc. Natl. Acad. Sci. U.S.A.*, 84, 6417, 1987.
49. **Zhou, J. A., McIndoe, A., Davies, H., Sun, X. Y., and Crawford, L.,** The induction of cytotoxic T-lymphocyte precursor cells by recombinant vaccinia virus expressing human papillomavirus type 16 L1, *Virology,* 181, 203, 1991.
50. **Beattie, E., Tartaglia, J., and Paoletti, E.,** Vaccinia-virus encoded eIF-2 alpha homolog abrogates the antiviral effect of interferon, *Virology,* 183, 419, 1991.
51. **Tsao, H., Liu, G. Q., Ruan, L., and Chu, C. M.,** Construction and application of plasmids containing bidirectional promoters of vaccinia virus, *J. Virol.*, 62, 4832, 1988.
52. **Mackett, M.,** Expression vectors using the vaccinia virus 19K promoter, unpublished data.
53. **Cochran, M. A., Mackett, M., and Moss, B.,** Eukaryotic transient expression system dependent on transcription factors and regulatory DNA sequences of vaccinia virus, *Proc. Natl. Acad. Sci. U.S.A.*, 82, 19, 1985.
54. **Venkatesan, S., Baroudy, B. M., and Moss, B.,** Distinctive nucleotide sequences adjacent to multiple initiation and termination sites of an early vaccinia virus gene, *Cell,* 25, 805, 1981.
55. **Andrew, M. E., Coupar, B. E. H., Ada, G. L., and Boyle, D. B.,** Cell-mediated immune responses to influenza virus antigens expressed by vaccinia virus recombinants, *Microb. Pathol.*, 1, 443, 1986.
56. **Townsend, A., Bastin, J., Gould, K., Brownlee, G., Andrew, A., Boyle, D. B., Chan, Y., and Smith, G.,** Defective presentation to class I-restricted cytotoxic T lymphocytes in vaccinia-infected cells is overcome by enhanced degradation of antigen, *J. Exp. Med.*, 168, 1211, 1988.
57. **Earl, P. L., Hugin, A. W., and Moss, B.,** Removal of cryptic poxvirus transcription termination signals from the human immunodeficiency virus type 1 envelope gene enhances expression and immunogenicity of a recombinant vaccinia virus, *J. Virol.*, 64, 2448, 1990.

58. **Condit, R. C. and Niles, E. G.,** Orthopoxvirus genetics, *Curr. Top. Microbiol. Immunol.,* 163, 1, 1990.
59. **Mackett, M. and Archard, L. C.,** Conservation and variation in *orthopoxvirus* genome structure, *J. Gen. Virol.,* 45, 683, 1979.
60. **Panicali, D., Davis, S. W., Mercer, S. R., and Paoletti, E.,**Two major variants present in serially propagated stocks of the WR strain of vaccinia virus, *J. Virol.,* 37, 1000, 1981.
61. **Moss. B., Winters, E., and Cooper, J. A.,** Deletion of a 9,000-base-pair segment of the vaccinia virus genome that encodes nonessential polypeptides, *J. Virol.,* 40, 387, 1981.
62. **Perkus, M. E., Panicali, D., Mercer, S., and Paoletti, E.,** Insertion and deletion mutants of vaccinia virus, *Virology,* 152, 285, 1986.
63. **Kotwal, G. J. and Moss, B.,** Analysis of a large cluster of nonessential genes deleted from a vaccinia virus terminal transposition mutant, *Virology,* 167, 524, 1988.
64. **Perkus, M. E., Piccini, A., Lipinskas, B. R., and Paoletti, E.,** Recombinant vaccinia virus: immunization against multiple pathogens, *Science,* 229, 981, 1985.
65. **Buller, R. M. L., Chakrabarti, S., Cooper, J. A., Twardzik, D. R., and Moss, B.,** Deletion of the vaccinia virus growth factor gene reduces virus virulence, *J. Virol.,* 62, 866, 1988.
66. **Kotwal, G. J., Hugin, A. W., and Moss, B.,** Mapping and insertional mutagenesis of a vaccinia virus gene encoding a 13,800-Da secreted protein, *Virology,* 171, 579, 1989.
67. **Patel, A. H., Gaffney, D. F., Subak-Sharpe, J. H., and Stow, N. D.,** DNA sequence of the gene encoding a major secreted protein of vaccinia virus, strain Lister, *J. Gen. Virol.,* 71, 2013, 1990.
68. **Perkus, M. E., Goebel, S. J., Davis, S. W., Johnson, G. P., Norton, E. K., and Paoletti, E.,** Deletion of 55 open reading frames from the termini of vaccinia virus, *Virology,* 180, 406, 1991.
69. **Hughes, S. J., Johnston, L. H., de Carlos, A., and Smith, G. L.,** Vaccinia virus encodes an active thymidylate kinase that complement a cdc8 mutant of *Saccharomyces cerevisiae, J. Biol. Chem.,* 266, 20,103, 1991.
70. **Child, S. J., Palumbo, G. J., Buller, R. M. L., and Hruby, D. E.,** Insertional inactivation of a large subunit of ribonucleotide reductase encoded by vaccinia virus is associated with reduced virulence *in vivo, Virology,* 174, 625, 1990.
71. **Colinas, R. J., Goebel, S. J., Davis, S. W., Johnson, G., Norton, E. K., and Paoletti, E.,** A DNA ligase gene in the Copenhagen strain of vaccinia virus is nonessential for virus replication and recombination, *Virology,* 179, 267, 1990.
72. **Kerr, S. M. and Smith, G. L.,** Vaccinia virus DNA ligase is nonessential for virus replication: recovery of plasmids from virus-infected cells, *Virology,* 180, 625, 1991.
73. **Moore, J. B. and Smith, G. L.,** Steroid hormone synthesis by a vaccinia enzyme: a new type of virus virulence factor, *EMBO J.,* in press.
74. **Niles, E. G. and Seto, J.,** Vaccinia virus gene D8 encodes a virion transmembrane protein, *J. Virol.,* 62, 3772, 1988.
75. **Zhou, J., McLean, L., Sun, X.-Y., Stanley, M., Almond, N., Crawford, L., and Smith, G. L.,** Increased antibody responses to human papillomavirus type 16 L1 protein expressed by recombinant vaccinia virus lacking serine protease inhibitor genes, *J. Gen. Virol.,* 71, 2185, 1990.
76. **Law, K. M. and Smith, G. L.,** A vaccinia serine protease inhibitor which prevents virus induced cell fusion, *J. Gen. Virol.,* 73, 549, 1992.
77. **Shida, H. Tochikura, T., Sato, T., Konno, T., Hirayoshi, K., Seki, M., Ito, Y., Hatanaka, M., Hinuma, Y., Sugimoto, M., Takahashi-Nishimaki, F., Maruyama, T., Miki, K., Suzuki, K., Morita, M., Sashiyama, H., and Hayami, M.,** Effect of the recombinant vaccinia viruses that express HTLV-I envelope gene on HTLV-I infection, *EMBO J.,* 6, 3379, 1987.

78. **Flexner, C., Hugin, A., and Moss, B.,** Prevention of vaccinia virus infection in immunodeficient mice by vector-directed IL-2 expression, *Nature,* 330, 259, 1987.
79. **Shida, H., Hinuma, Y., Hatanaka, M., Morita, M., Kidokoro, M., Suzuki, K., Maruyama, T., Takahashi-Nishimaki, F., Sugimoto, M., Kitamura, R., Miyazawa, T., and Hayami, M.,** Effects and virulences of recombinant vaccinia viruses derived from attenuated strains that express the human T-cell leukemia virus type I envelope gene, *J. Virol.,* 62, 4474, 1988.
80. **Duncan, S. A. and Smith, G. L.,** Vaccinia virus gene SalF2R is non-essential for virus replication in vitro, unpublished data.
80a. **Duncan, S. A. and Smith, G. L.,** Vaccinia virus gene SalF5R is non-essential for virus replication *in vitro* and *in vivo, J. Gen. Virol.,* in press.
81. **Engelstad, M., Howard, S. T., and Smith, G. L.,** A constitutively expressed vaccinia virus gene encodes a 42 kDa glycoprotein related to complement control factors that form part of the extra-cellular envelope, *Virology,* submitted for publication.
82. **Banham, A. H. and Smith, G. L.,** Vaccinia virus gene B12R is non-essential for virus replication in vitro, unpublished data.
83. **Buller, R. M. L., Smith, G. L., Cremer, K., Notkins, A. L., and Moss, B.,** Decreased virulence of recombinant vaccinia virus expression vectors is associated with a thymidine kinase-negative phenotype, *Nature,* 317, 813, 1985.
84. **Williamson, J. D., Reith, R. W., Jeffrey, L. J., Arrand, J. R., and Mackett, M.,** Biological characterization of recombinant vaccinia viruses in mice infected by the respiratory routes, *J. Gen. Virol.,* 71, 2761, 1990.
85. **Taylor, G., Stott, E. J., Wertz, G., and Ball, A.,** Comparison of the virulence of wild-type thymidine kinase (tk)-deficient and tk^+ phenotypes of vaccinia virus recombinants after intranasal inoculation of mice, *J. Gen. Virol.,* 72, 125, 1991.
86. **Kerr, S. M., Johnston, L. H., Odell, M., Duncan, S. A., Law, K. M., and Smith, G. L.,** Vaccinia DNA ligase complements *S. cerevisiae* cdc9, localizer in cytoplasmic factories, and affects virulence and virus sensitivity to DNA damaging agents, *EMBO J.,* 10, 4343, 1991.
87. **Smith, G L.,** Deletion of vaccinia virus thymidylate kinase gene causes virus attenuation, unpublished data.
88. **Itamura, S. Iinuma, H., Shida, H., Morikawa, Y., Nerome, K., and Oya, A.,** Characterization of antibody and cytotoxic T lymphocyte responses to human influenza virus H3 haemagglutinin expressed from the haemagglutinin locus of vaccinia virus, *J. Gen. Virol.,* 71, 2859, 1990.
89. **Smith, G. L., Chan, Y. S., and Howard, S. T.,** Nucleotide sequence of 42 kbp of vaccinia virus strain WR from near the right inverted terminal repeat, *J. Gen. Virol.,* 72, 1349, 1991.
90. **Hashizume, S., Yoshizawa, H., Morita, M., and Suzuki, K.,** Properties of attenuated mutant of vaccinia virus, LC16m8, derived from Lister strain, in *Vaccinia Viruses as Vectors for Vaccine Antigens,* Quinnan, G. V., Ed., Elsevier Science, New York, 1985, 87.
91. **Sugimoto, M., Yasuda, A., Miki, K., Morita, M., Suzuki, K., Uchida, N., and Hashizume, S.,** Gene structures of low-neurovirulent vaccinia virus LC16m0, LC16m8, and their Lister original (LO) strains, *Microbiol. Immunol.,* 29, 421, 1985.
92. **Takahashi-Nishimaki, F., Suzuki, K., Morita, M., Maruyama, T., Miki, K., Hashizume, S., and Sugimoto, M.,** Genetic analysis of vaccinia virus Lister strain and its attenuated mutant LC16m8: production of intermediate variants by homologous recombination, *J. Gen. Virol.,* 68, 2705, 1987.
93. **Morita, M., Suzuki, K., Yasuda, A., Kojima, A., Sugimoto, M., Watanabe, K., Kobayashi, H., Kajima, K., and Hashizume, S.,** Recombinant vaccinia virus LC16m0 or LC16m8 that expresses hepatitis B surface antigen while preserving the attenuation of the parental virus strain, *Vaccine,* 5, 65, 1987.

94. **Feng, G., Chongbai, L., Yao, Y., Ruan, L., and Zhu, J.,** Expression of hepatitis A virus proteins by recombinant vaccinia virus, *Chin. J. Virol.*, 5, 303, 1989.
95. **Rodriguez, D., Rodriguez, J.-R., Rodriguez, J. F., Trauber, D., and Esteban, M.,** Highly attenuated vaccinia virus mutants for the generation of safe recombinant viruses, *Proc. Natl. Acad. Sci. U.S.A.*, 86, 1287, 1989.
96. **Perkus, M. E., Limbach, K., and Paoletti, E.,** Cloning and expression of foreign genes in vaccinia virus, using a host range selection system, *J. Virol.*, 63, 3829, 1989.
97. **Chelyapov, N. V., Antonova, T. P., Yanova, N. N., and Chernos, V. L.,** Antigenic properties of vaccinia virus and of the virus recombinant strains expressing heterologous genes, *Acta Virol. Praha.*, 32, 409, 1988.
98. **Kotwal, G. J., Buller, R. M. L., Kapikian, A. Z., Moss, B., Stephens, E., and Compans, R. W.,** Analysis of recombinant vaccinia virions for envelope-associated foreign proteins, in *Vaccines 89, Modern Approaches to New Vaccine Including Prevention of AIDS,* Lerner, R. A., Ginsberg, H., Chanock, R. M., and Brown, F., Eds., Cold Spring Harbor Press, Cold Spring Harbor, NY, 1989, 389.
99. **Chambers, T. M., Essani, K., and Webster, R. G.,** Conditional expression of foreign genes by temperature-sensitive mutants virus, *Gene,* 95, 275, 1990.
100. **Morgan, A. J., Mackett, M., Finerty, S., Arrand, J. R., Scullion, F. T., and Epstein, M. A.,** Recombinant vaccinia virus expressing Epstein-Barr virus glycoprotein gp340 protects cottontop tamarins against EB virus-induced malignant lymphomas, *J. Med. Virol.*, 25, 189, 1988.
101. **Mackett, M., Smith, G. L., and Moss, B.,** The construction and characterization of vaccinia virus recombinants expressing foreign genes, in *DNA Cloning: A Practical Approach,* Vol. 2, Glover, D. M., Ed., IRL Press, Oxford, 1985, 191.
102. **Piccini, A., Perkus, M. E., and Paoletti, E.,** Vaccinia virus as an expression vector, *Methods Enzymol.*, 153, 545, 1987.
103. **Kieny, M.-P, Lathe, R., Drillien, R., Spehner, D., Skory, S., Schmitt, D., Wiktor, T. J., Korprowski, H., and Lecocq, J.-P.,** Expression of rabies virus glycoprotein from a recombinant vaccinia virus, *Nature,* 312, 163, 1984.
104. **de la Salle, H., Altenburger, W., Elkaim, R., Dott, K., Dieterle, A., Drillien, R., Cazenave, J. P., Tolstoshev, P., and Lecocq, J. P.,** Active gamma-carboxylated human factor IX expressed using recombinant DNA techniques, *Nature,* 316, 268, 1985.
105. **Fathi, Z., Sridhar, P., Pacha, R. F., and Condit, R. C.,** Efficient targeted insertion of an unselected marker into the vaccinia virus genome, *Virology,* 155, 97, 1986.
106. **Mackett, M.,** Vaccinia as a vector for delivering foreign antigens, in *Seminars in Virology,* Vol. 1, Brown, F., Ed., W. B. Saunders, Philadelphia, 1990, 39.
107. **Smith, G. L.,** Vaccinia: virus, vector and vaccine, in *Control of Virus Diseases (45th Symposium of the Society of General Microbiology),* Dimmock, N. J., Griffiths, P. D., and Madeley, C. R., Eds., Cambridge University Press, London, 1990, 77.
108. **Panicali, D., Grzelecki, A., and Huang, C.,** Vaccinia virus vectors utilizing the β-galactosidase assay for rapid selection of recombinant viruses and measurement of gene expression, *Gene,* 47, 193, 1986.
109. **Ramshaw, I. A., Andrew, M. E., Philips, S. M., Boyle, D. B., and Coupar, B. E. H.,** Recovery of immunodeficient mice from a vaccinia virus/IL-2 recombinant infection, *Nature,* 329, 545, 1987.
110. **Franke, C. A., Rice, C. M., Strauss, J. H., and Hruby, D. E.,** Neomycin resistance as a dominant selectable marker for selection and isolation of vaccinia virus recombinants, *Mol. Cell. Biol.*, 5, 1918, 1985.
111. **Boyle, D. B. and Coupar, B. E. H.,** A dominant selectable marker for the construction of recombinant poxviruses, *Gene,* 65, 123, 1988.
112. **Isaacs, S. N., Kotwal, G. J., and Moss, B.,** Reverse guanine phosphoribosyltransferase selection of recombinant vaccinia viruses, *Virology,* 178, 626, 1990.

113. **Dallo, S., Rodriguez, J. F., and Esteban, M.,** A 14K envelope protein of vaccinia virus with an important role in virus-host interactions is altered during virus persistence and determines the plaque size phenotype of the virus, *Virology,* 159, 423, 1987.
114. **Rodiguez, J. F. and Esteban, M.,** Plaque size phenotype as a selectable marker to generate vaccinia virus recombinants, *J. Virol.,* 63, 997, 1989.
115. **Payne, L. G. and Norrby, E.,** Presence of haemagglutinin in the envelope of extracellular vaccinia virus particles, *J. Gen. Virol.,* 32, 63, 1976.
116. **Shida, H.,** Nucleotide sequence of the vaccinia virus hemagglutinin gene, *Virology,* 150, 451, 1986.
117. **Ohishi, K., Maruyama, T., Shida, H., Nishimaki, J., Miki, K., Sagata, N., Ikawa, Y., and Sugimoto, M.,** Immunogenicity of a recombinant vaccinia virus expressing an envelope glycoprotein of bovine leukaemia virus, *Vaccine,* 6, 428, 1988.
118. **Gillard, S., Spehner, D., and Drillien, R.,** Mapping of a vaccinia virus host range sequence by insertion into the viral thymidine kinase gene, *J. Virol.,* 53, 316, 1985.
119. **Perkus, M. E., Goebel, S. J., Davis, S. W., Johnson, G. P., Limbach, K., Norton, E. K., and Paoletti, E.,** Vaccinia virus host range genes, *Virology,* 179, 276, 1990.
120. **Takahashi-Nishimaki, F., Funahasi, S.-I., Miki, K., Hashizume, S., and Sugimoto, M.,** Regulation of plaque size and host range by a vaccinia virus gene related to complement system proteins, *Virology,* 181, 158, 1991.
121. **Gillard, S., Spehner, D., Drillien, R., and Kirn, A.,** Localization and sequence of a vaccinia virus gene required for multiplication in human cells, *Proc. Natl. Acad. Sci. U.S.A.,* 83, 5573, 1986.
122. **Spehner, D., Drillien, R., and Lecocq, J.-P.,** Construction of fowlpox virus vectors with intergenic insertions: expression of the β-galactosidase gene and the measles virus fusion gene, *J. Virol.,* 64, 527, 1990.
123. **Falkner, F. G. and Moss, B.,** Transient dominant selection of recombinant vaccinia viruses, *J. Virol.,* 64, 3108, 1990.
124. **Ball, L. A.,** High-frequency homologous recombination in vaccinia virus DNA, *J. Virol.,* 61, 1788, 1987.
125. **Fuerst, T. R., Niles, E. G., Studier, F. W., and Moss, B.,** Eukaryotic transient expression system based upon recombinant vaccinia virus that synthesizes bacteriophage T7 RNA polymerase, *Proc. Natl. Acad. Sci. U.S.A.,* 83, 8122, 1986.
126. **Fuerst, T. R., Earl, P. L.., and Moss, B.,** Use of a hybrid vaccinia virus T7 RNA polymerase system for expression of target genes, *Mol. Cell. Biol.,* 7, 2538, 1987.
127. **Fuerst, T. R. and Moss, B.,** Structure and stability of mRNA synthesized by vaccinia virus encoded bacteriophage T7 RNA polymerase in mammalian cells. Importance of the 5′ untranslated leader, *J. Mol. Biol.,* 206, 333, 1989.
128. **Elroy-Stein, O., Fuerst, T. F., and Moss, B.,** Cap-independent translation of mRNA conferred by encephalomyocarditis virus 5′ sequence improves the performance of the vaccinia virus/bacteriophage T7 hybrid expression system, *Proc. Natl. Acad. Sci. U.S.A.,* 86, 6126, 1989.
129. **Li, Y., Luo, L. Z., Snyder, R. M., and Wagner, R. R.,** Expression of the M gene of vesicular stomatitis virus cloned in various vaccinia virus vectors, *J. Virol.,* 62, 776, 1988.
130. **Fuerst, T. R., Fernandez, M. P., and Moss, B.,** Transfer of the inducible lac repressor/operator system from *Escherichia coli* to a vaccinia virus expression vector, *Proc. Natl. Acad. Sci. U.S.A.,* 86, 2549, 1989.
131. **Rodriguez, J. F. and Smith, G. L.,** IPTG-dependent vaccinia virus: identification of a virus protein enabling virion envelopment of Golgi membrane and egress, *Nucleic Acids Res.,* 18, 5347, 1990.
132. **Zhang, Y. and Moss, B.,** Inducer-dependent conditional lethal mutant animal viruses, *Proc. Natl. Acad. Sci. U.S.A.,* 88, 1511, 1991.

133. **Coupar, B. E. H., Boyle, D. B., and Both, G. W.,** Effect of *in vitro* mutations in a vaccinia virus early promoter region monitored by herpes simplex virus thymidine kinase expression in recombinant vaccinia virus, *J. Gen. Virol.,* 68, 2299, 1987.
134. **Shuman, S., Golder, M., and Moss, B.,** Insertional mutagenesis of the vaccinia virus gene encoding a type I DNA topoisomerase: evidence that the gene is essential for virus growth, *Virology,* 170, 302, 1989.
135. **DeLange, A. M. and McFadden, G.,** Sequence-nonspecific replication of transfected plasmid DNA in poxvirus-infected cells, *Proc. Natl. Acad. Sci. U.S.A.,* 83, 614, 1986.
136. **Merchlinsky, M. and Moss, B.,** Resolution of linear minichromosomes with hairpin ends from circular plasmids containing vaccinia virus concatemer junctions, *Cell,* 45, 879, 1986.
137. **Vassef, A., Mars, M., Dru, A., Plucienniczak, A., Streeck, R. E., and Beaud, G.,** Isolation of cis-acting vaccinia virus DNA fragments promoting the expression of herpes simplex virus thymidine kinase by recombinant viruses, *J. Virol.,* 55, 163, 1985.
138. **Miner, J. N., Weinrich, S. L., and Hruby, D. E.,** Molecular dissection of cis-acting regulatory elements from 5′-proximal regions of a vaccinia virus late gene cluster, *J. Virol.,* 62, 297, 1988.
139. **de Magistris, L. and Stunnenberg, H. G.,** Cis-acting sequences affecting the length of the poly(A) head of vaccinia virus late transcripts, *Nucleic Acids Res.,* 16, 3141, 1988.
140. **Moss, B., Elroy-Stein, O., Mizukami, T., Alexander, W. A., and Fuerst, T.,** New mammalian expression vectors, (product review), *Nature,* 348, 91, 1990.
141. **Boulter, E. A. and Appleyard, G.,** Differences between extracellular and intracellular forms of poxviruses and their implications, *Prog. Med. Virol.,* 16, 86, 1973.
142. **VanSlyke, J. K. and Hruby, D. E.,** Posttranslational modification of vaccinia virus proteins, *Curr. Top. Microbiol. Immunol.,* 163, 185, 1990.
143. **Rice, C. M., Franke, C. A., Strauss, J. H., and Hruby, D. E.,** Expression of Sindbis virus structural proteins via recombinant vaccinia virus: synthesis, processing, and incorporation into mature Sindbis virions, *J. Virol.,* 56, 227, 1985.
144. **Edwards, R. H., Selby, M. J., Mobley, W. C. Weinrich, S. L., Hruby, D. E. and Rutter, W. J.,** Processing and secretion of nerve growth factor: expression in mammalian cells with a vaccinia virus vector, *Mol. Cell. Biol.,* 8, 2456, 1988.
145. **Gowda, S. D., Stein, B. S., and Engleman, E. G.,** Identification of protein intermediates in the processing of p55 HIV-1 gag precursor in cells infected with recombinant vaccinia virus, *J. Biol. Chem.,* 264, 8459, 1989.
146. **Lobigs, M., Zhao, H. X., and Garoff, H.,** Function of Semliki Forest virus E3 peptide in virus assembly: replacement of E3 with an artificial signal peptide abolishes spike heterodimerization and surface expression of E1, *J. Virol.,* 64, 4346, 1990.
147. **Eaton, B. T., Gould, A. R., Hyatt, A. D., Coupar, B. E., Martyn, J. C., and White, J. R.,** A bluetongue serogroup-reactive epitope in the amino terminal half of the major core protein VP7 is accessible on the surface of bluetongue virus particles, *Virology,* 180, 687, 1991.
148. **Mackett, M., Yilma, T., Rose, J. K., and Moses, B.,** Vaccinia virus recombinants: expression of VSV genes and protective immunization of mice and cattle, *Science,* 227, 433, 1985.
149. **Mackett, M. and Arrand, J. R.,** Recombinant vaccinia virus induces neutralising antibodies in rabbits against Epstein-Barr virus membrane antigen gp340, *EMBO J.,* 4, 3229,1985.
150. **Gregoriades, A., Guzman, G. G., and Paoletti, E.,** The phosphorylation of the integral membrane (M1) protein of influenza virus, *Virus Res.,* 16, 27, 1990.
151. **Connelly, P. A. and Stern, D. F.,** The epidermal growth factor receptor and the product of the neu protooncogene are members of a receptor tyrosine phosphorylation cascade, *Proc. Natl. Acad. Sci. U.S.A.,* 87, 6054, 1990.

152. **Smith, G. L., Levin, J. Z., Palese, P., and Moss, B.,** Synthesis and cellular location of the ten influenza virus polypeptides individually expressed by recombinant vaccinia viruses, *Virology,* 160, 336, 1987.
153. **Stamatos, N. M., Chakrabarti, S., Moss, B., and Hare, J. D.,** Expression of polyomavirus virion proteins by a vaccinia virus vector: association of VP1 and VP2 with the nuclear framework, *J. Virol.,* 61, 516, 1987.
154. **Siomi, H., Shida, H., Nam, S. H., Nosaka, T., Maki, M., and Hatanaka, M.,** Sequence requirements for nucleolar localization of human T cell leukemia virus type I pX protein, which regulates viral RNA processing, *Cell,* 55, 197, 1988.
155. **Sullivan, V. and Smith, G. L.,** The herpes simplex virus type 1 US7 gene product is a 66K glycoprotein and is a target for complement-dependent virus neutralization, *J. Gen. Virol.,* 69, 859, 1988.
156. **Cranage, M. P., Smith, G. L., Bell, S. E., Hart, H., Brown, C., Bankier, A. T., Tomlinson, P., Barrell, B. G., and Minson, A. C.,** Identification and expression of a human cytomegalovirus (HCMV) glycoprotein with homology to EBV BXLF2 gene product, VZV gpIII and HSV-1 glycoprotein H, *J. Virol.,* 62, 1416, 1988.
157. **Wiktor, T. J., Macfarlan, R. I., Reagan, K. J., Dietzschold, B., Curtis, P. J., Wunner, W. H., Kieny, M.-P., Lathe, R., Lecocq, J.-P., Mackett, M., Moss, B., and Koprowski, H.,** Protection from rabies by a vaccinia virus recombinant containing the rabies virus glycoprotein gene, *Proc. Natl. Acad. Sci. U.S.A.,* 81, 7194, 1984.
158. **Crise, B., Buonocore, L., and Rose, J. K.,** CD4 is retained in the endoplasmic reticulum by the human immunodeficiency virus type 1 glycoprotein precursor, *J. Virol.,* 64, 5585, 1990.
159. **Berger, E. A., Fuerst, T. R., and Moss, B.,** A soluble recombinant polypeptide comprising the amino-terminal half of the extracellular region of the CD4 molecule contains an active binding site for human immunodeficiency virus, *Proc. Natl. Acad. Sci. U.S.A.,* 85, 2357, 1988.
160. **Machamer, C. E., Mentone, S. A., Rose, J. K., and Farquhar, M. G.,** The E1 glycoprotein of an avian coronavirus is targeted to the cis Golgi complex, *Proc. Natl. Acad. Sci. U.S.A.,* 87, 6944, 1990.
161. **Zagouras, P., Ruusala, A., and Rose, J. K.,** Dissociation and reassociation of oligomeric viral glycoprotein subunits in the endoplasmic reticulum, *J. Virol.,* 65, 1976, 1991.
162. **Li, Y., Luo, L. Z., Snyder, R. M., and Wagner, R. R.,** Site-specific mutations in vectors that express antigenic and temperature-sensitive phenotypes of the M gene of vesicular stomatitis virus, *J. Virol.,* 62, 3729, 1988.
163. **Whitt, M. A., Chong, L., and Rose, J. K.,** Glycoprotein cytoplasmic domain sequences required for rescue of a vesicular stomatitis virus glycoprotein mutant, *J. Virol.,* 63, 3569, 1989.
164. **Pattnaik, A. K. and Wertz, G. W.,** Replication and amplification of defective interfering particle RNAs of vesicular stomatitis virus in cells expressing viral proteins from vectors containing cloned cDNAs, *J. Virol.,* 64, 2948, 1990.
165. **Pattnaik, A. K. and Wertz, G. W.,** Cells that express all five proteins of vesicular stomatitis virus from cloned cDNAs support replication, assembly, and budding of defective interfering particles, *Proc. Natl. Acad. Sci. U.S.A.,* 88, 1379, 1991.
166. **Ashorn, P. A., Berge, E. A., and Moss, B.,** Human immunodeficiency virus envelope glycoprotein/CD4-mediated fusion of nonprimate cells with human cells, *J. Virol.,* 64, 2149, 1990.
167. **Barrett, N., Mitterer, A., Mundt, W., Eibl, J. Eibl, M., Gallo, R. C., Moss, B., and Dorner, F.,** Large-scale production and purification of a vaccinia recombinant-derived HIV-1 gp160 and analysis of its immunogenicity, *AIDS Res. Hum. Retroviruses,* 5, 159, 1989.

168. **Gritz, L., Destree, A., Cormier, N., Day, E., Stallard, V., Caiazzo, T., Mazzara, G., and Panicali, D.,** Generation of hybrid genes and proteins by vaccinia virus-mediated recombination: application to human immunodeficiency virus type 1 env, *J. Virol.*, 64, 5948, 1990.
169. **Yilma, T., Ristow, S. S., Moss, B., and Jones, L.,** A novel approach for the production of monoclonial antibodies using infectious vaccinia virus recombinants, *Hybridoma*, 6, 329, 1987.
170. **Malvoisin, E. and Wild, F.,** Contribution of measles virus fusion protein in protective immunity: anti-F monoclonal antibodies neutralize infectivity and protect mice against challenge, *J. Virol.*, 64, 5160, 1990.
171. **McLean, C. S., Churcher, M. J., Meinke, J., Smith, G. L., Higgins, G., Stanley, M., and Minson, A. C.,** Production and characterisation of a monoclonial antibody to human papillomavirus type 16 using recombinant vaccinia virus, *J. Clin. Pathol.*, 43, 488, 1990.
172. **Boyer, V., Broly, H., Souche, S., Madaule, P., Rossier, J., Zagury, D., and Desgranges, C.,** Characterization and largescale production of human monoclonal antibodies against the HIV-1 envelope, *Clin. Exp. Immunol.*, 83, 452, 1991.
173. **Moss, B., Smith, G. L., Gerin, J. L., and Purcell, R. H.,** Live recombinant vaccinia virus protects chimpanzees against hepatitis B, *Nature*, 311, 67, 1984.
174. **Cranage, M. P., Kouzarides, T., Bankier, A. T., Satchwell, S., Weston, K., Tomlinson, P., Barrell, B. G., Bell, S. E., Minson, A. C., and Smith, G. L.,** Identification of the human cytomegalovirus glycoprotein B gene and induction of neutralizing antibodies via its expression in recombinant vaccinia virus, *EMBO J.*, 5, 3057, 1986.
175. **Hu, S.-L., Klaniecki, J., Sridhar, P., and Travis, B. M.,** Neutralizing antibodies against HIV-1 BRU and SF2 isolates generated in mice immunized with recombinant vaccinia virus expressing HIV-1 (BRU) envelope glycoproteins and boosted with homologous gp160, *AIDS Res. Hum. Retroviruses*, 7, 615, 1991.
176. **Bennink, J. R. and Yewdell, J. W.,** Recombinant vaccinia viruses as vectors for studying T lymphocyte specificity and function, *Curr. Top. Microbiol. Immunol.*, 163, 153, 1990.
177. **Bennink, J. R., Yewdell, J. W., Smith, G. L., Moller, C., and Moss, B.,** Recombinant vaccinia virus primes and stimulates influenza haemagglutinin-specific cytotoxic T cells, *Nature*, 311, 578, 1984.
178. **Yewdell, J. W., Bennink, J. R., Smith, G. L., and Moss, B.,** Influenza A virus nucleoprotein is a major antigen for influenza A virus cross-reactive cytotoxic T lymphocytes, *Proc. Natl. Acad. Sci. U.S.A.*, 82, 1785, 1985.
179. **Bennink, J. R., Yewdell, J. W., Smith, G. L., and Moss, B.,** Recognition of cloned influenza virus haemagglutinin gene products by cytotoxic T lymphocytes, *J. Virol.*, 57, 786, 1986.
180. **Bennink, J. R., Yewdell, J. W., Smith, G. L., and Moss, B.,** Anti-influenza cytotoxic T lymphocytes recognize the three viral polymerases and a nonstructural protein: responsiveness to individual virus antigens is major histocompatibility controlled, *J. Virol.*, 61, 1098, 1987.
181. **McMichael, A. J., Mitchie, C. A., Gotch, F. M., Smith, G. L., and Moss, B.,** Recognition of influenza nucleoprotein by human cytotoxic T lymphocytes, *J. Gen. Virol.*, 67, 719, 1986.
182. **Gotch, F. M., McMichael, A. J., Smith, G. L., and Moss, B.,** Identification of the viral molecules recognized by influenza specific human cytotoxic T lymphocytes, *J. Exp. Med.*, 165, 408, 1987.
183. **Andrew, M. E., Coupar, B. E. H., Boyle, D. B., and Ada, G. L.,** The roles of influenza virus haemagglutinin and nucleoprotein in protection: analysis using vaccinia virus recombinants, *Scand. J. Immunol.*, 25, 21, 1987.

184. **Gould, K., Gossins, J., Bastin, J., Brownlee, G. G., and Townsend, A. R. M.,** A 15 amino acid fragment of influenza virus nucleoprotein synthesised in the cytoplasm is presented to class 1-restricted cytotoxic T lymphocytes, *J. Exp. Med.*, 170, 1051, 1989.
185. **Whitton, J. L. and Oldstone, M. B.,** Class I MHC can present an endogenous peptide to cytotoxic T lymphocytes, *J. Exp. Med.*, 170, 1033, 1989.
186. **Sweetser, M. T., Morrison, L. A., Braciale, V. L., and Braciale, T. J.,** Recognition of pre-processed endogenous antigen by class I but not class II MHC-restricted T-cells, *Nature,* 342, 180,1989.
187. **del Val, M., Volkmer, H., Rothbard, J. B., Jonjic, S., Messerle, M., Schickedanz, J., Reddehase, M. J., and Koszinowski, U. H.,** Molecular basis for cytolytic T-lymphocyte recognition of the murine cytomegalovirus immediate-early protein pp89, *J. Virol.*, 62, 3965, 1988.
188. **Earl, P. L., Moss, B., Morrison, R. P., Wehrly, K., Nishio, J., and Chesebro, B.,** T-lymphocyte priming and protection against Friend leukemia by vaccinia-retrovirus env gene recombinant, *Science,* 234, 728, 1986.
189. **Lathe, R., Kieny, M. P., Gerlinger, P., Clertant, P., Guizanik, I., Cuzin, F., and Chambon, P.,** Tumour prevention and rejection with recombinant vaccinia, *Nature,* 326, 878, 1987.
190. **Bernards, R., Destree, A., McKenzie, S., Gordon, E., Weinberg, R. A., and Panicali, D.,** Effective tumor immunotherapy directed against an oncogone-encoded product using a vaccinia virus vector, *Proc. Natl. Acad. Sci. U.S.A.*, 84, 6854, 1987.
191. **Estin, C. D., Stevenson, U. S., Plowman, G. D., Hu, S.-L., Sridhar, P., Hellstrom, I., Brown, J. P., and Hellstrom, K. E.,** Recombinant vaccinia virus vaccine against the human melanoma antigen p97 for use in immunotherapy, *Proc. Natl. Acad. Sci. U.S.A.*, 85, 1052, 1988.
192. **Hu, S.-L, Plowman, G. D., Sridhar, P., Stevenson, U. S., Brown, J. P., and Estin, C. D.,** Characterization of a recombinant vaccinia virus expressing human melanoma-associated antigen p97, *J. Virol.*, 62, 176, 1988.
193. **Hareuveni, M., Gautier, C., Kieny, M.-P., Wreschner, D., Chambon, P., and Lathe, R.,** Vaccination against tumor cells expressing breast cancer epithelial tumor antigen, *Proc. Natl. Acad. Sci. U.S.A.*, 87, 9498, 1990.
194. **Meneguzzi, G., Kieny, M.-P., Lecocq, J.-P., Chambon, P., Cuzin, F., and Lathe, R.,** Vaccinia recombinants expressing early bovine papilloma virus (BPV1) proteins: retardation of BPV1 tumour development, *Vaccine,* 8, 199, 1990.
195. **Meneguzzi, G., Cerni, C., Kieny, M.-P., and Lathe, R.,** Immunization against human papillomavirus type 16 tumor cells with recombinant vaccinia viruses expressing E6 and E7, *Virology,* 181, 62, 1991.
196. **Kotwal, G. J., Isaacs, S. N., McKenzie, R., Frank, M. M., and Moss, B.,** Inhibition of the complement cascade by the major secretory protein of vaccinia virus, *Science,* 250, 827, 1990.
197. **Goebel, S. J., Johnson, G. P., Perkus, M. E., Davis, S. W., Winslow, J. P., and Paoletti, E.,** The complete DNA sequence of vaccinia virus, *Virology,* 179, 247, 1990.
198. **Mackett, M.,** pS1, a vaccinia insertion vector derived from pSC11, unpublished data.
199. **Mackett, M.,** Vaccinia virus co-insertion vectors using the 19K early promoter and *Eco*gpt selection, unpublished data.
200. **Smith, G. L., Howard, S. T., and Chan, Y. S.,** Vaccinia virus encodes a family of genes with homology to serine proteinase inhibitors, *J. Gen. Virol.*, 70, 2333, 1989.

Chapter 4

POXVIRUS RECOMBINANTS AS LIVE VACCINES

William I. Cox, James Tartaglia, and Enzo Paoletti

TABLE OF CONTENTS

I. CONSIDERATIONS FOR A SUCCESSFUL VACCINE

For any vaccine to be developed, marketed, and accepted for use by the general public, a variety of stringent ethical, economical, and practical criteria must be met. These considerations focus primarily, and rightfully so, on the risk-to-benefit issue (safety vs. efficacy), but also include the costs of development and production, stability in the field, and duration of immunity, as well as marketing considerations which often reflect desirable features of the vaccine such as ease of administration, among others.

The risk-to-benefit issue may be viewed as a flexible balance depending upon the seriousness of the target disease, current availability of safe and efficacious immunogens or therapies, and the perceptions of the threat to safety by the public, public health officials, and government regulatory agencies. For example, a vaccine, developed against a highly contagious, life-threatening disease agent, that demonstrates a high degree of efficacy while eliciting some adverse reactions (moderate risk, high benefit), may be considered at least as acceptable for public use as a second vaccine that is less efficacious, although clearly safe (low risk, moderate benefit). These two hypothetical vaccines may be viewed as possessing similar low risk-to-benefit ratios when considering the seriousness of the threat to those unvaccinated and at risk for the disease agent. In contrast, for a vaccine candidate for a disease with less severe consequences, such as transient viral enteritis, any adverse reactions elicited by the vaccine may be perceived as completely unacceptable. Thus, the acceptable degree of risk-to-benefit derived from a vaccine candidate is relative and must be evaluated for each disease.

A successful vaccine must also be practical to use. Stability is a necessary requirement for the vaccine to be successfully transported from the site of production or storage to the vaccinee while still retaining its efficacy, including potency and safety characteristics. Another practical consideration is the ease with which a vaccine may be administered. At the public level, this may affect compliance with vaccination programs, depending upon the frequency and number of immunizations necessary to achieve protection. No more than two immunizations that would yield adequate immunity lasting a year in duration is a realistic expectation for a practical vaccine.

Economic and associated marketing considerations also influence the availability of vaccines to the public. A vaccine must be relatively easy and inexpensive to produce in mass amounts, especially if competing vaccines are available or the vaccine is targeted to certain veterinary markets, such as poultry or livestock, where cost-to-benefit considerations routinely outweigh risk-to-benefit. In a society where pressures exist to recover research and development costs, marketing considerations must be realistically faced.

II. POXVIRUSES AS VACCINE VECTORS

A. VACCINIA VIRUS

Vaccinia virus as a live vector of vaccine antigens promises to meet many of the aforementioned criteria of safety, efficacy, practicality, and economy. After almost 200 years of use as a vaccine against smallpox, vaccinia virus is one of the best characterized viruses as a vaccine component. Its long history of use has yielded a wealth of information on its effects in man and other species. Economical production techniques capable of producing large numbers of doses have been developed. Its stability at room temperature following freeze drying has been clearly established. Vaccinia virus is easy to administer. Large, nonessential regions of its genome are available for the insertion of foreign DNAs into multiple cloning sites.[1] The vaccinia virus genome has been extensively mapped and, more recently, entirely sequenced.[2a,2b] It is also relatively easy to recover recombinants from transfected cells by marker transfer techniques (see Chapter 3). Furthermore, vaccinia virus displays a broad host range. These are all factors that enhance the attractiveness of vaccinia virus as a vaccine expression vector.

Another benefit of its long history as a vaccine is that it has been the subject of many basic research investigations characterizing its biological properties.[3,4] Since the first recombinant vaccinia viruses capable of expressing foreign gene products were constructed,[5,6] its usefulness as a potential vector of vaccine antigens was apparent. In the relatively short period of time since the first vaccinia virus recombinants were developed, reports have been commonplace in the scientific literature that illustrate the use of this important vaccine vector in eliciting both humoral and cellular immune mechanisms and, more importantly, in providing protective immunity in experimental models of disease.

Although the acceptability of the safety (risk-to-benefit) of vaccinia virus in the control and eventual eradication of smallpox is evident, the issue of the infrequent adverse reactions of wild type vaccinia virus is being reassessed.[7] Moreover, the safety of vaccinia virus altered by the introduction of foreign genes into its genome is currently being evaluated on a case by case basis. Eventually, attenuation of vaccinia virus vectors without compromising the potency of foreign gene expression and the associated induction of protective immunity may reduce risk to the point that any benefit gained would justify the use of this vector system, regardless of the lack of severity of the target disease entity (see Chapter 7).

An advantageous approach to attenuating recombinant vaccinia virus expression vectors is the deletion or inactivation of virulence genes or nonessential genes that influence the replication of the virus (for a review of pox virulence factors, see Reference 8). Introduction of various foreign encoding DNA sequences into the vaccinia virus thymidine kinase (TK) gene resulted in diminished virulence for this recombinant virus.[9] Similarly, the deletion of the vaccinia virus growth factor (VGF) gene which encodes a protein demonstrating amino acid sequence homology with epidermal growth factor (EGF) and transforming growth factor alpha (TGF-α) yielded a vaccinia virus recombinant with decreased virulence for mice and rabbits.[10] Other attenuated vaccinia virus expression vectors were derived by deleting a large portion of the left end of the viral genome as well as introducing point mutations in the gene encoding the fusion protein.[11] Others established temperature-sensitive (ts) recombinant vaccinia mutants with attenuated neurovirulence.[12] The identification of virus-encoded proteins that are related to plasma protein inhibitors of serine proteases[13,14] and complement control proteins[15] may also prove to be relevant targets of inactivation for the purpose of attenuation.

An intriguing strategy to vaccinia virus attenuation is the deletion of host range genes. First described in rabbitpox mutants, at least two host range genes have been identified in vaccinia virus.[16–26] Host range genes restrict virus growth in tissue culture cell lines of differing species origin. Such host range mutants as vectors would allow production of the vaccine entity while restricting growth in the species targeted by the vaccine, thereby limiting environmental spread. Expression of the foreign gene product in such a vector would undoubtedly be regulated by strong early promoters.

B. AVIPOXVIRUSES IN NONAVIAN SPECIES

A further extension of using highly attenuated poxviruses as vaccine vectors is to take advantage of poxviruses with naturally occurring restricted host ranges, such as the avipoxviruses (see Chapter 8). Not known to productively replicate in nonavian species, avipoxviruses are capable of undergoing incomplete replication in mammalian cells to the extent necessary to express inserted genes.[27–30] While providing the same advantages as recombinant vaccinia viruses: stability, ease of production, and the ability to provide relevant antigens to the immune system in a context that elicits not only humoral responses, but cell-mediated immunity as well, recombinant avipoxvirus vectors have the added benefit of nontransmissibility among nonavian hosts. Since avipoxviruses can only partially replicate in nonavian species, infectious virus is not shed from the vaccinee, thereby terminating any possibility of dispersal, even among avian species. Moreover, immunocompromised individuals who may be at risk from live vaccines or replicating recombinant vectors may be eligible for vaccination by recombinant avipoxvirus vectors. Pending demonstrations of efficacy, nonreplicating recombinant avipoxviruses may be the vectors of choice because they provide a safe and ecologically sound strategy to achieving recombinant virus vaccination programs.

III. RECOMBINANT POXVIRUS VECTORS AS POTENTIAL VACCINES

A. RABIES VIRUS

Rabies virus, a negative-sense, single-stranded RNA virus, causes a severe and often fatal disease of the central nervous system of warm blooded animals, including man. Of the five structural proteins comprising the rabies virion, only the glycoprotein spans the viral lipid envelope. Thus, antibodies to the glycoprotein are capable of neutralizing rabies virus infectivity. Moreover, virus neutralizing antibodies to the glycoprotein are correlated with protection against rabies.[31] Although effective vaccines against rabies virus currently exist, they are expensive to produce (usually in the brains of suckling mice or in tissue culture) and must be administerd almost exclusively by the intramuscular route. Rabies glycoprotein produced in recombinant bacteria has failed to adequately protect due to the absence of proper posttranslational modifications to the polypeptide. Utilizing a live virus vector system to express the glycoprotein in cells of the host species would overcome this shortcoming (see Chapter 5).

Rabies virus glycoprotein cDNA was cloned into the tk gene of vaccinia virus under the control of an early-late promoter.[32] Inoculation of rabbits with the recombinant vaccinia virus elicited high titers of rabies virus-neutralizing antibody, demonstrating the faithful expression of the glycoprotein. Following immunization with the same recombinant virus, mice were protected against a lethal experimental challenge with street rabies virus. A vaccinia recombinant containing a codon for leucine instead of proline at the position for amino acid 8 in the glycoprotein gene failed to elicit neutralizing antibodies or protect mice from rabies challenge.[33] With proline in the same position, virus neutralizing antibodies were detected in mice following immunization and provided protection against challenge. Furthermore, vaccination with the recombinant virus primed mice to produce virus specific cytotoxic T lymphocytes following the *in vitro* culture of spleen cells with rabies virus.

Following these promising experimental results, recombinant vaccinia viruses containing the rabies glycoprotein gene were tested in a variety of vaccine target species. Foxes (*Vulpes vulpes* L.) developed rabies virus-neutralizing antibody titers and completely resisted experimental infection with a street strain of rabies virus following subcutaneous, intradermal, and oral immunization.[34] In another study of 13 fox cubs orally administered vaccinia virus-rabies glycoprotein, 12 developed virus neutralizing titers, of which 11 were protected from experimental infection.[35] Virus-neutralizing antibodies and protection were apparent as early as 28 and 33 d post-vaccination, respectively. Although virus-neutralizing antibodies were not detected 1 year after immunization, the cubs survived rabies challenge infection, demonstrating that a single oral immunization with the recombinant vaccinia virus rabies glycoprotein vaccine primed a potent anamnestic response. Aside from the efficacy aspects, residual pathogenicity of the vaccine vector was not observed, nor could horizontal transmission of the vector be demonstrated as evidenced by the absence of relevant virus-neutralizing antibody titers and the absence of acquisition of protective immunity to challenge in unvaccinated foxes kept in close contact with the vaccinates.

Studies of immunity and protection elicited by live vaccinia virus containing the rabies glycoprotein gene have yielded similar results in wild species that serve as the major reservoir of rabies virus, as well as in dogs. Oral immunization of racoons (*Procyon lotor*) and skunks (*Mephitis mephitis*) with a rabies glycoprotein recombinant vaccinia virus stimulated virus-neutralizing antibody titers and afforded protection against virulent rabies virus challenge, as were dogs following classical intramuscular immunization.[36–39] Successful oral rabies vaccination of racoons with a racoon recombinant poxvirus expressing the rabies virus glycoprotein has also been demonstrated.[40]

Initial application to the U.S. Department of Agriculture to field test a live vaccinia vectored rabies vaccine in South Carolina and Virginia was met with resistance, based primarily upon ecological concerns, despite the issuance of a "Finding of No Significant Impact".[41] This report concluded that (1) the expression of the rabies glycoprotein gene inserted into the tk gene of vaccinia virus cannot induce rabies, (2) the vaccine virus causes no adverse clinical signs or lesions, (3) horizontal transmission to nonvaccinated contact animals cannot be demonstrated, (4) the gene insertion is stable, (5) the recombinant vaccinia virus does not cause cancer, and (6) the recombinant vaccinia virus vaccine is safe, efficacious, and nonpathogenic in many animal species. Subsequently, the State of Virginia allowed limited field trials.[42] Preliminary results from these open field trials confirmed the positive results of more extensive field trials performed in a number of European countries over a several year period.[43,44] Recombinant vaccinia virus containing the rabies glycoprotein gene may be the first live virus vaccine licensed for use in baits directed against wildlife rabies.

Recently, a fowlpoxvirus recombinant expressing the rabies glycoprotein was constructed and tested for expression and ability to protect several non-avian species from challenge with virulent rabies virus.[29] After infection with the fowlpox recombinant, both permissive (avian) and nonpermissive (non-avian) cells expressed the rabies glycoprotein as a membrane-associated antigen, despite the inability of the virus to replicate in the nonpermissive cells. Following immunization with the fowlpox recombinant, mice produced rabies virus-neutralizing antibodies and were protected from a lethal challenge of virulent rabies virus. When rabbits, rats, cats, dogs, and cattle were vaccinated by one of several different routes (intradermal, subcutaneous, or intramuscular) with the recombinant fowlpox, each developed significant antibody titers to rabies glycoprotein by 14 d postinoculation. The dogs and cats which were administered the fowlpox recombinant were able to withstand an experimentally induced fatal infection with rabies virus. When the cattle were revaccinated with the fowlpox recombinant, rabies glycoprotein antibody titers increased. Furthermore, when the recombinant virus was inactivated by β-propiolactone and injected into rabbits, antibodies were produced which reacted only with fowlpox antigens, but not with rabies glycoprotein, indicating that *de novo* expression of the foreign gene was essential. These preliminary studies indicated that fowlpox recombinants, though protective against rabies challenge, were less efficient than vaccinia rabies recombinants. Further work has demonstrated that other members of the avipoxvirus genus (canarypox), when engineered to express the rabies glycoprotein, are as efficacious in protection studies as vaccinia recombinants.[45]

B. HEPATITIS B VIRUS

With over 200 million acute or chronic cases, some progressing to hepatocellular carcinoma, hepatitis B virus infections represent a health problem of global proportions.[46] Vaccination with the hepatitis B virus surface antigen (HBsAg) confers immunity; however, it is difficult to obtain as a vaccine since the virus is unable to grow in tissue culture. Antibodies elicited by HBsAg have been described as virus-neutralizing and protective, although the role of cell-mediated immunity has yet to be evaluated.[47] HBsAg, derived from the plasma of chronically infected individuals as 22-nm noninfectious lipoprotein particles, has been helpful, but is in limited supply, requiring purification and extensive safety testing prior to use. HBsAg has been produced as 22-nm particles in yeast and although this expression system fails to secrete the particle,[48–50] significant amounts of the expressed efficacious product are available.

The major protein located on the surface of the 22-nm particle is the S protein, along with the minor proteins, MS (middle surface protein) and LS (large surface protein). While the hepatitis B virus surface structural unit is composed of dimers of S proteins, the LS and MS proteins are thought to be responsible for mediating attachment to hepatocytes.[51,52] Each of the three proteins share the same carboxyl terminus.[53–55] Furthermore, a single open reading frame is divided into S, pre-S1, and pre-S2 regions.[46] The S protein is encoded by the S region, whereas the MS protein is encoded by the S and pre-S regions. All three regions encode the LS protein.

Recombinant vaccinia virus vectors were constructed that expressed HBsAg as secreted 22-nm particles that were indistinguishable from native HBsAg as determined by buoyancy in CsCl density gradients, detection using a commercially available radioimmunoassay (RIA), and by electron microscopy.[56,57] Moreover, rabbits produced high levels of circulating antibodies to authentic HBsAg following a single immunization with the recombinant vaccinia vectors. Following intradermal inoculation with a HBsAg-expressing vaccinia vector, chimpanzees failed to produce detectable levels of circulating anti-HBsAg antibodies.[58] Nonetheless, they were protected from clinical signs of hepatitis after challenge with hepatitis B virus. Following the challenge, a rapid and strong anti-HBsAg response was detected in the vaccinates that was not observed in the control animal, suggesting that the immunization primed a protective anamnestic antibody response.

As a further extension of these studies, recombinant vaccinia viruses were constructed containing the hepatitis B virus coding sequences for the LS or MS antigens.[59,60] Cells infected with the LS-vaccinia virus (containing the full open reading frame) produced polypeptides that corresponded to both the unglycosylated and glycosylated forms of LS. However, LS was not found to be secreted, despite glycosylation. Nonetheless, the recombinant virus stimulated the production of circulating antibodies that recognized epitopes encoded by S and pre-S regions. In contrast, cells infected with recombinant vaccinia virus carrying only the S and pre-S regions expressed glycosylated MS as particles that resembled human hepatitis B surface antigen as well as particles comprised entirely of S antigen. The MS particles were distinguishable from S-containing particles by their ability to bind to polymerized human albumin, suggesting the presence of the portion of the polypeptide encoded by the pre-S2 region. Additionally, rabbit antiserum to the MS particles inhibited the binding of human HBsAg derived from human plasma from binding polymerized human albumin, whereas antiserum against the S recombinant vaccinia virus failed to do so.

C. HERPES VIRUSES

1. Herpes Simplex Virus

Members of the herpes virus family are large, complex DNA viruses that produce a variety of human diseases that exhibit complex clinical, and sometimes fatal, symptomology. Herpes simplex virus (HSV) causes localized, acute primary infections and persists through latent infection of the nerve ganglia. Periodically, the latent virus is reactivated and produces cutaneous lesions despite high levels of circulating antibodies to the virus. Thus, the morbidity associated with HSV infections is characterized by recurrent episodes that can endure for life. The rapid increase in incidence of HSV infection has heightened the desirability for an effective vaccine.

HSV expresses a number of antigenic glycoproteins on the surface of the envelope, several of which may serve as pertinent targets for vaccine strategies. Cell penetration by the virus appears to be mediated by gB while gC may serve as a receptor for the complement component, C3b.[61–63] Glycoprotein D, one of the best characterized surface determinants of HSV, has been implicated in cell fusion and elicits virus-neutralizing antibodies.[64,65] Virus-neutralizing antibodies also target gG and gH.[66,67]

Since HSV infections and associated latency are quite complicated, it is not surprising that immunity to this virus is also complex, involving not only the evocation of virus-neutralizing antibodies but also the inclusion of cell-mediated immunity. Recombinant vaccinia viruses expressing HSV gD illustrate the complexity of immune responses mediating protection. A vaccinia virus vector containing the cloned gene for gD expressed the authentic, membrane-bound glycoprotein in tissue culture cells.[57] When injected into rabbits, antibodies were produced which neutralized the virus. Moreover, mice immunized with the recombinant virus were protected against a lethal, experimental challenge with HSV. Similar results, as well as a reduction in latency in mice, were obtained with another vaccinia virus recombinant expressing gD.[68] In these studies, gD from HSV type 1 elicited virus-neutralizing antibodies and protected mice from fatal infection with HSV types 1 and 2. Furthemore, the majority of vaccinated mice were also protected from latent infection since virus could not be recovered from the trigeminal ganglia of mice challenged by the lip route.[68]

A number of studies have focused on the role of cell-mediated immunity in protection or recovery from HSV infections. Immunization of mice with vaccinia virus expressing gD led to a state of split tolerance in that mice produced T-lymphocyte dependent virus-neutralizing antibodies, but could effectively clear only a low-dose virus challenge and were unable to demonstrate virus-specific delayed-type hypersensitivity.[69] The extent of protection afforded mice by the recombinant vaccinia virus was less than that elicited by inactivated HSV which also was capable of evoking delayed-type hypersensitivity. Furthermore, passive transfer of lymph node cells treated with antibodies reactive against cytotoxic T cells and with complement failed to adequately protect recipient mice from HSV infection. In other studies, vaccinia virus vectors expressing gD elicited T suppressor cells capable of inhibiting the development of class I MHC restricted, but not Class II restricted, cytotoxic T cells (see Chapter 6).[70] In guinea pigs, protection against primary, recurrent, and latent infection following HSV type 2 challenge was provided by a recombinant vaccinia virus expressing type 1 gD, but incomplete protection was obtained by a vaccinia recombinant expressing type 2 gD.[71] Protection correlated with the appearance of virus-specific lymphoproliferative responses and lymphokine elaboration that stimulated natural killer cells, but not with virus-neutralizing antibodies. The results of these studies suggested an important role of cell-mediated immunity in complete protection against HSV infections.

When the temporal regulation of expression of HSV type 1 gD in antigen-presenting cells was examined, it was found that lasting protection in mice was affected when the expression of gD was under the control of vaccinia early promoters.[72,73] Early promoter control led to gD expression on the cell surface, which elicited significantly higher virus-specific delayed-type hypersensitivity and T-cell proliferative responses than when gD was regulated by a late vaccinia promoter. Although both recombinants, with gD regulated by early or late promoters, induced virus-neutralizing antibodies, only early expression of gD provided long-term protection against challenge. When tested in the guinea pig model of recurrent disease, similar results were noted.[74] Immunized guinea pigs produced virus-neutralizing antibodies, but only those immunized with a vaccinia virus vector expressing gD under the control of an early promoter produced significant virus-specific T-cell responses and were protected from primary and recurrent herpes lesions following challenge. This clearly indicated that proper expression of gD in accessory antigen presenting cells by recombinant vaccinia virus vectors has consequences upon the quality of the immune response and associated protection.

Cell-mediated immunity may also be important in protection or recovery from HSV infections in man. Human cytotoxic T-cell clones have been established which lyse autologous target cells infected with HSV type 1 or 2 following *in vitro* stimulation with a vaccinia virus recombinant expressing type 1 gD.[75] Additionally, the T-cell clones killed targets infected with the gD-expressing vaccinia virus, demonstrating that gD contains a target structure for human cytotoxic T lymphocytes.

The role of immunity of other surface glycoproteins against HSV infections has also been investigated. A recombinant vaccinia virus expressing gB was shown to induce MHC-restricted cytotoxic T lymphocytes in mice.[76] Expression of gB also elicited virus-neutralizing antibodies in mice, as well as protection against a lethal HSV 1 infection.[77] HSV-neutralizing antibodies were also raised in rabbits and mice which were shown to be complement-dependent.[66]

The effectiveness of revaccination with recombinant vaccinia virus vectors expressing HSV gD seems controversial. While enhanced virus-neutralizing antibody titers following booster immunizations were observed in rabbits[78] and cattle,[79] declining titers have been reported in mice.[80] Nonetheless, mice that had recovered from experimental HSV infections produced virus-neutralizing antibodies following immunization with a gD recombinant vaccinia virus.[81] These studies demonstrated the feasibility of stimulating an anamnestic response with vaccinia vectors despite preexisting immunity to the vector or the foreign gene product.

2. Cytomegalovirus

Another member of the herpes virus family which may cause morbidity and mortality in man is cytomegalovirus (CMV). As a ubiquitous agent in human populations, CMV infection of the immunocompetent individual occasionally leads to disease. However, the consequences of CMV infection in the immunocompromised host are serious and often life threatening. CMV infection is a frequent complication in organ transplantation patients.[82,83] More recently, CMV has emerged as an important cause of disease in patients with acquired immunodeficiency syndrome.[84–86] Furthermore, CMV is the most frequent cause of intrauterine infections, often resulting in severe central nervous system pathology in the developing fetus.[87]

Immunity against CMV infection is thought to be directed against the components of the virion envelope. The envelope of CMV contains at least eight distinct glycoproteins of which several are likely targets of protective responses.[88] Monoclonal antibodies and CMV-reactive human sera neutralize infectious virus.[89–91] Furthermore, immunoaffinity-purified CMV gB or specific peptides can induce virus-neutralizing antibodies in experimental animals.[92,93] Glycoprotein B of CMV, also termed gp55-116 complex, is located within the mature virion as a heterodimer of proteins of Mr 116,000 and 55,000. Intracellular forms of the two mature proteins exist as well as glycosylated intermediates of Mr 150,000 and 160,000.

Recombinant vaccinia virus harboring the gene for gB of CMV after infecting human MRC-5 cells, expresses a glycoprotein product recognized by monoclonal antibodies with specificity for CMV-derived gB.[94] In addition, the recombinant gene product is immunoprecipitated by human sera from individuals known to be chronically infected with CMV. Following immunization with the recombinant vaccinia virus, rabbits produced antisera which also immunoprecipitated authentic gB as well as neutralizing infectious virus *in vitro*. Others have also cloned the gB gene into vaccinia virus as well as into *Escherichia coli*.[95] The recombinant *E. coli* produced a protein the size of the nonglycosylated precursor found in infected cells which reacted with anti-CMV gp55 monoclonal antibodies. Mice immunized with the *E. coli*-gB product produced virus-neutralizing antibodies that were not dependent upon complement. In contrast, mice immunized with the recombinant vaccinia virus, expressing the Mr 55,000, 150,000, and 160,000 intracellular forms, produced antisera which neutralized infectious CMV only in the presence of complement. The results of studies such as these implicate CMV gB as a useful component of a CMV vaccine.

Cell-mediated immunity has also been implicated in the control of CMV infections. Cytotoxic T lymphocytes have been suggested as important participants, since the inability to mount virus-specific cytotoxic T lymphocyte responses correlates with disease progression. This contrasts sharply to the poor correlation between virus-specific antibody titers and recovery from infection.

As a model of CMV infection of humans, murine cytomegalovirus (MCMV) has been used to investigate the role of cell-mediated immunity in protection and recovery. Unlike studies focusing on humoral responses to CMV, a nonstructural protein of MCMV, pp89, has been found to be the major antigenic target recognized by cytotoxic T lymphocytes.[96] The first viral-encoded protein produced following infection, pp89, a nonstructural regulator of the immediate-early phase of replication, is localized in the nucleus of infected cells. Adoptively transferred pp89-specific cytotoxic T lymphocytes of the $CD8^+$ subset limit experimentally induced viral infection while preventing tissue damage and protect against lethal disease. Infection of mice with vaccinia virus-expressing pp89 protected against a lethal infection of MCMV.[97] Protection was entirely mediated by $CD8^+$ T lymphocytes since their depletion by anti-CD8 and complement abolished protective immunity. Furthermore,

$CD8^+$ T lymphocytes from donors immunized with the vaccinia virus-pp89 recombinant protected passively transferred donor mice by limiting MCMV replication. No protective effect was demonstrated by $CD4^+$ T lymphocytes or antiserum to pp89. By the introduction of mutations in the pp89 gene and expression of the fragments by recombinant vaccinia virus, it was found that a region of the protein spanning amino acids 154 to 249 was sufficient to induce cytotoxic T lymphocytes as well as serve as the target determinant recognized by cytotoxic T lymphocytes.[98] Peptide-pulse experiments confirmed that cytotoxic T lymphocytes recognize this epitope in association with class I major histocompatibility antigens.

A similar immune phenomenon has been reported regarding human CMV.[99] Shortly following the infection of permissive cells by CMV, several nonstructural virus-encoded proteins appear in the nucleus. The most predominant is the 72 kD IE (immediate-early) protein.[100] Using clonal analysis of precursors of cytotoxic T lymphocytes from persistently infected donors and autologous target B cells or fibroblasts infected with recombinant vaccinia viruses expressing either gB or the 72 kDa IE protein of CMV, the effector populations were characterized. Only a small proportion of cytotoxic T lymphocytes recognized gB in contrast to the 72 kDa IE protein. This suggests, as demonstrated in the MCMV model, that a nonstructural IE protein serves as a major recognition antigen in chronically infected individuals. Since CMV grows only slowly in permissive diploid cells, using attenuated vaccinia viruses as vectors to express gB and the 72 kDa IE protein to elicit virus-neutralizing antibodies and cytotoxic T lymphocytes, respectively, may prove helpful in the intervention of CMV related diseases.

3. Epstein-Barr Virus

Epstein-Barr virus (EBV), another herpes virus, is the etiological agent of human infectious mononucleosis and is associated with the development of Burkitt's lymphoma and nasopharyngeal carcinoma, the most prevalent cause of cancer in males of southern China (reviewed in Reference 101). More recently, EBV has been associated with the development of lymphomas in the immunocompromised.[102] The need for a vaccine has been recognized for some time, but has not been feasible due to the low production of virions and antigens of interest by classical tissue culture techniques.

EBV synthesizes a membrane antigen (MA) on its envelope composed of at least two glycoproteins, gp220 and gp340. Both glycoproteins are encoded by the same open reading frame within the same gene and, not surprisingly, share antigenic determinants.[103,104] Monoclonal antibodies and MA-specific antisera neutralize infectious EBV.[105–107] Furthermore, cottontop tamarins (*Saguinus oedipus oedipus*) are protected from EBV-induced lymphomas by vaccination with purified gp340.[108]

The gene for MA has been cloned into two strains of vaccinia virus under the regulation of the vaccinia early-late 7.5K promoter.[109] The gene product, expressed on the surface of infected cells, appeared to be authentic gp340 as determined by glycosylation characteristics, molecular weight, and recognition by rabbit polyclonal anti-gp340 and a virus-neutralizing monoclonal antibody with specificity toward gp340. Following vaccination with the two recombinant vaccinia viruses, rabbits produced antibodies which reacted with EBV-transformed cell lines that expressed gp340, but not with Raji cells which, although infected with EBV, do not produce MA. Moreover, these rabbit antisera were capable of neutralizing EBV *in vitro*. When cottontop tamarins were immunized with one of the two recombinant vaccinia virus strains, only the laboratory strain, derived from WR, provided protection against the development of lymphomas in three of the four animals tested.[110] No significant protection was elicited from the vaccine-derived recombinant strain, although one animal underwent complete remission ten weeks after challenge, following a delay in the development of lymphoma. Although significant antibody titers against vaccinia virus appeared to correlate with immunity and protection, circulating EBV-neutralizing antibody titers could not be detected in any of the animals. In the absence of virus-neutralizing antibodies, the mechanism of protection was not identified and may have been associated with gp340-specific cell-mediated immunity, which was not assessed in this study.

D. INFLUENZA VIRUS

Influenza viruses pose a serious health threat not only to humans, but to many animal species. Although influenza viruses may infect the entire respiratory tract, it is the pathological effects upon the lower respiratory tract that are significant. Following infection, clinical symptoms may abruptly appear. Primary viral pneumonia can rapidly progress to fatal proportions. Influenza virus spreads by the aerosol route in droplets formed by sneezing and coughing, which enhances dispersal throughout a nonimmune population.

Influenza virus types A, B, and C are serologically distinguished by antigenic differences found in the internal structural protein, the nucleoprotein (NP). The genomes of types A and B are contained on eight distinct RNA segments (seven segments for type C) encoding ten proteins, including a hemagglutinin (HA), neuraminidase (NA), NP, two matrix proteins (M_1 and M_2), polymerase subunits, and several nonstructural proteins. It is the reassortment of the genomic segments during coinfections and amino acid substitutions in HA that allow a high rate of natural variation among influenza viruses.

During infection, antibodies are produced against HA, NA, NP, and M proteins.[111,112] However, it is the humoral responses to HA and NA that are associated with resistance.[113] Neutralizing antibody directed solely at the HA, which contains the cell receptor and mediates attachment to the cell, also correlates with protection, but is not cross-rective with other HA subtypes.[104] The high degree of variability of the HA strongly contributes to the constant progression of influenza epidemics and our inability to control the virus through vaccination. Cell-mediated immune responses appear to play an important role in recovery from infection.[115,116]

Since the influenza HA is the target of virus-neutralizing antibodies, it serves as a primary candidate for a vaccine component. Recombinant vaccinia viruses containing the cloned influenza virus HA gene expressed authentic HA in infected cells.[117] The recombinant protein reacted with specific anti-HA antiserum. When rabbits were infected with the recombinant vaccinia virus, antibodies were produced which reacted with influenza HA in radioimmunoassays, inhibited HA-mediated agglutination of erythrocytes, and neutralized infectivity of influenza virus. Recombinant vaccinia viruses expressing authentic influenza HA could also protect hamsters from experimentally induced respiratory infection following intradermal inoculation.[118] After infection with recombinant vaccinia viruses expressing the swine influenza HA, cattle, sheep, and poultry produced antibodies which reacted with authentic HA.[119] Chickens could be protected from lethal influenza challenge by immunization with vaccinia-HA recombinants despite low levels of antibodies capable of hemagglutination inhibition and neutralization of influenza virus.[120] Furthermore, unimmunized chickens were protected from challenge following passive immunization with the low-titer serum. Protection elicited by vaccinia virus recombinants expressing influenza HA was also extended to suckling offspring of inoculated ferrets.[121] Despite low hemagglutination inhibiting titers, milk-derived anti-influenza IgG was identified as responsible for protection. However, adult ferrets, and consequently, their offspring, were susceptible to influenza infection following immunization with vaccinia virus recombinants expressing influenza NA, polymerase subunits, matrix proteins, NP, or nonstructural proteins.

A different approach to producing an avian influenza vaccine was to immunize chickens with lysates from cells infected with a recombinant vaccinia virus expressing the avian H5 influenza hemagglutinin.[122] Immunized chickens were protected from challenge. Similar results were obtained using a fowlpox virus recombinant expressing the avian influenza H5 hemagglutinin.[123] In these studies, chickens and turkeys were inoculated with a conventional inactivated H5 avian influenza vaccine or with the H5 hemagglutinin fowlpox recombinant. Despite the absence of significant levels of HA-inhibition or virus-neutralizing antibodies, only the vaccinates survived homologous as well as heterologous (a related but distinct H5 avian influenza) challenge.

In a side by side comparison of the efficacy of live or inactivated influenza viruses and recombinant vaccinia encoding the HA, the vaccinia-based vaccine proved to be as effective in reducing the amount of influenza in the lungs of mice following challenge as the conventional vaccines.[124] In this study, mice inoculated intranasally with any of the vaccine preparations produced poor levels of circulating antibodies that could inhibit influenza-mediated agglutination of erythrocytes or that could neutralize infectious influenza virus.

Nonetheless, vaccines administered by this route were demonstrated to be highly effective in controlling influenza infection.

Inoculation with recombinant vaccinia viruses also produced cell-mediated immune responses. Cytotoxic T lymphocytes capable of killing influenza-infected target cells were detected, following stimulation *in vitro*, in the spleens of mice previously inoculated with vaccinia viruses harboring and expressing the influenza HA gene.[125] These mice also produced antibodies against HA and were protected from challenge with virulent influenza virus. The effect of expression of the influenza HA under the regulation of early, early-late, and late promoters in recombinant vaccinia viruses was compared in mice.[126] Whereas all three recombinant viruses stimulated antibody responses to HA, only recombinant vaccinia viruses expressing HA-utilizing promoters with early function (pre-DNA replication) elicited cytotoxic T lymphocytes and were able to stimulate memory cytotoxic T cells *in vitro*. This indicated that T-lymphocyte effector involvement required synthesis of HA early in the immune process and is reminiscent of the data for HSV gD temporal expression and guinea pig protection studies.[74]

The role of cell-mediated immune responses to the NP and its correlation to immunity were also examined.[127] Immunization of mice with vaccinia recombinants expressing either the influenza HA or the NP stimulated cytotoxic T lymphocytes. It was found that cytotoxic T lymphocytes against HA eptiopes were subtype specific whereas T-cell recognition of NP eptitopes was cross-reactive. The dominant T-cell response was directed toward NP as determined by limiting dilution analysis. Despite these results, mice immunized with the NP-vaccinia recombinant were not as well protected as were mice immunized with the recombinant expressing HA. Subsequent studies demonstrated that mice passively transferred with *in vitro* stimulated, NP-specific or HA-specific cytotoxic T lymphocytes were almost totally protected against lethal influenza challenge.[128,129] However, only immunization with vaccinia recombinants expressing HA, but not NP, provided protection. When mice were immunized with a 42 kD component of the NP or a recombinant vaccinia virus expressing NP, high levels of monospecific antibody and cytotoxic T lymphocytes were detected.[130] In spite of these vigorous cellular and humoral responses, mice were not protected from virulent influenza infection.

The role of the influenza NA in eliciting protection was investigated in chickens.[131] When administered in adjuvant, purified NA prevented the spread of influenza and prevented death, but failed to deter infection. Vaccinia virus recombinants expressing NA were less effective. The NA produced by vaccinia recombinants retains its enzymatic activity as evidenced by its ability to cleave sialic acid from glycoconjugates on the surface of infected cells.[132]

A single influenza virus protein, the HA, when expressed in recipients administered recombinant vaccinia viruses, confers protection against infection by homologous influenza virus. However, both humoral and cell-mediated responses directed toward the HA are subtype specific and, therefore, immunity does not extend to antigenically unrelated influenza viruses. Since a high degree of variation in the HA exists in nature, combined with genetic reassortment of the influenza virus genomic segments, eliciting cross-reactive immune responses that protect against heterologous influenza viruses remains the key to successfully preventing and controlling influenza outbreaks.

E. MEASLES VIRUS

Measles is an acute, highly contagious respiratory disease primarily affecting children.[133–136] Typical measles virus infections are of the nasopharynx and conjunctivae that over the course of days disseminate through the reticuloendothelial system and to the skin, respiratory tract, and other distant sites.[137] Early symptoms of fever, malaise, conjunctivitis, and cough precede the development of erythematous maculopapular rash that spreads from the head to the extremities.[137–140] In industrialized nations, the most frequent complications that may arise are otitis media, pneumonia, encephalitis, and death.[133,138,139,141–150] Subacute sclerosing panencephalitis (SSPE), a chronic, progressive cerebral degeneration that manifests 4 to 10 years after a normal measles infection, is also a frightening potential consequence of measles infections.[148] In developing countries, malnutrition and rapid decline of maternal antibodies accelerate the susceptibility of children at an earlier age.[151,152] The consequence of measles infection in developing countries is increased mortality rates with an estimated 2.1 million measles-related deaths reported annually, worldwide.[153]

Measles is transmitted person-to-person by respiratory droplets or by aerosol nuclei,[133,154–156] Transmission occurs mainly among young children, but during outbreaks, older individuals may become involved. Measles is so contagious that it can rapidly disseminate among large aggregates of susceptible individuals.

The measles virus virion contains an unsegmented, single-stranded RNA genome.[157] As a member of the *Morbillivirus* genus of the Paramyxoviridae, measles is closely related to canine distemper virus (CDV).[158] Six structural proteins are produced; of these, two (fusion protein and HA) are glycosylated and are located on the external side of the envelope.[157,159] The HA is responsible for adsorption of the virus to cells and mediates agglutination of erythrocytes. The fusion protein (F) promotes fusion of the virus envelope with cell membranes initiating virus penetration and spread. The F protein also effects hemolysis. Neutralizing antibodies are directed against epitopes found on both the F and HA proteins, although those reacting with F appear to provide better protection.[160] Infection can be precluded by passive transfer of immune gammaglobulin.[161–163]

A number of live, attenuated measles vaccines are currently available, derived from the Edmonston B vaccine which was originally licensed in 1963. Continuous improvements to the early vaccine have reduced the untoward effects: high rate of fever and rash. The live attenuated vaccines are highly effective in preventing measles. Following immunization, antimeasles IgM, IgG, and IgA can be detected in both serum and nasal secretions.[164,165] Cell-mediated immunity is induced as well.[166–168]

Administration of high doses of vaccine is recommended due to possible deterioration of the vaccine as a result of improper storage or use.[169] Measles virus is highly labile, being sensitive to both heat and light. The attenuated vaccine strains are no exception. Although the measles vaccines are highly stable at low temperatures (−70 to −20°C), they are increasingly unstable at higher temperatures. Lyophilized vaccine remains stable for 2 years at 2°C to 8°C, 2 months at room temperature, but only 2 weeks at 37°C. After reconstitution, the vaccine is stable for only 2 days at room temperature.

Even though the frequency of many of the adverse effects observed with use of the early vaccine has been reduced, mild fever and rash are still ordinary occurrences.[170–178] Although vaccination-associated central nervous system disorders remain controversial, acute Guillain-Barré syndrom, Reye's syndrome, sight and hearing disorders, arthritis, allergic reactions, and skin and soft tissue reactions continue to be reported following vaccination.[179–190]

Although considered safe, the achievement of worldwide eradication of measles, a goal worthy of increased attention, will be difficult using the current measles vaccine, considering its relative instability under conditions common to developing countries where it will be most utilized. The development of highly thermostable vaccines using recombinant virus vectors, such as poxviruses, expressing the measles F and HA proteins may hasten the reality of eradication.

Vaccinia virus recombinants carrying the measles gene for either the F or HA protein were constructed.[191] Infected cells synthesized proteins that migrated in electrophoretic gels identically to authentic F or HA and could be detected using specific antibodies. Following infection with the recombinant expressing the HA protein, mice produced antibodies which neutralized infectivity of measles virus and inhibited measles virus mediated hemagglutination. Infection with the F recombinant induced antibodies which also neutralized measles infectivity. Vaccination with either recombinant vaccinia virus achieved protection against a lethal intracerebral challenge of an SSPE strain of measles. Other vaccinia virus recombinants containing sequences encoding the measles virus F or HA proteins have also been developed.[27] Cells infected with the recombinants expressed authentic F or HA as determined by immunoprecipitation, using antibodies specific for the appropriate measles antigens. Autoradiography, following electrophoresis, revealed the presence of the fusion precursor molecule, F_0, and the two cleavage products, F_1 and F_2, in the immunoprecipitates of the F recombinant infected cells. Similar techniques revealed the glycosylated form of HA. Both proteins were expressed on the surface of infected cells as demonstrated by immunofluorescence. When cells were infected with the vaccinia recombinant to test the functionality of the expressed F protein, no syncytia formation resulting from cell fusion was observed. However, when cells were infected with both recombinants expressing F or HA, efficient cell fusing activity was apparent. Dogs immunized with the HA recombinant or a combination of recombinants expressing HA or F developed antibodies capable of neutralizing measles virus as well as measles virus vaccinated dogs, but not the dogs immunized with the recombinant expressing solely F. Since dogs vaccinated with measles virus are protected from lethal canine distemper virus by one-way cross neutralizing antibodies,[192–203] a challenge study was performed on the vaccinates. Although surviving, dogs immunized with recombinant vaccinia expressing F showed mild clinical signs of transient infection by canine distemper virus. On the other hand, dogs that had been vaccinated with recombinants expressing HA, or a combination of recombinants expressing HA or F, resisted the lethal challenge with no significant symptomology as well as those dogs immunized with measles virus or a vaccine strain of canine distemper virus.

The measles F gene was cloned into and expressed in fowlpoxvirus vectors, but was not tested for its ability to protect mammals from challenge infections.[204] Recently, the genes encoding the measles HA and F proteins were inserted into canarypox.[27] Although, like other avipox, canarypox is not known to replicate in mammalian species, immunization of dogs with the canarypox recombinants expressing measles HA or both HA and F elicited antibody responses which neutralized infectious measles virus as well as vaccinia recombinants expressing similar genes or live measles vaccine. Dogs immunized with a canarypox recombinant expressing only F failed to produce neutralizing antibodies. Nonetheless, all dogs immunized with canarypox and vaccinia recombinants were protected to the same extent from a lethal challenge of canine distemper virus as measles vaccinated dogs, although those immunized with recombinants expressing only F exhibited mild clinical symptoms before recovery. These results would seem to suggest that vaccination of humans with recombinant vaccinia or canarypox viruses expressing HA or a combination of F and HA would be capable of affording protection against measles virus infection.

F. FLAVIVIRUSES

One of the most medically important groups of viruses are the flaviviruses. Of some 60 viruses of the arthropod-borne group, 28 are known to cause disease in man, while most of the remainder are of veterinary importance.[205] Three members of the flaviviruses, dengue virus, Japanese encephalitis virus (JEV), and yellow fever virus, cause serious health concerns on a global basis. The flaviviruses cause a wide range of symptomology and pathology reflecting neurological and hemorrhagic disorders.

Classic dengue, epidemic in many tropical and subtropical sections of the world, is normally self-limiting and is typified by fever, headaches, rash and arthralgia. Dengue appeared for the first time in epidemic form in the western hemisphere (Cuba) in 1981. Dengue hemorrhagic fever (DHF), a more serious consequence of infection with an immunological basis, is characterized by increased blood concentration via fluid loss and hemorrhage into various tissues. The immunopathology is related to antibody enhancement of infection of cells bearing Fc receptors for IgG. At its extreme, DHF develops into dengue shock syndrome (DSS), a state of shock resulting from the loss of fluid proteins. DSS has a 5 to 10% mortality rate.

Yellow fever has affected nearly a quarter of a million individuals in West Africa alone over the past 20 years. Fever, hemorrhage, and renal, hepatic, and myocardial dysfunction with a 20% fatality rate characterize this disease.

Japanese encephalitis is endemic and epidemic in many parts of Asia. In Thailand, the disease incidence among children is estimated to be as high as 50 cases per 100,000.

Flaviviruses encode a capsid protein (C), matrix protein (M), envelope protein (E), and several nonstructural proteins from a single large open reading frame (reviewed in Reference 206). The individual proteins are proteolytically cleaved from a single precursor polyprotein. Using passively administered monoclonal antibodies, protection from virus infection in experimental animal models has been associated with immunity to each of the three structural proteins and usually correlates with the ability to neutralize infectious virus.[207–210] However, a nonstructural glycoprotein, NS1, has been implicated as a protective antigen despite the absence of virus-neutralizing proficiency of antibodies targeting it.[207,211–215] The protective effects of the nonneutralizing antibodies may involve complement fixation.[211,216]

Effective live, attenuated vaccines are currently available against yellow fever and inactivated vaccines successfully immunize against Japanese encephalitis and tick-borne encephalitis viruses.[217,218] However, the availability of inactivated dengue vaccines has been hampered by low virus yields in certified cell lines. Moreover, there are concerns about a dengue vaccine containing virion proteins that could potentially elicit antibodies associated with Fc receptor mediated antibody enhancement of infection, thought to be fundamental in the pathogenesis of DHF and DSS. Therefore, the development of subunit or recombinant vaccines involving only the relevant protective antigens would overcome the production problems and immunopathology associated with some conventional flavivirus vaccines, and most important, allow the evolution of an effective dengue vaccine.

A recombinant vaccinia virus was constructed by inserting a cDNA copy of the portion of dengue 2 virus genome encoding the C, M, and E proteins.[219] Regulation of the expression of these foreign genes was controlled by a late vaccinia promoter. Expression of the envelope protein was detected by immunofluorescence in infected cells using five different dengue 2 E-specific monoclonal antibodies. Proteins from cells infected with the recombinant vaccinia virus were immunoprecipitated with polyclonal antiserum to dengue 2 virus and migrated in polyacrylamide gels identically to E from purified dengue 2 and E from dengue 2 infected cells. Neither C nor pre-M were detected. Neither the mice nor the *Cynomolgus* monkeys intradermally administered the recombinant vaccinia virus produced dengue neutralizing antibodies although vaccinia neutralizing antibodies were present 4 weeks post-inoculation. Following challenge with dengue 2 virus, monkeys vaccinated with recombinant vaccinia, similar to unvaccinated controls, became viremic, whereas monkeys which had received inactivated dengue 2 virus and had developed dengue 2 neutralizing antibodies remained free of virus. The failure of the vaccinia recombinant to provide protection was attributed to lack of expression of E on the surface of the infected cell.

The cDNA fragments encoding dengue 4 virus structural proteins C, M (or pre-M), and E, as well as the nonstructural proteins NS1 and NS2a were inserted into vaccinia virus, which expressed genuine pre-M, E, and NS1 under the control of an early-late promoter.[220] Upon administration to cotton rats (*Sigmodon fulviventer*), low levels of anti-NS1 were detected, but not antibodies to pre-M or E. When mice were vaccinated intraperitoneally with the vaccinia recombinant and challenged with mouse brain adapted dengue 4 virus, they were protected from fatal encephalitis after developing antibodies to NS1, but not M or E.[221] When mice were immunized with recombinant vaccinia viruses expressing authentic C, pre-M and E, or solely E, they also were protected from fatal disease following challenge, although only a portion of the mice produced detectable antibody titers. In contrast to mice immunized against C, pre-M, and E, mice immunized against only E expressed by the recombinant developed mild, transient signs of encephalitis, but nonetheless survived the challenge. To evaluate the role of NS1 in protection, recombinant vaccinia viruses were constructed containing the entire coding sequences of NS1 and NS2a which express authentic NS1 and recombinants containing NS1 and incomplete NS2a, expressing abnormal NS1.[222] Recombinants faithfully expressing NS1 elicited high antibody titers in immunized mice that reacted with NS1 and were capable of activating complement. Furthermore, these sera protected mice from dengue 4 encephalitis following passive immunization. Only low antibody titers were induced by the recombinants containing the truncated form of NS2a. Mice immunized with the full NS1-NS2a recombinant were protected from challenge, whereas those immunized against the aberrant form were only partially protected. Additionally, two vaccinia recombinants encoding NS1, NS2a, 84% of NS2b, and E of differing expanse from dengue 2 virus were partially protective for mice against fatal encephalitis.

In a comparable approach as the result of previous successes with NS1, a vaccinia virus recombinant was constructed containing the coding sequence for the yellow fever virus nonstructural proteins NS1, NS2a, and NS2b.[223] NS1, whose expression was regulated by a vaccinia late promoter, was detected on the surface of infected cells by immunofluorescence and complement-mediated lysis using a monoclonal antibody against native NS1. NS1 was also detected in secreted form in the medium of recombinant vaccinia infected cells. Mice immunized with the recombinant vaccinia virus produced antibodies which lysed cells infected with yellow fever virus in the presence of complement, although to a lesser extent than yellow fever virus-infected cells. Nonetheless, only moderate protection, although reproducible, was afforded to mice vaccinated with the recombinant vaccinia virus after challenge with virulent yellow fever virus.

Reflecting the varied host-pathogen relationships of different flaviviruses, the success of utilizing other antigens for protection against JEV was explored. The cDNA sequences representing the structural proteins of JEV were inserted into the TK gene of vaccinia virus and tested for immunogenicity and its ability to protect mice.[224] Under the control of an early-late promoter, proteins were produced in infected cells that could be immunoprecipitated with antiserum to JEV and comigrated with authentic pre-M and E. E was detected on the surface of infected cells by fluorescence-activated flow cytometry and specific monoclonal antibodies. Following intradermal inoculation of guinea pigs with the recombinant virus, JEV antigens were detected in histological sections of dermis by immunofluorescence. The expression and distribution of fluorescence was similar to that observed when antiserum to vaccinia virus was used, indicating that the recombinant vaccinia virus produced JEV antigens *in vivo*. Rabbits inoculated with the recombinant viruses generated antibodies that immunoprecipitated pre-M and E from JEV-infected cell lysates. These antisera were capable also of inhibiting JEV mediated hemagglutination as well as neutralizing infectious JEV. Mice vaccinated with the recombinant rapidly produced neutralizing antibodies and were protected from lethal encephalitis after a low dose challenge with virulent JEV.

In another study, four different recombinant vaccinia viruses were constructed containing JEV cDNA sequences encoding E, NS1, and NS2a under the control of an early-late promoter.[225] These four recombinants differed in the M sequences (M in its entirety or a small portion of M) and the presence or absence of NS2b. All four recombinants faithfully expressed glycosylated NS1 and E. Those recombinants containing the entire sequence for M also produced a protein (pre-M) which was reactive with a monoclonal antibody to authentic JEV-derived M protein. Interestingly, cells infected with the recombinants expressing pre-M secreted a slow sedimenting particle with hemagglutinating activity as well as the fully processed form of M. The hemagglutinating activity of the particles was sensitive to anti-JEV and pH conditions similar to authentic empty JEV envelopes.[226] Although preliminary studies indicated that all four of the recombinants could provide some protection to mice from the lethal encephalitis of the Beijing strain of JEV, two recombinants were selected for detailed study. Following a single inoculation, the recombinant expressing the extracellular form of E (particles) provided better protection than did the recombinant containing only a portion of M and the entire NS2b. After the second administration, mice that received the particle-producing recombinant were completely protected from a severe challenge of 10,000 LD_{50} (50% lethal doses) of JEV. Mice that were vaccinated with the second recombinant, incapable of producing extracellular E, did not fare nearly as well; only about one third of the animals survived. Protection was clearly correlated with the production of antibodies that immunoprecipitated genuine E protein, inhibited JEV hemagglutination, and neutralized infectious JEV. Thus, a recombinant vaccinia virus was constructed which was capable of directing the synthesis, processing, and secretion of immunogenic particles resembling empty JEV envelopes that provided a high degree of protection against JEV infection.

The use of recombinant vaccinia viruses in combination with advances in our understanding of the role of various flavivirus antigens in evoking protective immunity brings the development of safe and effective vaccines against this group of important pathogenic viruses closer to fruition. Notable is the realization that nonstructural proteins, such as the dengue virus NS1 glycoprotein, may be highly relevant antigens. The ability to induce protection by recombinant viruses directing the synthesis and secretion of highly immunogenic virus-like particles following accurate processing of antigens increases optimism in achieving success. Utilizing the tremendous capacity of poxviruses to accept large foreign inserts into its genome may allow the expression of sufficiently numerous antigens to allow future vaccines to surmount the difficulties of immunizing against the various virus serotypes typical of this group of pathogens.

G. HUMAN IMMUNODEFICIENCY VIRUS

In recent years much attention within the field of medical virology has been focused on the escalating incidence of AIDS caused by human immunodeficiency virus (HIV). This viral system has been scrutinized with respect to its molecular biology, immunobiology, and pathogenesis in an effort to develop safe and efficacious vaccines and antiviral therapies. Vaccinia virus has been used extensively to express a number of HIV gene products to investigate their biochemical, functional, and immunological properties.

With regard to HIV vaccinology, recombinant vaccinia virus expressing HIV-derived gene products has been used as a reagent to characterize HIV-specific immune responses, particularly cell-mediated responses. Using these specific reagents, cell-mediated cytotoxic activities have been demonstrated in seropositive individuals toward the envelope,[227–229] core proteins (*gag*),[228–230] *pol*,[231,232] and *nef* gene products.[228] Additionally, vaccinia virus recombinants containing HIV genes have been shown to elicit both cell-mediated and humoral immune responses in small laboratory animals,[233–236] macaques,[237] chimpanzees,[238] and significantly, humans.[239,240]

To date, limited vaccination/protection studies in primates have been completed with vaccinia virus recombinants expressing gene products from HIV or related viruses. An initial *in vivo* challenge study described by Hu et al. was not very encouraging.[238] In these studies, a vaccinia virus recombinant expressing the HIV-1 envelope glycoprotein was used to vaccinate chimpanzees, followed by a subsequent inoculation with either 3×10^5 $TCID_{50}$ or 100 $TCID_{50}$ units of infectious HIV. With either challenge dose, no evidence of protection was attained.

Other studies, however, have indicated that vaccination of primates with a recombinant vaccinia virus expressing the envelope glycoprotein from AIDS-causing retroviruses can elicit humoral and cell-mediated immune responses and, more significantly, protect against simian immunodeficiency virus (SIV) infection.[241] In the results described by Hu et al., a vaccinia virus recombinant expressing the envelope glycoprotein from simian AIDS retrovirus (SRV-2) was used to vaccinate pig-tailed macaques (*Macaca nemestrina*).[241] All immunized animals developed both SRV-specific cell-mediated and humoral immune responses, including neutralizing antibodies and antibodies which mediate antibody-dependent cellular cytotoxicity (ADCC) toward SRV-2 infected cells. Animals were challenged intravenously with 5×10^3 $TCID_{50}$ of SRV-2/W. The challenged control animals (nonvaccinated) became viremic by 2 weeks postchallenge and those that did not seroconvert died by 7 weeks postchallenge. Significantly, all of the challenged animals which were previously inoculated with the vaccinia virus-*env* recombinant remained healthy, virus-free, and seropositive exclusively against the envelope antigen.

In a pilot experiment in humans, HIV seronegative individuals were vaccinated with a vaccinia virus/HIV-1 envelope recombinant. This primary inoculation resulted in weak immunological responses.[240] These primary responses were subsequently boosted with various protocols. The use of an intravenous injection of paraformaldehyde-fixed autologous cells infected *in vitro* with the vaccinia virus HIV recombinant, however, provided the most significant booster effect. With this immunization protocol, anamnestic immune responses were acheived against the envelope antigen consisting of group-specific neutralizing antibodies, cytotoxic T-lymphocytes, and delayed-type hypersensitivity.[240] In clinical trials performed in the U.S. using a recombinant vaccinia virus expressing the HIV *env* gene (HIVAC-le; Bristol-Meyers), individuals receiving the recombinant mounted T-cell proliferative responses to HIV.[239] The intensity of these responses, however, was affected by prior exposure to vaccinia virus. Consequently, individuals immune to vaccinia virus mounted a weak and transient T-cell response whereas in individuals not immune to vaccinia virus, a strong response was observed toward the HIV envelope antigen (reviewed in Reference 239). These results are encouraging and have provided evidence that an immune state can be obtained in man prior to HIV exposure using a poxvirus based immunization vehicle.

An intriguing potential in terms of HIV-1 vaccinology is provided by expression of the HIV-1 *gag* products by vaccinia virus either alone or in combination with the envelope glycoprotein. Several laboratories have demonstrated that expression of *gag* or *gag/pol* sequences by vaccinia virus results in the production of noninfectious particles.[242–244] Analysis of these vaccinia virus recombinant infected cells by electron microscopy revealed retrovirus-like particles budding from the plasma membrane and extracellular forms which were indistinguishable from particles observed in HIV-1 infected cells. Biochemically, these particles are also similar to native HIV-1 particles. Moreover, rodents and chimpanzees inoculated with a vaccinia virus-HIV *gag* recombinant generated both humoral and cell-mediated immune responses to the HIV-1 core antigens.[242]

The coexpression of the HIV-1 envelope glycoprotein with the core proteins in mammalian cells by vaccinia virus also resulted in the assembly and release of HIV-1 particles.[245] These typical type C-retrovirus particles contained both the envelope glycoproteins (gp120 and gp41) and the *gag* proteins (p24, p17, p55, and p39). These proteins were also described as being present in these recombinant-made particles in the same ratio as observed in HIV-1 virions. The production of these noninfectious HIV particles either in purified form or synthesized in a vaccinee upon inoculation with a multivalent vaccinia virus recombinant may provide valuable ''whole-virus'' vaccinating agents against HIV. This is especially appealing since protection against the lentivirus, SIV, has been demonstrated using whole inactivated virus preparations.[246,247]

The above reviewed examples demonstrate the potential use of vaccinia virus recombinants expressing HIV antigens to induce pertinent immunological responses necessary for protection. Worthy of reiteration, especially in the context of HIV and many of those at risk for HIV, a major concern about the use of live viral vaccines is the issue of safety. Rare complications from immunization with vaccinia virus have been documented, particularly in immunocompromised individuals, and have raised some objections for their acceptance as vaccine candidates.[248,249] Recently, however, significant strides have been made in understanding viral virulence factors with the hope of modifying strains for use as immunization vehicles.[250] With more relevance to the development of vaccinia virus-based HIV vaccine candidates for the immunoprophylaxis and immunotherapy of AIDS, the genes responsible for the productive replication of vaccinia virus in human cell systems have been identified.[25,26] Potential use of vaccinia virus vectors devoid of these genes provide a nonreplicating vector vaccine candidate that may present appropriate HIV antigens in a fashion that elicits a protective immune response.

H. OTHER PATHOGENS

A variety of other disease agents, primarily pathogenic viruses of humans or other animals, have been targeted for disease intervention by recombinant vaccinia viruses. When cDNA copies encoding the glycoprotein (G) of vesicular stomatitis virus (VSV), causative agent of a contagious disease of horses, cattle, and pigs, were inserted into the TK locus of vaccinia virus, a glycoprotein was produced which resembled authentic G.[251] Mice and cattle produced VSV-neutralizing antibodies following inoculation with the recombinant virus and were protected against experimental infection. One of these laboratory investigators was accidentally infected with a recombinant expressing the N protein of VSV and produced antibodies specific for N.[252] This clearly demonstrated that humans could respond to recombinant vaccinia viruses expressing foreign gene products.

Respiratory syncytial virus (RSV) is the primary cause of lower respiratory tract infections in children. The major glycoprotein (G) and the fusion protein (F) of human RSV have been cloned into vaccinia virus TK.[253,254] The recombinants directed syntheses of proteins indistinguishable from the native proteins. Immunization of cotton rats with the F recombinant induced higher levels of neutralizing antibodies than did the recombinant expressing G, as well as eliciting substantial protection against challenge. Similar results were obtained when mice were immunized with a vaccinia recombinant expressing F.[255] Mice produced antibodies that reacted with F and were afforded protection from infectious RSV.

Similarly, the hemagglutinin-neuraminidase (HN) and F of human parainfluenza virus (PIV), second only to RSV in frequency of severe lower respiratory tract infections of children, were cloned into and expressed by recombinant vaccinia viruses.[256,257] The HN recombinant evoked higher neutralizing antibody titers than did the F recombinant in immunized cotton rats. Furthermore, the recombinant expressing HN induced substantially better protection than did the F recombinant in that PIV replication in the respiratory tract of these animals was reduced over 3000-fold as compared to a 6-fold reduction. Further extending this investigation, patas monkeys (*Eryphrocebus patas*) were immunized once with both of the vaccinia recombinants. PIV-neutralizing antibody titers increased following immunization. After intratracheal challenge, PIV replication was considerably reduced in both the upper and lower respiratory tracts, as was the duration of virus shedding.

Recombinant vaccinia virus expressing the Lassa virus glycoprotein also afforded protection to guinea pigs.[258] Although these animals developed a transient low-grade fever and became viremic, they survived the lethal challenge. Similar success has been reported with rinderpest virus, the causative agent of cattle plague.[259] Recombinants expressing the hemagglutinin and fusion glycoproteins elicited neutralizing antibodies and protected cattle from a severe challenge. Pseudorabies virus (PRV) glycoprotein gp50 (gD) protected mice when expressed by a recombinant vaccinia virus.[260] Although not tested directly in swine, the natural host of PRV, vaccinia recombinant gp50 produced *in vitro* protected pigs from a virulent PRV challenge.

IV. CONCLUSIONS

The strategy for developing recombinant poxviruses as vaccine vectors should now be clear. Initially, the pertinent proteins of the disease-causing agent must be identified. In many cases, recombinant vaccinia viruses have proved invaluable in this phase of development by providing the expression system for potentially important antigens. The contribution of a single protein from a pathogenic agent as it relates to protective immunity can be evaluated once the gene has been identified and inserted into vaccinia virus. Expression of the gene product, usually under the control of a strong promoter with early function, must be verified to be nearly indistinguishable from the native protein. Usually, this may be accomplished *in vitro* by infection of a permissive cell line. Protein expression is then evaluated by immunofluorescence techniques, immunoprecipitation, probing Western blots, RIA, or by ELISA using specific antisera or monoclonal antibodies. After confirmation of accurate protein expression, immunogenicity is assessed by analyzing sera from rabbits or rodents inoculated with the recombinant virus for immunoreactivity with the native antigen or by the ability to inhibit a biological function associated with the protein, such as neutralization of infectivity or the inhibition of fusion or hemagglutination. The ability of the recombinant to protect against the pathogen is then tested in challenge studies using appropriate animal model systems. In the case of rodents, immunogenicity and protection may be evaluated simultaneously. Similarly, if the recombinant vaccinia virus vaccine has veterinary application, immunogenicity and efficacy may be tested directly in the target species.

As this review reveals, this strategy has been productively utilized in the research of potential vaccines for humans and other animals against a wide variety of pathogens, primarily viruses. A live recombinant rabies vaccine is currently being field tested. Vaccines against hepatitis B virus, herpes simplex, cytomegalovirus, Epstein-Barr virus, influenza, measles, dengue, and JEV, employing recombinant vaccinia virus technology, are only the beginning and will, most likely, be entering the development phase within the next few years. Second generation recombinant poxvirus vaccines may implement nonreplicating vectors, such as the avipoxes or highly attenuated vaccinia viruses restricted in replication in the target species, yet be able to express multiple foreign antigens that may successfully provide protection against many unrelated diseases.

REFERENCES

1. **Fenner, F.,** Poxviruses, in *Virology,* Vol. 2, 2nd ed., Fields, B. N. and Knipe, D. M., Eds., Raven Press, New York, 1990, 2113.

2a. **Goebel, S. J., Johnson, G. P., Perkus, M. E., Davis, S. W., Winslow, J. P., and Paoletti, E.,** The complete DNA sequence of vaccinia virus, *Virology,* 179, 247, 1990.

2b. **Goebel, S. J., Johnson, G. P., Perkus, M. E., Davis, S. W., Winslow, J. P., and Paoletti, E.,** The complete DNA sequence of vaccinia virus, *Virology,* 179, 517 (appendix), 1990.

3. **Moss, B.,** Poxviridae and their replication, in *Virology,* Vol. 2, 2nd ed., Fields, B. N. and Knipe, D. M., Eds., Raven Press, New York, 1990, 2079.
4. **Fenner, F.,** The poxviruses, in *Portraits of Viruses. A History of Virology,* Fenner, F. and Gibbs, A., Eds., S. Karger, Basel, 1988, 1.
5. **Panicali, D. and Paoletti, E.,** Construction of poxviruses as cloning vectors: insertion of the thymidine kinase gene from herpes simplex virus into the DNA of infectious vaccinia virus, *Proc. Natl. Acad. Sci. U.S.A.,* 79, 4927, 1982.
6. **Mackett, M., Smith, G. L., and Moss, B.,** Vaccinia virus: a selectable eukaryotic cloning and expression vector, *Proc. Natl. Acad. Sci. U.S.A.,* 79, 7415, 1982.
7. **Lane, J. M. and Millar, J. D.,** Risks of smallpox vaccination complications in the United States, *Am. J. Epidemiol.,* 93, 238, 1971.
8. **Joklik, W. K., Pickup, D. J., Patel, D. D., and Moody, M. D.,** Virulence genes of poxviruses and reoviruses, *Vaccine,* 6, 123, 1988.
9. **Buller, R. M. L., Smith, G. L., Cremer, K., Notkins, A. L., and Moss, B.,** Decreased virulence of recombinant vaccinia virus expression vectors is associated with a thymidine kinase-negative phenotype, *Nature,* 317, 813, 1985.
10. **Buller, R. M. L., Chakrabarti, S., Cooper, J. A., Twardzik, D. R., and Moss, B.,** Deletion of the vaccinia virus growth factor gene reduces virus virulence, *J. Virol.,* 62, 866, 1988.
11. **Dallo, S., Rodriguez, J. F., and Esteban, M.,** A 14K envelope protein of vaccinia virus with an important role in virus-host cell interactions is altered during virus persistence and determines the plaque size phenotype of the virus, *Virology,* 159, 423, 1987.
12. **Morita, M., Suzuki, K., Yasuda, A., Kojima, A., Sugimoto, M., Watanabe, K., Kobayashi, H., Kajima, K., and Hashizume, S.,** Recombinant vaccinia virus LC16m0 or LC16m8 that expresses hepatitis B surface antigen while preserving the attenuation of the parental virus strain, *Vaccine,* 5, 65, 1987.
13. **Pickup, D. J., Ink, B. S., Hu, W., Ray, C. A., and Joklik, W. K.,** Hemorrhage in lesions caused by cowpox virus is induced by a viral protein that is related to plasma protein inhibitors of serine proteases, *Proc. Natl. Acad. Sci. U.S.A.,* 83, 7698, 1986.
14. **Smith, G. L., Howard, S. T., and Chan, Y. S.,** Vaccinia virus encodes a family of genes with homology to serine proteinase inhibitors, *J. Gen. Virol.,* 70, 2333, 1989.
15. **Kotwal, G. J. and Moss, B.,** Vaccinia virus encodes a secretory polypeptide structurally related to complement control proteins, *Nature,* 335, 176, 1988.
16. **Gemmel, A. and Fenner, F.,** Genetic studies with mammalian poxviruses. III. White (μ) mutants of rabbitpox virus, *Virology,* 11, 219, 1960.
17. **McClain, M. E.,** The host range and plaque morphology of rabbitpox virus (RPμ+) and its μ mutants on chick fibroblast, PK-2a, and L929 cells, *Aust. J. Exp. Biol. Med. Sci.,* 43, 31, 1965.
18. **McClain, M. E. and Greenland, R. M.,** Recombination between rabbitpox virus mutants in permissive and nonpermissive cells, *Virology,* 25, 516, 1965.
19. **Fenner, F. and Sambrook, J. F.,** Conditional lethal mutants of rabbitpox virus. II. Mutants *p* that fail to multiply in PK-2a cells, *Virology,* 28, 600, 1966.
20. **Tagaya, I., Kitamura, T., and Sano, Y.,** A new mutant of dermovaccinia virus, *Nature,* 192, 381, 1961.
21. **Mayr, A., Hochstein-Mintzel, V., and Stickl, H.,** Abstammung, Eigenschaften und Verwendung des attenuierten vacciniastammes MVA, *Infection,* 3, 6, 1975.
22. **Drillien, R. Spehner, D., and Kirn, A.,** Host range restriction of vaccinia virus in Chinese hamster ovary cells: relationship to shutoff of protein synthesis, *J. Virol.,* 28, 843, 1978.
23. **Hruby, D. E., Lynn, D. L., Condit, R. C., and Kates, J. R.,** Cellular differences in the molecular mechanisms of vaccinia virus host range restriction, *J. Gen. Virol.,* 47, 485, 1980.
24. **Drillien, R., Koehren, F., and Kirn, A.,** Host range deletion mutant of vaccinia virus defective in human cells, *Virology,* 111, 488, 1981.

25. **Perkus, M. E., Goebel, S. J., Davis, S. W., Johnson, G. P., Limbach, K., Norton, E. K., and Paoletti, E.,** Vaccinia virus host range genes, *Virology,* 179, 276, 1990.
26. **Gillard, S., Spehner, D., Drillien, R., and Kirn, A.,** Localization and sequence of a vaccinia virus gene required for multiplication in human cells, *Proc. Natl. Acad. Sci. U.S.A.,* 83, 5573, 1986.
27. **Taylor, J., Pincus, S., Tartaglia, J., Richardson, C., Alkhatib, G., Briedis, D., Appel, M., Norton, E., and Paoletti, E.,** Vaccinia virus recombinants expressing either the measles fusion or hemagglutinin glycoprotein protect dogs against canine distemper virus challenge, *J. Virol.,* 65, 4263, 1991.
28. **Taylor, J. and Paoletti, E.,** Fowlpox virus as a vector for non-avian species., *Vaccine,* 6, 466, 1988.
29. **Taylor, J., Weinberg, R., Languet, B., Desmettre, P., and Paoletti, E.,** Recombinant fowlpox virus inducing protective immunity in non-avian species, *Vaccine,* 6, 497, 1988.
30. **Taylor, J., Weinberg, R., Tartaglia, J., Richardson, C., Alkhatib, G., Briedis, D., Appel, M., Norton, E., and Paoletti, E.,** Nonreplicating viral vectors as potential vaccines: recombinant canarypox virus expressing measles virus fusion (F) and hemagglutinin (HA) glycoproteins, *Virology,* in press.
31. **Wunner, W. H., Dietzschold, B., Curtis, P. J., and Wiktor, T. J.,** Rabies subunit vaccines, *J. Gen. Virol.,* 64, 1649, 1983.
32. **Kieny, M.-P., Lathe, R., Drillien, R., Spehner, D., Skory, S., Schmitt, D., Wiktor, T., Koprowski, H., and Lecocq, J.-P.,** Expression of rabies glycoprotein from a recombinant vaccinia virus, *Nature,* 312, 163, 1984.
33. **Wiktor, T. J., Macfarlan, R. I., Reagan, J. K., Dietzschold, B., Curtis, P. J., Wunner, W. H., Kieny, M.-P., Lathe, R., Lecocq, J.-P., Mackett, M., Moss, B., and Koprowski, H.,** Protection from rabies by a vaccinia virus recombinant containing the rabies virus glycoprotein gene, *Proc. Natl. Acad. Sci. U.S.A.,* 81, 7194, 1984.
34. **Blancou, J., Kieny, M.-P., Lathe, R., Lecocq, J.-P., Pastoret, P.-P., Soulebot, J. P., and Desmettre, P.,** Oral vaccination of the fox against rabies using a live recombinant vaccinia virus, *Nature,* 322, 373, 1986.
35. **Brochier, B. M., Languet, B., Blancou, J., Kieny, M.-P., Lecocq, J.-P., Costy, F., Desmettre, P., and Pastoret, P.-P.** Use of recombinant vaccinia-rabies virus for oral vaccination of fox cubs (*Vulpes vulpes,* L.) against rabies, *Vet. Microbiol.,* 18, 103, 1988.
36. **Rupprecht, C. E., Wiktor, T. J., Johnston, D. H., Hamir, A. N., Dietzschold, B., Wunner, W. H., Glickman, L. T., and Koprowski, H.,** Oral immunization and protection of racoons (*Procyon lotor*) with a vaccinia-rabies glycoprotein recombinant virus vaccine, *Proc. Natl. Acad. Sci. U.S.A.,* 83, 7947, 1986.
37. **Rupprecht, C. E., Hamir, A. N., Johnston, D. H., and Koprowski, H.,** Efficacy of a vaccinia-rabies glycoprotein recombinant virus vaccine in racoons *(Procyon lotor), Rev. Inf. Dis.* 4, (Suppl. 10), S803, 1988.
38. **Tolson, N. D., Charlton, K. M., Stewart, R. B., Campbell, J. B., and Wiktor, T. J.,** Immune response in skunks to a vaccinia virus recombinant expressing the rabies glycoprotein, *Can. J. Vet. Res.,* 51, 363, 1987.
39. **Esposito, J., Brechling, K., Baer, G. M., and Moss, B.** Vaccinia virus recombinants expressing rabies virus glycoprotein protect against rabies, *Virus Genes,* 1, 7, 1987.
40. **Esposito, J. J., Knight, J. C., Shaddock, J. H., Novembre, F. J., and Baer, G. M.,** Successful oral rabies vaccination of racoons with racoon poxvirus recombinants expressing rabies virus glycoprotein, *Virology,* 165, 313, 1988, 87.
41. **Shibley, G. and Langston, A.,** An application to field test a live vaccinia vectored rabies vaccine, Abstract 87, in *Modern Approaches to New Vaccines Including Prevention of AIDS,* Cold Spring Harbor Laboratory, Cold Spring Harbor, NY, 1989.
42. **Anon.,** *Science,* 249, 982, 1990.

43. **Pastoret, P.-P., Brochier, B., Languet, B., Thomas, I., Paquot, A., Bauduin, B., Kieny, M.-P., Lecocq, J.-P., De Bruyn, J., Costy, F., Antoine, H., and Desmettre, P.,** First field trial of fox vaccination against rabies using a vaccinia-rabies recombinant virus, *Vet. Rec.,* 123, 481, 1988.
44. **Brochier, B., Thomas, I., Baudin, B., Leveau, T., Pastoret, P.-P., Languet, B., Chappuis, G., Desmettre, P., Blancou, J., and Artois, M.,** Use of a vaccinia-rabies recombinant virus for the oral vaccination of foxes against rabies, *Vaccine,* 8, 101, 1990.
45. **Taylor, J., Trimarchi, C., Weinberg, R., Languet, B., Riviere, M., Desmettre, P., and Paoletti, E.,** Efficacy studies on a canarypox-rabies recombinant virus, *Vaccine,* 9, 190, 1991.
46. **Tiollais, P., Pourcel, C., and Dejean, A.,** The hepatitis B virus, *Nature,* 317, 489, 1985.
47. **Moss, B. and Flexner, C.,** Vaccinia virus expression vectors, *Annu. Rev. Immunol.,* 5, 305, 1987.
48. **Valenzuela, P., Medina, A., Rutter, W. J., Ammerer, G., and Hall, B. D.,** Synthesis and assembly of hepatitis B virus surface antigen particles in yeast, *Nature,* 298, 347, 1982.
49. **Miyanohara, A., Toh-e, A., Nozaki, C., Hamada, F., Ohtomo, N., and Matsubara, K.,** Expression of hepatitis B surface antigen in yeast, *Proc. Natl. Acad. Sci. U.S.A.,* 80, 1, 1983.
50. **Valenzuela, P., Coit, D., and Kuo, C. H.,** Synthesis and assembly in yeast of hepatitis B surface antigen particles containing the polyalbumin receptor, *Bio/Technology,* 3, 317, 1985.
51. **Machida, A., Kishimoto, S., Ohnuma, H., Baba, K., Ito, Y., Miyamoto, H., Funatsu, G., Oda, K., Usuda, S., Togami, S., Nakamura, T., Miyakawa, Y., and Miyuma, M.,** A polypeptide containing 55 amino acid residues coded by the pre-S region of hepatitis B virus deoxyribonucleic acid bears the receptor for polymerized human as well as chimpanzee albumins, *Gastroenterology,* 86, 910, 1984.
52. **Neurath, A. R., Kent, S. B. H., Strick, N., Taylor, P., and Stevens, C. E.,** Hepatitis B virus contains pre-S gene-encoded domains, *Nature,* 315, 154, 1985.
53. **Heermann, K. H., Goldmann, U., Schwartz, W., Seyffarth, T., Baumgarten, H., and Gerlich, W. H.,** Large surface proteins of hepatitis B virus containing the pre-s sequence, *J. Virol.,* 52, 396, 1984.
54. **Machida, A., Kishimoto, S., Ohnuma, H., Miyamoto, H., Baba, K., Oda, K., Nakamura, T., Miyakawa, Y., and Miyuma, M.,** A hepatitis B surface antigen polypeptide (P31) with the receptor for polymerized human as well as Chimpanzee albumins, *Gastroenterology,* 85, 268, 1983.
55. **Stibbe, W. and Gerlich, W. H.,** Structural relationships between minor and major proteins of hepatitis B surface antigens, *J. Virol.,* 46, 626, 1983.
56. **Smith, G. L., Mackett, M., and Moss, B.,** Infectious vaccinia virus recombinants that express hepatitis B virus surface antigen, *Nature,* 302, 490, 1983.
57. **Paoletti, E., Lipinskas, B. R., Samsonoff, C., Mercer, S., and Panicali, D.,** Construction of live vaccines using genetically engineered poxviruses: biological activity of vaccinia virus recombinants expressing the hepatitis B virus surface antigen and the herpes simplex virus glycoprotein D., *Proc. Natl. Acad. Sci. U.S.A.,* 81, 193, 1984.
58. **Moss, B., Smith, G. L., Gerin, J. L., and Purcell, R. H.,** Live recombinant vaccinia virus protects chimpanzees against hepatitis B, *Nature,* 311, 67, 1984.
59. **Cheng, K.-C., Smith, G. L., and Moss, B.,** Hepatitis B virus large surface protein is not secreted but is immunogenic when selectively expressed by recombinant vaccinia virus, *J. Virol.,* 60, 337, 1986.
60. **Cheng, K.-C. and Moss, B.,** Selective synthesis and secretion of particles composed of the hepatitis B virus middle surface protein directed by a recombinant vaccinia virus: induction of antibodies to pre-S and S epitopes, *J. Virol.,* 61, 1286, 1987.

61. **Little, S. P., Jofre, J. T., Courtney, R. J., and Schaffer, P. A.,** A virion-associated glycoprotein essential for infectivity of herpes simplex virus type 1, *Virology,* 115, 149, 1981.
62. **Sarmiento, M., Haffey, M., and Spear, P. G.,** Membrane proteins specified by herpes simplex viruses. III. Role of glycoprotein VP7(B2) in virion infectivity, *J. Virol.,* 29, 1149, 1979.
63. **Friedman, H. M., Cohen, G. H., Eisenberg, R. J., Seidel, C. A., and Cines, D. B.,** Glycoprotein C of herpes simplex virus 1 acts as a receptor for the C3b complement component on infected cells, *Nature,* 309, 633, 1984.
64. **Noble, A. G., Lee, G. T.-Y., Sprague, R., Parish, M. L., and Spear, P. G.,** Anti-gD monoclonal antibodies inhibit cell fusion induced by herpes simplex virus type 1, *Virology,* 129, 218, 1983.
65. **Para, M. F., Parish, M. L., Noble, A. G., and Spear, P. G.,** Potent neutralizing activity associated with anti-glycoprotein D specificity among monoclonal antibodies selected for binding to herpes simplex virions, *J. Virol.,* 55, 483, 1985.
66. **Sullivan, V. and Smith, G. L.,** Expression and characterization of herpes simplex type 1 (HSV-1) glycoprotein G (gG) by recombinant vaccinia virus: neutralization of HSV-1 infectivity with anti-G antibody, *J. Gen. Virol.,* 68, 2587, 1987.
67. **Gompels, U. and Minson A.,** The properties and sequence of glycoprotein H of herpes simplex virus type 1, *Virology,* 153, 230, 1986.
68. **Cremer, K. J., Mackett, M., Wohlenberg, C., Notkins, A. L., and Moss, B.,** Vaccinia virus recombinant expressing herpes simplex virus type 1 glycoprotein D prevents latent herpes in mice, *Science,* 228, 737, 1985.
69. **Martin, S. and Rouse, B. T.,** The mechanisms of antiviral immunity induced by a vaccinia virus recombinant expressing herpes simplex virus type 1 glycoprotein D: clearance of local infection, *J. Immunol.,* 138, 3431, 1987.
70. **Martin, S., Moss, B., Berman, P. W., Laskey, L. A., and Rouse, B. T.,** Mechanisms of antiviral immunity induced by a vaccinia virus recombinant expressing herpes simplex virus type 1 glycoprotein D: cytotoxic T cells, *J. Virol.,* 61, 726, 1987.
71. **Wachsman, M., Aurelian, L., Smith, C. C., Lipinskas, B. R., Perkus, M. E., and Paoletti, E.,** Protection of guinea pigs from primary and recurrent herpes simplex virus (HSV) type 2 cutaneous disease with vaccinia virus recombinants expressing HSV glycoprotein D., *J. Inf. Dis.,* 155, 1188, 1987.
72. **Wachsman, M., Aurelian, L., Hunter, J. C. R., Perkus, M. E., and Paoletti, E.,** Expression of herpes simplex virus glycoprotein D on antigen presenting cells infected with vaccinia recombinants and protective immunity, *Bio. Sci. Rep.,* 8, 323, 1988.
73. **Wachsman, M., Luo, J. H., Aurelian, L., Perkus, M. E., and Paoletti, E.,** Antigen-presenting capacity of epidermal cells infected with vaccinia virus recombinants containing the herpes simplex virus glycoprotein D, and protective immunity, *J. Gen. Virol.,* 70, 2513, 1989.
74. **Wachsman, M., Aurelian, L., Smith, C. C., Perkus, M. E., and Paoletti, E.,** Regulation of expression of herpes simplex virus (HSV) glycoprotein D in vaccinia recombinants affects their ability to protect from cutaneous HSV-2 disease, *J. Inf. Dis.,* 159, 625, 1989.
75. **Zarling, J. M., Moran, P. A., Lasky, L. A., and Moss, B.,** Herpes simplex virus (HSV)-specific human T-cell clones recognize HSV glycoprotein D expressed by a recombinant vaccinia virus, *J. Virol.,* 59, 506, 1986.
76. **McLaughlin-Taylor, E., Willey, D. E., Cantin, E. M., Eberle, R., Moss, B., and Openshaw, H.,** A recombinant vaccinia virus expressing herpes simplex virus type 1 glycoprotein B induces cytotoxic T lymphocytes in mice, *J. Gen. Virol.,* 69, 1731, 1988.
77. **Cantin, E. M., Eberle, R., Baldick, J. L., Moss, B., Willey, D. E., Notkins, A. L., and Openshaw, H.,** Expressions of herpes simplex virus 1 glycoprotein B by a recombinant vaccinia virus and protection of mice against a lethal herpes simplex virus 1 infection, *Proc. Natl. Acad. Sci. U.S.A.,* 84, 5908, 1987.

78. **Perkus, M. E., Piccini, A., Lipinskas, B. R., and Paoletti, E.,** Recombinant vaccinia virus: immunization against multiple pathogens, *Science,* 229, 981, 1985.
79. **Gillespie, J. H., Geissinger, C., Scott, F. W., Higgins, W. P., Holmes, D. F., Perkus, M., Mercer, S., and Paoletti, E.,** Response of dairy calves to vaccinia viruses that express foreign genes, *J. Clin. Microbiol.,* 23, 283, 1986.
80. **Rooney, J. F., Wohlenberg, C., Cremer, K. J., Moss, B., and Notkins, A. L.,** Immunization with a vaccinia virus recombinant expressing herpes simplex virus type 1 glycoprotein D: long-term protection and effect of revaccination, *J. Virol.,* 62, 1530, 1988.
81. **Flexner, C., Murphy, B. R., Rooney, J. F., Wohlenberg, C., Yuferov, V., Notkins, A. L., and Moss, B.,** Successful vaccination with a polyvalent live vector vaccine despite existing immunity to an expressed antigen, *Nature,* 335, 259, 1988.
82. **Meyers, J. D., Fluornoy, N., and Thomas, E. D.,** Nonbacterial pneumonia after allogeneic transplantation: a review of ten years' experience, *Rev. Infect. Dis.,* 4, 1119, 1982.
83. **Meyers, J. D.,** Cytomegalovirus infection following marrow transplantation: risk, treatment, and prevention, *Birth Defects Orig. Article Ser.,* 20, 101, 1986.
84. **Quinnan, G. V., Jr., Rook, A. H., Frederick, W. R., Manischewitz, J. F., Epstein, J. S., Siegel, J., Masur, H., Macher, A. M., Mitchell, C., Armstrong, G., et al.,** Prevalence, clinical manifestations, and immunology of herpesvirus infections in the acquired immunodeficiency syndrome, *Ann. N.Y. Acad. Sci.,* 437, 200, 1984.
85. **Moskovitz, L., Hensley, G. T., Chan, J. C., and Adams, K.,** Immediate causes of death in acquired immunodeficiency syndrome, *Arch. Pathol. Lab. Med.,* 109, 735, 1985.
86. **Levin, A. V., Zeichner, S., Duker, J. S., Starr, S. E., Augsberger, J. J., and Kronwith, S.,** Cytomegalovirus retinitis in an infant with acquired immunodeficiency syndrome, *Pediatrics,* 84, 683, 1989.
87. **Alford, C. A., Stagno, S., and Pass, R.,** Natural history of perinatal cytomegalovirus infection, in *Perinatal Infections,* Elliot, K., O'Connor, M., and Whelan, J., Eds., Elsevier/North-Holland, Amsterdam, 1980, 125.
88. **Stinski, M. F.,** Human cytomegalovirus: glycoproteins associated with virions and dense bodies, *J. Virol.,* 19, 594, 1976.
89. **Britt, W. J.,** Neutralizing antibodies detect a disulfide-linked glycoprotein complex within the envelope of human cytomegalovirus, *Virology,* 135, 369, 1984.
90. **Kari, B., Lussenhop, N., Goertz, R., Wabuke-Bunoti, M., Radeke, R., and Gehrz, R.,** Characterization of monoclonal antibodies reactive to several biochemically distinct human cytomegalovirus glycoprotein complexes, *J. Virol.,* 60, 345, 1986.
91. **Law, K. M., Wilton-Smith, P., and Farrar, G. H.,** A murine monoclonal antibody recognizing a single glycoprotein within a human cytomegalovirus virion envelope complex, *J. Med. Virol.,* 17, 255, 1985.
92. **Gonczol, E., Hudecz, F., Ianacone, J., Dietzschold, B., Starr, S., and Plotkin, S. A.,** Immune response to isolated human cytomegalovirus envelope proteins, *J. Virol.,* 58, 661, 1986.
93. **Rasmussen, L., Mullenax, J., Nelson, M., and Merigan, T. C.,** Human cytomegalovirus polypeptides stimulate virus neutralizing antibody *in vivo, Virology,* 145, 186, 1985.
94. **Cranage, M. P., Kouzarides, T., Bankier, A. T., Satchwell, S., Weston, K., Tomlinson, P., Barrell, B., Hart, H., Bell, S. E., Minson, A. C., and Smith, G. L.,** Identification of human cytomegalovirus glycoprotein B gene and induction of neutralizing antibodies via its expression in recombinant vaccinia virus, *EMBO J.,* 5, 3057, 1986.
95. **Britt, W. J., Vugler, L., and Stephens, E. B.,** Induction of complement-dependent and -independent neutralizing antibodies by recombinant-derived cytomegalovirus gp55-116 (gB), *J. Virol.,* 62, 3309, 1988

96. **Reddehase, M. J., Mutter, W., Munch, K., Buhring, H.-J., and Koszinowski, U. H.,** CD8-positive T lymphocytes specific for murine cytomegalovirus immediate-early antigens mediate protective immunity, *J. Virol.*, 61, 3102, 1987.
97. **Jonjic, S., del Val, M., Keil, G. M., Reddehase, M. J., and Koszinowski, U. H.,** A nonstructural viral protein expressed by a recombinant vaccinia virus protects against lethal cytomegalovirus infection, *J. Virol.*, 62, 1653, 1988.
98. **del Val, M., Volkmer, H., Rothbard, J. B., Jonjic, S., Messerle, M., Schickedanz, J., Reddehase, M. J., and Koszinowski, U. H.,**Molecular basis for cytotoxic T-lymphocyte recognition of the murine cytomegalovirus immediate-early protein pp89, *J. Virol.*, 62, 3965, 1988.
99. **Borysiewicz, L. K., Hickling, J. K., Graham, S., Sinclair, J., Cranage, M. P., Smith, G. L., and Sissons, J. G. P.,** Human cytomegalovirus-specific cytotoxic T cells: relative frequency of stage-specific CTL recognizing the 72-kDa immediate-early protein and glycoprotein B expressed by recombinant vaccinia viruses, *J. Exp. Med.*, 168, 919, 1988.
100. **Stenberg, R. M., Thomsen, D. R., and Stinski, M. F.,** Structural analysis of the major immediate early gene of human cytomegalovirus, *J. Virol.*, 49, 190, 1984.
101. **Epstein, M. A. and Achong, B. G., Eds.,** *The Epstein-Barr Virus: Recent Advances,* Heineman, London, 1986.
102. **Clearly, M. L., Dorfman, R. F., and Sklar, J.,** Failure in immunological control of the virus infection: post-transplant lymphomas, in *The Epstein-Barr Virus: Recent Advances,* Epstein, M. A. and Achong, B. G., Eds., Heineman, London, 1986, 163.
103. **Beisel, C., Tanner, J., Matsuo, T., Thorley-Lawson, D., Kezdy, F., and Kieff, E.,** Two major outer envelope glycoproteins of Epstein-Barr virus are encoded by the same gene, *J. Virol.*, 54, 665, 1985.
104. **Qualtiere, L. F., and Pearson, G. R.,** Epstein-Barr virus-induced membrane antigens: immunochemical characterization of Triton X-100 solubilized viral membrane antigens from EBV-superinfected Raji cells, *Int. J. Cancer,* 23, 808, 1979.
105. **Hoffman, G. J., Lazarowitz, S. G., and Hayward, S. D.,** Monoclonal antibody against a 250,000-dalton glycoprotein of Epstein-Barr virus identifies a membrane antigen and a neutralizing antigen, *Proc. Natl. Acad. Sci. U.S.A.*, 77, 2879, 1980.
106. **Thorley-Lawson, D. A. and Geilinger K.,** Monoclonoal antibodies against the major glycoprotein (gp350/220) of Epstein-Barr virus neutralize infectivity, *Proc. Natl. Acad. Sci. U.S.A.*, 77, 5307, 1980.
107. **Franklin, S. M., North, J. R., Morgan, A. J., and Epstein, M. A.,** Antigenic differences between the membrane antigen polypeptides determined by different EB virus isolates, *J. Gen. Virol.*, 53, 371, 1981.
108. **Epstein, M. A., Morgan, A. J., Finerty, S., Randle, B. J., and Kirkwood, J. K.,** Protection of cottontop tamarins against Epstein-Barr virus-induced malignant lymphoma by a prototype subunit vaccine, *Nature,* 318, 287, 1985.
109. **Mackett, M. and Arrand, J. R.,** Recombinant vaccinia virus induces neutralizing antibodies in rabbits against Epstein-Barr virus membrane antigen gp340, *EMBO J.*, 4, 3229, 1985.
110. **Morgan, A. J., Mackett, M., Finerty, S., Arrand, J. R., Scullion, F. T., and Epstein, M. A.,** Recombinant vaccinia virus expressing Epstein-Barr virus glycoprotein gp340 protects cottontop tamarins against EB virus-induced malignant lymphomas, *J. Med. Virol.*, 25, 189, 1988.
111. **Cretescu, L., Beare, A. S., and Schild, G. C.,** Formation of antibody to matrix protein in experimental human influenza A virus infections, *Infect. Immun.*, 22, 322, 1978.
112. **Potter, C. W. and Oxford, J. S.,** Determinants of immunity to influenza infection in man, *Br. Med. Bull.*, 35, 69, 1979.
113. **Askonas, B. A., McMichael, A. J., and Webster, R. J.,** The immune response to influenza viruses and the problem of protection against infection, in *Basic and Applied Influenza Research,* Beare, A. S., Ed., CRC Press, Boca Raton, FL, 1982, 159.

114. **Clements, M. L., Betts, R. F., Tierney, E. L., and Murphy, B. R.,** Serum and nasal wash antibodies associated with resistance to experimental challenge with influenza A wild-type virus, *J. Clin. Microbiol.,* 24, 157, 1986.
115. **Yewdell, J. W. and Hackett, C. J.,** The specificity and function of T lymphocytes induced by influenza A viruses, in *The Influenza Viruses,* Krug, R., Ed., Plenum Press, New York, 1989, 361.
116. **McMichael, A. J., Gotch, F. M., Noble, G. R., and Beare, P. A. S.,** Cytotoxic T-cell immunity to influenza, *N. Engl. J. Med.,* 309, 13, 1983.
117. **Panicali, D., Davis, S. W., Weinberg, R. L., and Paoletti, E.,** Construction of live vaccines by using genetically engineered poxviruses: biological activity of recombinant vaccinia virus expressing influenza virus hemagglutinin, *Proc. Natl. Acad. Sci. U.S.A.,* 80, 5364, 1983.
118. **Smith, G. L., Murphy, B. R., and Moss, B.,** Construction and characterization of an infectious vaccinia virus recombinant that expresses the influenza hemagglutinin gene and induces resistance to influenza virus infection in hamsters, *Proc. Natl. Acad. Sci. U.S.A.,* 80, 1755, 1983.
119. **Boyle, D. B., Coupar, B. E. H., Parsonson, I. M., Bagust, T. J., and Both, G. W.,** Responses of cattle, sheep, and poultry to a recombinant vaccinia virus expressing a swine influenza hemagglutinin, *Res. Vet. Sci.,* 41, 40, 1986.
120. **Chambers, T. M., Kawaoka, Y., and Webster, R. G.,** Protection of chickens from lethal influenza infection by vaccinia-expressed hemagglutinin, *Virology,* 167, 414, 1988.
121. **Jakeman, K. J., Smith, H., and Sweet, C.,** Mechanism of immunity to influenza: maternal and passive neonatal protection following immunization of adult ferrets with a live vaccinia-influenza virus haemagglutinin recombinant but not recominants containing other influenza virus proteins, *J. Gen. Virol.,* 70, 1523, 1989.
122. **De, B. K., Shaw, M. W., Rota, P. A., Harmon, M. W., Esposito, J. J., Rott, R., Cox, N. J., and Kendal, A. P.,** Protection against virulent H5 avian influenza virus infection in chickens by an inactivated vaccine produced with recombinant vaccinia virus, *Vaccine,* 6, 257, 1988.
123. **Taylor, J., Weinberg, R., Kawaoka, Y., Webster, R. G., and Paoletti, E.,** Protective immunity against avian influenza induced by a fowlpox virus recombinant, *Vaccine,* 6, 504, 1988.
124. **Rota, P. A., De, B. K., Shaw, M. W., Black, R. A., Gamble, W. C., and Kendal, A. P.** Comparison of inactivated, live and recombinant DNA vaccines against influenza virus in a mouse model, *Virus Res.,* 16, 83, 1990.
125. **Bennink, J. R., Yewdell, J. W., Smith, G. L., Moller, C., and Moss, B.,** Recombinant vaccinia virus primes and stimulates influenza haemagglutinin-specific cytotoxic T cells, *Nature,* 311, 578, 1984.
126. **Coupar, B. E. H., Andrew, M. E., Both, G. W., and Boyle, D. B.,** Temporal regulation of influenza hemagglutinin expression in vaccinia virus recombinants and effects on the immune response, *Eur. J. Immunol.,* 16, 1479, 1986.
127. **Andrew, M. E., Coupar, B. E. H., Ada, G. L., and Boyle, D. B.,** Cell mediated immune responses to influenza virus antigens expressed by vaccinia virus recombinants, *Microb. Pathol.,* 1, 443, 1986.
128. **Andrew, M. E., Coupar, B. E. H., Boyle, D. B., and Ada., G. L.,** The roles of influenza virus haemagglutinin and nucleoprotein in protection: analysis using vaccinia virus recombinants, *Scand. J. Immuol.,* 15, 21, 1987.
129. **Andrew, M. E. and Coupar, B. E. H.,** Efficacy of influenza haemagglutinin and nucleoprotein as protective antigens against influenza virus infection in mice, *Scand. J. Immunol.,* 28, 81, 1988.
130. **Stitz, L., Schmitz, C., Binder, D., Zinkernagel, R., Paoletti, E., and Becht, H.,** Characterization and immunological properties of influenza A virus nucleoprotein (NP): cell-associated NP isolated from infected cells or viral NP expressed by vaccinia recombinant virus do not confer protection, *J. Gen. Virol.,* 71, 1169, 1990.

131. **Webster, R. G., Reay, P. A., and Laver, W. G.,** Protection against lethal influenza with neuraminidase, *Virology,* 164, 230, 1988.
132. **Els, M. C., Laver, W. G., and Air, G. M.,** Sialic acid is cleaved from glycoconjugates at the cell surface when influenza virus neuraminidases are expressed from recombinant vaccinia viruses, *Virology,* 170, 346, 1989.
133. **Babbott, F. L., Jr. and Gordon, J. E.,** Modern measles, *Am. J. Med. Sci.,* 228, 334, 1954.
134. **Langmuir, A. D.,** Medical importance of measles, *Am. J. Dis. Child.,* 103, 224, 1962.
135. **Black, F. L.,** A nationwide screen survey of United States military recruits, 1962. III. Measles and mumps antibodies, *Am. J. Hyg.,* 80, 304, 1962.
136. **Chou, T., Weil, D., and Arnow, P. M.,** Prevalence of measles antibody in hospital personnel, *Infect. Contr.,* 7, 309, 1986.
137. **Pap, K.,** Experiences prouvant que la voie d'infection de la rougeole est la contamination de la muqueuse conjunctival, *Rev. Immunol. Ther. Antimicrob.,* 20, 27, 1956.
138. **Robbins, F. C.,** Measles: clinical features. Pathogenesis, pathology, and complications, *Am. J. Dis. Child.,* 103, 266, 1962.
139. **Kempe, C. J. and Fulginiti, V. A.,** The pathogenesis of measles virus infection, *Arch. Virusforsch.,* 16, 103, 1965.
140. Centers for Disease Control, Classification of Measles Cases and Categorization of Measles Elimination Programs, Morbidity and Mortality Weekly Rep. No. 31, Centers for Disease Control, Public Health Service, U.S. Department of Health, Education and Welfare, Atlanta, 1983, 707.
141. **Appelbaum, E. Dolgopol, V. B., and Dolgin, J.,** Measles encephalitis, *Am. J. Dis. Child.,* 77, 25, 1949.
142. **Miller, H. G., Stanton, J. B., and Gibbons, J. L.,** Para-infections encephalomyelitis and related syndromes: a critical review of the neurologic complications of certain specifc fevers, *Q. J. Med.,* 25, 247, 1956.
143. **Gibbs, F. A., Gibbs, E. L., Carpenter, E. R., and Spies, H. W.,** Electroencephalographic abnormality in ''uncomplicated'' childhood diseases, *J. Am. Med. Assoc.,* 72, 1050, 1959.
144. **Katz, S. L.,** Measles — its complications, treatment and prophylaxis, *Med. Clin. North Am.,* 46, 1163, 1962.
145. **Bech, V.,** Measles epidemic in Greenland, *Am. J. Dis. Child.,* 103, 252, 1962.
146. **Mclean, D. M., Best, J. M., Smith, P. A., Larke, R. P. B., and McNaughton, G. A.,** Viral infections of Toronto children during 1965. II. Measles encephalitis and other complications, *Can. Med. Assoc. J.,* 94, 905, 1966.
147. **Barkin, R. M.,** Measles mortality. Analysis of the primary cause of death, *Am. J. Dis. Child.,* 129, 307, 1975.
148. **Morgan, E. M. and Rapp, F.,** Measles virus and its associated diseases, *Bacteriol. Rev.,* 41, 636, 1977.
149. **Modlin, J. F., Jabbour, J. T., Witte, J. J., and Halsey, N. A.,** Epidemiological studies of measles, measles vaccine, and subacute sclerosing panencephalitis, *Pediatrics,* 59, 505, 1977.
150. **Gremillion, D. H. and Crawford, G. E.,** Measles pneumonia in young adults: an analysis of 106 cases, *Am. J. Med.,* 71, 539, 1981.
151. **Halsey, N. A., Boulos, R., Mode, F., Andre, J., Bowman, L., Yaeger, R. G., Toureau, S., Rohde, J., and Boulos, C.,** Responses to measles vaccine in Haitian infants 6 to 12 months old. Influence of maternal antibodies, malnutrition, and concurrent illnesses, *N. Engl. J. Med.,* 313, 544, 1985.
152. **Black, F. L., Berman, L. L., Borgono, J. M., Capper, R. A., Carvalho, A. A., Collins, C. Glover, O., Hijazi, Z., Jacobson, D. L., Lee. Y. L., et al.,** Geographic variation in infant loss of maternal measles antibody and in prevalence of rubella antibody, *Am. J. Epidemiol.,* 124, 442, 1986.

153. **Anon.,** Issues in world health: immunizing the world's children, *Pop. Rep.,* 14, L153, 1986.
154. **Riley, E. C., Murphy, G., and Riley, R. L.,** Airborne spread of measles in a suburban elementary school, *Am. J. Epidemiol.,* 107, 421, 1978.
155. **Bloch, A. B., Orenstein, W. A., Ewing, W. M., Spain, W. H., Mallison, G. F., Herrmann, K. L., and Hinman, A. R.,** Measles outbreak in a pediatric practice: airborne transmission in an office setting, *Pediatrics,* 75, 676, 1985.
156. **Remington, P. L., Hall, W. N., Davis, I. H., Herald, A., and Gunn, R. A.,** Airborne transmission of measles in a physician's office, *J. Am. Med. Assoc.,* 253, 1574, 1985.
157. **Norrby, E. and Oxman, M. N.,** Measles, in *Virology,* Vol. 1, 2nd ed., Fields, B. N. and Knipe, D. M., Eds., Raven Press, New York, 1985, 1013.
158. **Imagawa, D. T.,** Relationships among measles, canine distemper and rinderpest viruses, *Prog. Med. Virol.,* 10, 160, 1968.
159. **Diallo, A.,** Morbillivirus group: genome organisation and proteins, *Vet. Microbiol.,* 23, 155, 1990.
160. **Norrby, E., Enders-Ruckle, G., and ter Meulen, V.,** Differences in the appearance of antibodies to structural components of measles virus after immunization with inactivated and live virus, *J. Infect. Dis.,* 132, 262, 1975.
161. **Janeway, C. A.,** Use of concentrated human serum γ-globulin in the prevention and attenuation of measles, *Bull. N.Y. Acad. Med.,* 21, 202, 1945.
162. **Black, F. L. and Yannet, H.,** Inapparent measles after gamma globulin administration, *J. Am. Med. Assoc.,* 173, 1183, 1960.
163. **Brody, J. A. and Bridenbraugh, E.,** Prophylactic γ-globulin and live measles vaccine in an island epidemic of measles, *Lancet,* 2, 811, 1964.
164. **Bellanti, J. A., Sanga, R. L., Klutinis, B., Brandt, B., and Artenstein, M. S.,** Antibody responses in serum and nasal secretions of children immunized with inactivated and attenuated measles-virus vaccines, *N. Engl. J. Med.,* 280, 628, 1969.
165. **Friedman, M., Hadari, I., Goldstein, V., and Sarov, I.,** Virus-specific secretory IgA antibodies as a means of rapid diagnosis of measles and mumps infection, *Isr. J. Med. Sci.,* 19, 881, 1983.
166. **Krause, P. J., Cherry, J. D., Naiditch, M. J., Deseda-Tous, J., and Walbergh, E. J.,** Revaccination of previous recipients of killed measles vaccine: clinical and immunologic studies, *J. Pediatr.,* 93, 565, 1978.
167. **Krause, P. J., Cherry, J. D., Carney, J. M., Naiditch, M. J., and O'Connor, K.,** Measles-specific lymphocyte reactivity and serum antibody in subjects with different measles histories, *Am. J. Dis. Child.,* 134, 567, 1980.
168. **Gallagher, M. R., Welliver, R., Yamanaka, T., Eisenberg, B., Sun, M., and Ogra, P. L.,** Cell-mediated immune responsiveness to measles. Its occurrence as a result of naturally acquired or vaccine-induced infection and in infants of immune mothers, *Am. J. Dis. Child.,* 135, 48, 1981.
169. **Preblud, S. R. and Katz, S. L.,** Measles vaccine, in *Vaccines,* Plotkin, S. A. and Mortimer, E. A., Eds., Harcourt Brace Jovanovich, Philadelphia, 1988, 182.
170. **Krugman, S., Giles, J. P., Jacobs, A. M., and Friedman, H.,** Studies with a further attenuated live measles-virus vaccine, *Pediatrics,* 31, 919, 1963.
171. **Crawford, G. E. and Gremillion, D. H.,** Epidemic measles and rubella in Air Force recruits: impact of immunization, *J. Inf. Dis.,* 144, 403, 1981.
172. Centers for Disease Control, Recommendations of the Immunizations Practice Advisory Committee. New Recommended Schedule for Active Immunization of Normal Infants and Children, Morbidity and Mortality Weekly Rep. No. 35, Centers for Disease Control, Public Health Service, U.S. Department of Health, Education and Welfare, Atlanta, 1986, 577.
173. **Arbeter, A. M., Baker, L., Starr, S. E., Levine, B. L., Books, E., and Plotkin, S. A.,** Combination measles, mumps, rubella, and varicella vaccine, *Pediatrics,* 78 (Suppl.), 742, 1986.

174. **Arbeter, A. M., Baker, L., Starr, S. E., and Plotkin, S. A.,** The combination measles, mumps, rubella and varicella vaccine in healthy children, *Dev. Biol. Stand.,* 65, 89, 1986.
175. **Just, M., Berger, R, and Just, V.,** Evaluation of a measles-mumps-rubella-chickenpox vaccine, *Dev. Biol. Stand.,* 65, 85, 1986.
176. **Byrne, E. B., Rosenstein, B. J., Jaworski, A. A., and Jaworski, R. A.,** A statewide mass measles immunization program, *J. Am. Med. Assoc.,* 199, 619, 1967.
177. Centers for Disease Control, Measles Vaccination Reactions Among College Students — North Carolina; Massachusetts, Morbidity and Mortality Weekly Rep. No. 29, Centers for Disesase Control, Public Health Service, U.S. Department of Health, Education and Welfare, Atlanta, 1980, 549.
178. Centers for Disease Control, Measles Vaccination Reactions Among College Students — Wisconsin, Morbidity and Mortality Weekly Rep. No. 29, Centers for Disease Control, Public Health Service, U.S. Department of Health, Education and Welfare, Atlanta, 1980, 21.
179. **Trump, R. C. and White, T. R.,** Cerebellar ataxia presumed due to live, attenuated measles virus vaccine, *J. Am. Med. Assoc.,* 199, 129, 1967.
180. **Grose, C. G. and Spigland, I.,** Guillian-Barré syndrome following administration of live measles vaccine, *Am. J. Med.,* 60, 441, 1976.
181. **Kazarian, E. L. and Gager, W. E.,** Optic neuritis complicating measles, mumps, and rubella vaccination, *Am. J. Ophthalmol.,* 86, 544, 1978.
182. **Morens, D. M., Halsey, N. A., Schonberger, L. B., and Baublis, J. V.,** Reye syndrome associated with vaccination with live virus vaccines. An exploration of possible etiologic relationships, *Clin. Pediatr.,* 18, 42, 1979.
183. **Chan, C. C., Sogg, R. L., and Steinman, L.,** Isolated oculomotor palsy after measles immunization, *Am. J. Ophthalmol.,* 89, 446, 1980.
184. **Halsey, N. A., Weiner L. B., Meyers, M. G., Herrmann, K. L., and Hinman, A. R.,** Clinical evaluation of a new live measles vaccine derived in chick chorioallantoic membranes, *J. Biol. Stand.,* 9, 507, 1981.
185. **Marshall, G. S., Wright, P. F., Fenichel, G. M., and Karzon, D. T.,** Diffuse retinopathy following measles, mumps, and rubella vaccination, *Pediatrics,* 76, 989, 1985.
186. **Brodsky, L. and Stanievich, J.,** Sensorineural hearing loss following live measles virus vaccination, *Int. J. Pediatr. Otorhinol.,* 10, 159, 1985.
187. **Wilhelm, D. J. and Paegle, R. D.,** Thrombocytopenic purpura following live measles vaccine, *Am. J. Dis. Child.,* 113, 534, 1967.
188. **Zwemer, R., Hodge, S., Owen, L. G., and Fliegelman, M. T.,** Persistent toxic erythema and chronic urticaria. Possible association with the use of measles virus vaccine, *Arch. Dermatol.,* 104, 390, 1971.
189. **Vessal, S. and Kravis, L. P.,** Immunologic mechanisms responsible for adverse reactions to routine immunizations in children, *Clin. Pediatr.,* 15, 688, 1976.
190. **Van Aspersen, P. P., McEniery, J., and Kemp, A. S.,** Immediate reactions following live attenuated measles vaccine, *Med. J. Austr.,* 2, 330, 1976.
191. **Drillien, R., Spehner, D., Kirn, A., Giraudon, P., Buckland, R., Wild, F., and Lecocq, J.-P.,** Protection of mice from fatal measles encephalitis by vaccination with vaccinia virus recombinants encoding either the hemagglutinin or the fusion protein, *Proc. Natl. Acad. Sci. U.S.A.,* 85, 1252, 1988.
192. **Adams, J. M. and Imagawa, D. T.,** Immunological relationship between measles and distemper viruses, *Proc. Soc. Exp. Biol. Med.,* 96, 240, 1957.
193. **Imagawa, D. T., Goret, P., and Adams, J. M.,** Immunological relationships among measles, distemper, and rinderpest viruses, *Proc. Nat. Acad. Sci. U.S.A.,* 46, 1119, 1960.
194. **Gillespie, J. H. and Karzon, D. T.,** A study of the relationship between canine distemper and measles in the dog, *Proc. Soc. Exp. Biol. Med.,* 105, 547, 1960.

195. **Warren, J., Nadel, M. K., Slater, E., and Millian, S. J.,** The canine distemper measles complex. I. Immune response of dogs to canine distemper and measles viruses, *Am. J. Vet. Res.,* 21, 111, 1960.
196. **Moura, R. A. and Warren, J.,** Subclinical infection of dogs by canine adapted measles virus evidenced by their subsequent immunity to canine distemper virus, *J. Bacteriol.,* 82, 702, 1961.
197. **Delay, P. B., Stone, S. S., Karzon, D. T., Katz, S., and Enders, J.,** Clinical and immune response of alien hosts to inoculation with measles, rinderpest and canine distemper viruses, *Am. J. Vet. Res.,* 26, 1359, 1965.
198. **Roberts, J. A.,** A study of the antigenic relationships between human measles virus and canine distemper virus, *J. Immunol.,* 94, 622, 1965.
199. **Orvell, C. and Norrby, E.,** Further studies on the immunologic relationships among measles, distemper and rinderpest viruses, *J. Immunol.,* 113, 1850, 1974.
200. **Sheshberadaran, H., Norrby, E., McCullough, K. C., Carpenter, W. C., and Orvell, C.,** The antigenic relationship between measles, canine distemper and rinderpest viruses studied with monoclonal antibodies, *J. Gen. Virol.,* 67, 1381, 1986.
201. **Stephenson, J. R. and ter Meulen, V.,** Antigenic relationships between measles and canine distemper viruses: comparison of immune response in animals and humans to individual virus specific polypeptides, *Proc. Natl. Acad. Sci. U.S.A.,* 76, 6601, 1979.
202. **Hall, W. W., Lamb, R. A., and Choppin, P. W.,** The polypeptides of canine distemper virus: synthesis in infected cells and relatedness to the polypeptides of other Morbilliviruses, *Virology,* 100, 433, 1980.
203. **Orvell, C. and Norrby, E.,** Immunological relationships between homologous structural polypeptides of measles and canine distemper virus, *J. Gen. Virol.,* 50, 231, 1980.
204. **Spehner, D., Drillien, R., and Lecocq, J.-P.,** Construction of fowlpox virus vectors with intergenic insertions: expression of the beta-galactosidase gene and the measles virus fusion gene, *J. Virol.,* 64, 527, 1990.
205. **Monath, T. P.,** Pathobiology of the flaviviruses, in *The Togaviridae and Flaviviridae,* Schlesinger, S. and Schlesinger, M. J., Eds., Plenum Press, New York, 1986, 375.
206. **Rice, C. M., Ellen, G. S., and Strauss, J. H.,** Structure of the flavivirus genome, in *The Togaviridae and Flaviviridae,* Schlesinger, S. and Schlesinger, M. J., Eds., Plenum Press, New York, 1986, 279.
207. **Guld, E. A., Buckley, A., Barrett, A. D. T., and Cammack, N.,** Neutralizing (54K) and non-neutralizing (54K and 48K) monoclonal antibodies against structural and nonstructural yellow fever virus proteins confer immunity in mice, *J. Gen. Virol.,* 67, 591, 1986.
208. **Kaufman, B. M., Summers, P. L., Dubois, D. R., and Eckels, K. H.,** Monoclonal antibodies against dengue 2 virus E-glycoprotein protect mice against lethal dengue infection, *Am. J. Trop. Med. Hyg.,* 36, 427, 1987.
209. **Kimura-Kuroda, J. and Yasui, K.,** Protection of mice against Japanese encephalitis virus by passive administration with monoclonal antibodies, *J. Immunol.,* 141, 3606, 1988.
210. **Mason, P. W., Dalrymple, J. M., Gentry, M. K., McCown, J. M., Hoke, C. H., Burke, D. S., Fournier, M. J., and Mason, T. L.,** Molecular characterization of a neutralizing domain of the Japanese encephalitis virus structural glycoprotein, *J. Gen. Virol.,* 70, 2037, 1989.
211. **Schlesinger, J. J., Brandriss, M. W., and Walsh, E. E.,** Protection against 17D yellow fever encephalitis in mice by passive transfer of monoclonal antibodies to the nonstructural glycoprotein gp48 and by active immunization with gp48, *J. Immunol.,* 135, 2805, 1985.
212. **Schlesinger, J. J., Brandriss, M. W., Cropp, C. B., and Monath, T. P.,** Protection against yellow fever in monkeys by immunization with yellow fever virus nonstructural protein NS1, *J. Virol.,* 60, 1153, 1986.

213. **Schlesinger, J. J., Brandriss, M. W., and Walsh, E. E.,** Protection of mice against dengue 2 virus encephalitis by immunization with the dengue 2 virus non-structural glycoprotein NS1, *J. Gen. Virol.,* 68, 853, 1987.
214. **Henchal, E. A., Henchal, L. S., and Schlesinger, J. J.,** Synergystic interactions of anti-NS1 monoclonal antibodies protect passively immunized mice from lethal challenge with dengue 2 virus, *J. Gen. Virol.,* 69, 2101, 1988.
215. **Winkler, G., Randolph, V. B., Cleaves, G. R., Ryan, T. E., and Stollar, V.,** Evidence that the mature form of the flavivirus nonstructural protein NS1 is a dimer, *Virology,* 162, 187, 1988.
216. **Smith, G. W. and Wright, P. J.,** Synthesis of proteins in dengue type 2-infected Vero and *Aedes albopictus* cells, *J. Gen. Virol.,* 66, 559, 1985.
217. **Brandt, W. E.,** Current approaches to the development of dengue vaccines and related aspects of the molecular biology of flaviviruses, *J. Infect. Dis.,* 157, 1105, 1988.
218. **Gibson, C. A., Schlesinger, J. J., and Barrett, A. D. T.,** Prospects for a virus non-structural protein as a subunit vaccine, *Vaccine,* 6, 7, 1988.
219. **Deubel, V., Kinney, R. M., Esposito, J. J., Cropp, C. B., Vorndam, A. V., Monath, T. P., and Trent, D. W.,** Dengue 2 virus envelope protein expressed by a recombinant vaccinia virus fails to protect monkeys against dengue, *J. Gen. Virol.,* 69, 1921, 1988.
220. **Zhao, B. T., Prince, G., Horswood, R., Eckels, K., Summers, P., Chanock, R., and Lai, C.-J.,** Expression of dengue virus structural proteins and nonstructural protein NS1 by a recombinant vaccinia virus, *J. Virol.,* 61, 4019, 1987.
221. **Bray, M., Zhao, B. T., Markoff, L., Eckels, K. H., Chanock, R. M., and Lai, C.-J.,** Mice immunized with recombinant vaccinia virus expressing dengue 4 virus structural proteins with or without nonstructural protein NS1 are protected against fatal dengue 4 virus encephalitis, *J. Virol.,* 63, 2853, 1989.
222. **Falgout, B., Bray, M., Schlesinger, J. J., and Lai, C.-J.,** Immunization of mice with recombinant vaccinia virus expressing authentic dengue virus nonstructural protein NS1 protects against lethal dengue virus encephalitis, *J. Virol.,* 64, 4356, 1990.
223. **Putnak, J. R. and Schlesinger, J. J.,** Protection of mice against yellow fever virus encephalitis by immunization with a vaccinia virus recombinant encoding the yellow fever virus non-structural proteins, NS1, NS2a, and NS2b, *J. Gen. Virol.,* 71, 1697, 1990.
224. **Yasuda, A., Kimura-Kuroda, J., Ogimoto, M., Miyamoto, M., Sata, T., Sato, T., Takamura, C., Kurata, T., Kojima, A., and Yasui, K.,** Induction of protective immunity in animals vaccinated with recombinant vaccinia viruses that express new preM and E glycoproteins of Japanese encephalitis virus, *J. Virol.,* 64, 2788, 1990.
225. **Mason, P. W., Pincus, S., Fournier, M. J., Mason, T. L., Shope, R. E. and Paoletti, E.,** Japanese encephalitis virus-vaccinia recombinants produce particulate forms of the structural membrane proteins and induce high levels of protection against lethal JEV infection, *Virology,* 180, 294, 1991.
226. **Eckels, K. H., Ketrick, F. M., and Russell, P. K.,** Virion and soluble antigens of Japanese encephalitis virus, *Infect. Immun.,* 11, 1053, 1975.
227. **Walker, B. D., Chakrabarti, S., Moss, B., Paradis, T. J., Flynn, T., Durno, A. G., Blumberg, R. S., Kaplan, J. C., Hirsch, M. S., and Schooley, R. T.,** HIV-specific cytotoxic T lymphocytes in seropositive individuals, *Nature,* 328, 345, 1987.
228. **Riviere, Y., Tanneau-Salvadori, F., Regnault, A., Lopez, O., Sansonetti, P., Guy, B., Kieny, M.-P., Fournel, J.-J., and Montagnier, L.,** Human immunodeficiency virus-specific cytotoxic responses of seropositive individuals: distinct types of effector cells mediate killing of targets expressing *gag* and *env* proteins, *J. Virol.,* 63, 2270, 1989.
229. **Koup, R. A., Sullivan, J. L., Levine, P. H., Brettler, D., Mahr, A., Mazzara, G., McKenzie, S., and Panicali, D.,** Detection of major histocompatiblity complex class I-restricted HIV-specific cytotoxic T lymphocytes in the blood of infected hemophiliacs, *Blood,* 73, 1909, 1989.

230. **Nixon, D. F., Townsend, A. R. M., Elvin, J. G., Rizza, C. R., Gallwey, J., and McMichael, A. J.,** HIV-1 *gag*-specific cytotoxic T lymphocytes defined with recombinant vaccinia virus and synthetic peptides, *Nature,* 336, 484, 1988.
231. **Walker, B. D., Flexner, C., Paradis, T. J., Fuller, T. C., Hirsch, M. S., Schooley, R. T., and Moss, B.,** HIV-1 reverse transcriptase is a target for cytotoxic T lymphocytes in infected individuals, *Science,* 240, 64, 1988.
232. **Walker, B. D., Flexner, C., Birch-Limberger, K., Fisher, L., Paradis, T. J., Aldovini, A., Young, R., Moss, B., and Schooley, R. T.,** Long-term culture and fine specificity of human cytotoxic T-lymphocyte clones reactive with human immunodeficiency virus type 1, *Proc. Natl. Acad. Sci. U.S.A.,* 86, 9514, 1989.
233. **Michel, F., Hoffenbach, A., Langlade-Demoyen, P., Guy, B., Lecocq, J.-P., Wain-Hobson, S., Kieny, M.-P., and Plata, F.,** HIV-specific T lymphocyte immunity in mice immunized with a recombinant vaccinia virus, *Eur. J. Immunol.,* 18, 1917, 1988.
234. **Rautmann, G., Kieny, M.-P., Brandely, R., Dott, K., Girard, M., Montagnier, L., and Lecocq, J.-P.,** HIV-1 core proteins expressed from recombinant vaccinia viruses, *AIDS Res. Human Retrovir.,* 5, 147, 1989.
235. **Zarling, J. M., Morton, W., Moran, P. A., McClure, J., Kosowski, S. G., and Hu, S.-L.,** T-cell responses to human AIDS virus in macaques immunized with recombinant vaccinia virus, *Nature,* 323, 344, 1986.
236. **Hu, S.-L., Kosowski, S. G., and Dalrymple, J. M.,** Expression of the AIDS virus envelope gene in recombinant vaccinia viruses, *Nature,* 320, 537, 1986.
237. **Chakrabarti, S., Robert-Gurhoff, M., Wong-Staal, F., Gallo, R., and Moss, B.,** Expression of the HTLV III envelope gene by a recombinant vaccinia virus, *Nature,* 320, 535, 1986.
238. **Hu, S.-L, Fultz, P. N., McClure, H. M., Eichberg, J. W., Thomas, E. K., Zarling, J., Singhal, M. C., Kosowski, S. G., Swenson, R. B., Anderson, D. C., and Todaro, G.,** Effect of immunization with a vaccinia-HIV env recombinant on HIV infection in chimpanzees, *Nature,* 328, 721, 1987.
239. **Koff, W. C. and Fauci, A. S.,** Human trials of AIDS vaccines: current status and future directions, *AIDS,* 3, S125, 1989.
240. **Zagury, D., Bernard, J., Cheynier, R., Desportes, I., Leonard, R., Fouchard, M., Reveil, B., Ittele, D., Lurhuma, Z., Mbayo, K., Wane, J., Salaun, J.-J., Goussard, B., Dechazal, L., Burny, A., Nara, P., and Gallo, R. C.,** A group-specific anamnestic immune reaction against HIV-1 induced by a candidate vaccine against AIDS, *Nature,* 332, 728, 1988.
241. **Hu, S.-L., Zarling, J. M., Chinn, J., Travis, B. M., Moran, P. A., Sias, J., Kuller, L., Morton, W. R., Heidecker, G., and Benveniste, R. E.,** Protection of macaques against simian AIDS by immunization with a recombinant vaccinia virus expressing the envelope glycoproteins of simian type D retrovirus, *Proc. Nat. Acad. Sci. U.S.A.,* 86, 7213, 1989.
242. **Hu, S.-L., Travis, B. M., Garrigues, J., Zarling, J. M., Eichberg, J. W., and Alpers, C. E.,** Processing, assembly, and immunogenicity of HIV-1 core antigens expressed by recombinant vaccinia virus, in *VACCINE 90,* Chanock, R. M., Lerner, R. A., Brown, F., and Ginsberg, H., Eds., Cold Spring Harbor Press, Cold Spring Harbor, NY, 1990, 231.
243. **Shioda, T. and Shibuta, H.,** Production of human immunodeficiency virus (HIV)-like particles from cells infected with recombinant vaccinia viruses carrying the *gag* gene of HIV, *Virology,* 175, 139, 1990.
244. **Karacostas, V., Nagashima, K., Gonda, M. A., and Moss, B.,** Human immunodeficiency virus-like particles produced by a vaccinia virus expression vector, *Proc. Natl. Acad. Sci. U.S.A.,* 86, 8964, 1989.
245. **Haffar, O., Garrigues, J., Travis, B., Moran, P., Zarling, J., and Hu, S.-L.,** Human immunodeficiency virus-like, nonreplicating, *gag-env* particles assemble in a recombinant vaccinia virus expression system, *J. Virol.,* 64, 2653, 1990.

246. **Murphy-Corb, M., Martin, L. N., Davison-Fairburn, B., Montelaro, R. C., Miller, M., West, M., Ohkawa, S., Baskin, G. B., Zhang, J.-Y., Putney, S. D., Allison, A. C., and Eppstein, D. A.,** A formalin-inactivated whole SIV vaccine confers protection in macaques, *Science,* 246, 1293, 1989.
247. **Desrosiers, R. C., Wyand, M. S., Kodama, T., Ringler, D. J., Arthur, L. O., Sehgal, P. K., Letvin, N. L., King, N. W., and Daniel, M. D.,** Vaccine protection against simian immunodeficiency virus infection, *Proc. Natl. Acad. Sci. U.S.A.,* 86, 6353, 1989.
248. **Behbehani, A. M.,** The smallpox story: life and death of an old disease, *Microbiol. Rev.,* 47, 455, 1983.
249. **Lane, J. M., Ruben, F. L., Neff, J. M., and Millar, J. D.,** Complications of smallpox vaccination, *New Engl. J. Med.,* 281, 1201, 1969.
250. **Tartaglia, J. and Paoletti, E.,** Live recombinant viral vaccines, in *Immunochemistry of Viruses. II. The Basis for Serodiagnosis and Vaccines,* van Regenmortel, M. H. V. and Neurath, A. R., Eds., Elsevier, Amsterdam, 1990, 125.
251. **Mackett, M., Yilma, T., Rose, J. K., and Moss, B.,** Vaccinia virus recombinants: expression of VSV genes and protective immunization of mice and cattle, *Science,* 227, 433, 1985.
252. **Jones, L., Ristow, S., Yilma, T., and Moss, B.,** Accidental human vaccination with vaccinia virus expressing nucleoprotein gene, *Nature,* 319, 543, 1986.
253. **Ball, L. A., Young, K. K. Y., Anderson, K., Collins, P. L., and Wertz, G. W.,** Expression of the major glycoprotein G of human respiratory syncytial virus from recombinant vaccinia virus vectors, *Proc. Natl. Acad. Sci. U.S.A.,* 83, 246, 1986.
254. **Olmsted, R. A., Elango, N., Prince, G. A., Murphy, B. R., Johnson, P. R., Moss, B., Chanock, R. M., and Collins, P. L.,** Expression of the F glycoprotein of respiratory syncytial virus by a recombinant vaccinia virus: comparison of the individual contributions of the F and G glycoproteins to host immunity, *Proc. Natl. Acad. Sci. U.S.A.,* 83, 7462, 1986.
255. **Wertz, G. W., Stott, E. J., Young, K. K.-Y., Anderson, K., and Ball, L. A.,** Expression of the fusion protein of human respiratory syncytial virus from recombinant vaccinia virus vectors and protection of vaccinated mice, *J. Virol.,* 61, 293, 1987.
256. **Spriggs, M. K., Murphy, B. R., Prince, G. A., Olmsted, R. A., and Collins, P. L.,** Expression of the F and HN glycoproteins of human parainfluenza virus type 3 by recombinant vaccinia viruses: contributions of the individual proteins to host immunity, *J. Virol.,* 62, 3146, 1987.
257. **Spriggs, M. K., Collins, P. L., Tierney, E., London, W. T., and Murphy, B. R.,** Immunization with vaccinia virus recombinants that express the surface glycoproteins of human parainfluenza virus type 3 (PIV3) protects patas monkeys against PIV3 infection, *J. Virol.,* 62, 1293, 1988.
258. **Auperin, D. D., Esposito, J. J., Lange, J. V., Bauer, S. P., Knight, J., Sasso, D. R., and McCormick, J. B.,** Construction of a recombinant vaccinia virus expressing the Lassa virus glycoprotein gene and protection of guinea pigs from a lethal Lassa virus infection, *Virus Res.,* 9, 233, 1988.
259. **Yilma, T., Hsu, D., Jones, L., Owens, S., Grubman, M., Mebus, C., Yamanaka, M., and Dale, B.,** Protection of cattle against rinderpest with vaccinia virus recombinants expressing the HA or F gene, *Science,* 242, 1058, 1988.
260. **Marchioli, C. C., Yancey, R. J., Jr., Petrovskis, E. A., Timmins, J. G., and Post, L. E.,** Evaluation of pseudorabies virus glycoprotein gp50 as a vaccine for Aujesky's disease in mice and swine: expression by vaccinia virus and Chinese hamster ovary cells, *J. Virol.,* 61, 3977, 1987.

Chapter 5

DEVELOPMENT AND DELIBERATE RELEASE OF A VACCINIA-RABIES RECOMBINANT VIRUS FOR THE ORAL VACCINATION OF FOXES AGAINST RABIES

Paul-Pierre Pastoret, Bernard Brochier, Jean Blancou, Marc Artois, Michel Aubert, Marie-Paule Kieny, Jean-Pierre Lecocq, Bernard Languet, Gilles Chappuis, and Philippe Desmettre

TABLE OF CONTENTS

I. INTRODUCTION

The main vector of the current epidemic of sylvatic rabies in Europe is the red fox. Here, the term vector means the animals most susceptible to rabies in a region at a given time, solely responsible for maintaining the infection. All other species are victims, even if they are able to transmit rabies. Consequently, their destruction or immunization has no effect on the disease cycle. Whereas rabies among domestic animals can be controlled by appropriate prophylactic measures, it poses a bigger problem in wildlife, and, until 1960, the only possible available means of control was considered to be the reduction of vector populations.

As in most other European countries, control measures reducing fox population were only temporarily effective and did not stop the spread of the disease. For this reason, other methods, such as oral immunization of foxes against rabies, needed to be assessed in order to disrupt the cycle of sylvatic rabies.[1] Research focused on oral vaccination, the only procedure appropriate for the immunization of a sufficient number (75%) of wild foxes (by distribution of vaccine baits). There have been numerous experiments in which procedures varied according to the type of vaccine and the type of bait used as a vaccine vehicle. The method of vaccinating wild animals against rabies was developed first in the U.S.,[1] then in Europe,[2] and was used for the first time in the field in October 1978 in Switzerland.[3] Since 1978, several European countries have conducted, at different times, large-scale field trials of oral vaccination of foxes using the Street Alabama Dufferin (SAD), standard or B19-modified, attenuated strain of rabies virus.[4–8] The promising results obtained with these vaccination campaigns attest to the feasibility and efficacy of the method. However, the use of attenuated rabies virus still remains controversial as far as innocuity and stability are concerned, since these virus strains retain pathogenicity for laboratory and wild rodents and are heat sensitive.[9,10] Moreover, attenuated strains of rabies virus may revert to virulence[11] and may still be pathogenic for man; humans exposed to those strains must be treated using standard procedures.[57] Therefore, recent work has focused on technical adjustments, particularly the type of vaccine used. In order to improve both the safety and the stability of the vaccine used, a recombinant vaccinia virus which expresses the immunizing glycoprotein of rabies virus has been developed and tested in the field.

II. CONSTRUCTION OF THE VACCINIA-RABIES RECOMBINANT VIRUS

Rabies viral particles contain five virus-encoded proteins: L (large, transcriptase), G (glycoprotein), M (membrane), NS (nonstructural, phosphoprotein), and N (nucleocapsid).[12–14] These proteins are separately encoded in the genome (gene order N, NS, M, G, and L) and are translated from five separate polyadenylated RNAs transcribed from the negative-sense genome. N (about 62 kDa), with the minor component NS (about 42 kDa), serves to condense the viral RNA into the nucleoprotein core. L protein (about 190 kDa) is responsible for the transcriptase activity present in the virion and is associated with the viral nucleocapsid. M (about 25 kDa) is found in association with the inner surface of the viral envelope G protein (about 67 kDa), which traverses the membrane surrounding the virus where it self-associates to form spikes.[12,15,16]

Physical studies have demonstrated that G is the sole viral protein present on the external surface of the membrane.[17] G is the only viral antigen capable of eliciting the formation of virus-neutralizing antibodies (VNA) and has been shown to confer immunity to rabies.[18–22] More recent studies have demonstrated that vaccination of animals with the N-protein is also capable of conferring immunity to the disease, through the induction of cellular immunity.[23] For the construction of a new vaccine against rabies, most attention has nevertheless been focused on the G-protein.

A. CHARACTERIZATION OF G-CODING SEQUENCES

Anilionis et al.[24] have cloned the cDNA corresponding to the G-protein and nucleotide sequence analysis revealed an open reading frame of 524 amino acids. Nineteen predominantly hydrophobic amino acids precede the N-terminus (sequence LysPheProIle) of the mature protein,[25] indicating the presence of a signal sequence removed from the precursor during maturation and membrane transport. Towards the carboxy-terminus of the encoded protein, a second stretch of 22 hydrophobic amino acids is thought to represent a transmembrane "anchor" zone serving to tether G in the envelope surrounding the virus.

It is worth mentioning that the eighth amino acid of native ERA strain G is proline[26] rather than leucine,[24] which appears to be a spontaneous mutation in the G-coding sequence. We surmised that this mutation could be of importance and accordingly, oligonucleotide-directed mutagenesis was used to correct the triplet corresponding to the eighth amino acid of mature G from CTA (leucine) to CCA (proline).

B. CONSTRUCTION OF A LIVE RECOMBINANT VACCINIA VIRUS BEARING THE RABIES G-CODING SEQUENCE

Live attenuated vaccines are particularly effective immunogens since expression of viral antigens occurs in conjunction with host histocompatibility determinants, ensuring an effective cellular immune response (see Chapter 6). The vaccine potential of live recombinant viruses bearing foreign antigens is thus clear. Upon inoculation of the live virus, it is to be expected that expression of the foreign antigen, within cells of the vaccinated host, will also lead to optimal cell-surface presentation of the antigen. A live apathogenic virus, whether attenuated or recombinant, is particularly appropriate in cases such as rabies, where the induction of cytotoxic T lymphocytes is of central importance to defense against the disease. Vaccinia virus (VV), a member of the *Orthopoxvirus* genus, has been used successfully as a live vaccine for the prevention and the elimination of a serious human disease, smallpox. The eradication of smallpox rendered the vaccine obsolete and was considered a strong contraindicator to any further immunization with VV. Recent technical advances have permitted the development of VV as a cloning and expression vector[27] and vaccinia recombinants bearing foreign antigen-coding sequences have been used in experiments to immunize against cognate diseases.[28,29] Expression of exogenous protein-coding sequences in VV involves essentially two steps. First, the exogenous coding sequence is aligned with a VV promoter and inserted *in vitro* at a site within a (nonessential) segment of vaccinia DNA cloned into a suitable bacterial plasmid replicon. Second, the flanking VV sequences now permit homologous recombination *in vivo* between the plasmid and the viral genome. Double-reciprocal recombination results in the transfer of the DNA insert from the plasmid to the viral genome, wherein it is propagated and expressed. Thus, the tailored cDNA (leu8 → pro8) was linked to a promoter sequence derived from an early VV gene encoding a 7.5-kDa protein. This block was inserted into a cloned segment of the VV genome bearing the TK gene, and recombinational exchange was used to transfer the expression block to the genome of VV, generating a selectable TK^- virus, VVTGgRAB 26D3.[30–32] This recombinant virus has also been called V-RG in Wiktor et al.[33]

C. IMMUNOGENIC PROPERTIES OF VVTGgRAB

Infection of tissue-culture cells with the recombinant elicited the production of a correctly processed rabies G-protein. Indeed, the profile of reactivity of the recombinant G with a panel of monoclonal antibodies was largely identical to that obtained with native ERA virus, attesting to the authenticity of the recombinant G (Figure 1).

1. Immunogenicity in Mice

Vaccination of mice by scarification with VVTGgRAB induced high levels of rabies-neutralizing antibodies, and these animals resisted severe intracerebral challenge infection (2400 mouse LD_{50} units) with street rabies virus (Table 1). The minimum dose of VVTGgRAB capable of protecting 50% of recipient mice was 10^4 plaque forming units (pfu).

The leucine (leu) to proline (pro) alteration at position 8 of the mature glycoprotein (above) is of particular interest. Animals vaccinated with the recombinant bearing leu instead of pro at position 8 (VVTGgRAB-leu) were not protected against rabies.[32] We have shown that substituting pro by leu dramatically affects the structure of G and folding, glycosylation, and processing are abnormal (unpublished data) although removal of the N-terminal signal sequence appears to take place.[58]

Further examination of the VVTGgRAB (pro) revealed that CTL induced by the recombinant were primarily directed against VV determinants. Nevertheless, lymphocytes from mice inoculated with VVTGgRAB were effectively primed for a secondary CTL response upon culture in the presence of inactivated ERA-strain rabies virus.[33] This result indicated that the recombinant VVTGgRAB is a powerful immunogen able to prime the animal for both the antibody and CTL responses against rabies.

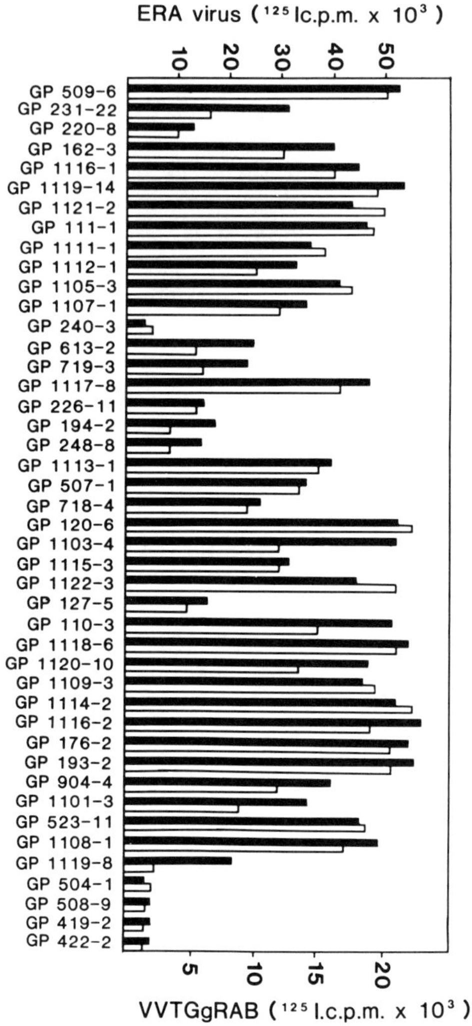

FIGURE 1. Comparison of VVTGgRAB and rabies ERA-strain antigens for reaction with a panel of monoclonal antibodies. Lysates of infected cells were adsorbed to microtiter plates and incubated with 1/1000 dilutions of ascites fluid. Bound antibody was detected by the addition of ^{125}I-labeled goat anti-mouse immunoglobin. Solid bars, ERA-strain rabies virus; open bars, VVTGgRAB. (Reproduced from Kieny, M.-P. et al., *Nature*, 312, 163, 1984.)

2. Immunogenicity in Rabbits

Intradermal inoculation of rabbits with VVTGgRAB virus also resulted in rapid induction of rabies VNA and in protection from severe intracerebral challenge with large doses of street rabies virus. Rabies VNA titers at 5, 11, and 14 d after inoculation were 800, 10,000, and greater than 30,000, respectively. Anti-vaccinia VNA titers after 14 d were substantially lower;

TABLE 1
Protection of Mice from Rabies by VVTGgRAB Administration

Route of inoculation	Vaccine	VNA on day 15	Protection
Intradermal	VVTGgRAB	>30,000	12/12
	VV — wild type	<10	0/12
Footpad	VVTGgRAB	>30,000	12/12
	VV — wild type	<10	0/12

Note: VNA, virus-neutralizing antibody; VV, vaccinia virus.

15/16 of the rabbits vaccinated with 10^8 pfu of VVTGgRAB resisted a challenge with an infectious dose of 2,400 mouse LD_{50} units. Sera from rabbits immunized with VVTGgRAB (day 14) neutralized between $10^{5.3}$ and $10^{6.6}$ tissue culture LD_{50} both of ERA rabies virus and three street rabies virus isolates previously shown to differ from the ERA strain in their reactivity with a panel of anti-glycoprotein monoclonal antibodies. The sera also neutralized the Duvenhage strain of rabies-related virus but not Mokola and Lagos bat viruses.

Since intradermal inoculation may not be a practical route of administration for a rabies vaccine, we compared the immune response to VVTGgRAB virus administered by the intradermal, intramuscular, subcutaneous, and oral routes. Pairs of rabbits were injected with, or induced to swallow, $10^{7.8}$ pfu of VVTGgRAB. Levels of VNA were evaluated and on day 21 all rabbits and two unimmunized controls were challenged intracerebrally with street rabies virus as above. VNA could be detected on day 3 in rabbits injected by intradermal, intramuscular, or subcutaneous route. All animals showed high levels of antibodies, reaching titers of 100,000 to 120,000 on day 21 and resisted challenge with street rabies virus.

3. Booster Response to VVTGgRAB Inoculation

It is important to know whether previous vaccination interferes with the immune response to a second inoculation with VVTGgRAB. Three rabbits immunized intradermally with $10^{7.6}$ pfu of VVTGgRAB virus and showing a VNA titer >30,000 15 d after vaccination were inoculated intradermally 6 months later with the same dose of virus. Animals were bled before and after the booster inoculation. Three weeks after the booster, they received 24,000 mouse LD_{50} of street rabies virus intracerebrally. As shown in Table 2, the level of rabies virus-neutralizing antibodies on the day of the booster inoculation was between 8000 and 12,000. Levels of VNA in all three animals increased drastically, starting on day 3 (titers 24,000) and reached titers of 70,000 or higher by day 15. As expected, all animals resisted challenge with street rabies virus. These results indicate that the immunity induced by

TABLE 2
Booster Response to VVTGgRAB in Rabbits

Rabbit	Rabies VNA Day 0	Day 3	Day 15
A	8,000	24,000	>70,000
B	8,000	24,000	>70,000
C	12,000	24,000	>70,000

Note: VNA, virus-neutralizing antibody.

TABLE 3
Induction of Rabies Virus-Neutralizing Antibodies (VNA) and Protection from Rabies by Inactivated VVTGgRAB

Vaccine	Rabies VNA titers	Protection
VV-WT cell lysate	<10	0/12
VVTGgRAB cell lysate	8,000	12/12
VVTGgRAB virus (purified)	4,000	12/12
VVTGgRAB G-protein	15,000	12/12

VVTGgRAB did not interfere with the response to the same immunogen given 6 months later.[34]

4. Immunogenicity of Inactivated VVTGgRAB Preparations

Live VV has a long history of safe use as a vaccine for humans, despite a low incidence of serious complications. Nevertheless, the reintroduction of VV-based vaccines for humans may be controversial. The immunogenicity of wild-type VV and VVTGgRAB-infected inactivated cell extracts, purified inactivated VVTGgRAB, and the rabies glycoprotein isolated from VVTGgRAB-infected cell extracts using an affinity column prepared with anti-rabies virus glycoprotein antibody, was therefore evaluated. Mice were vaccinated and challenged intracerebrally with rabies virus.[34] All preparations, with the exception of the control cell extract, induced high levels of rabies VNA and protected against rabies virus challenge (Table 3). The possibility of using the inactivated VV recombinant in place of inactivated rabies virus is thus open.

5. Postexposure Protection in Hamsters by Live or Inactivated VVTGgRAB

An experimental vaccine was prepared from VVTGgRAB-infected cells by centrifugation through a 36% sucrose cushion. A portion of the resuspended

TABLE 4
Postexposure Protection of Hamsters with VVTGgRAB

Vaccine	Percentage protection	
	Vaccination − 24 h	Vaccination + 1 h and 3 d
Live VVTGgRAB	50	83
Inactivated VVTGgRAB	42	50
Rabies vaccine	33	17
None	0	—

virus was inactivated with β-propiolactone, yielding a rabies vaccine harboring an antigenic value of 3.8 IU as determined by the standard National Institutes of Health (NIH) potency test for inactivated rabies vaccines. A group of 12 hamsters was inoculated intramuscularly with $10^{7.7}$ pfu of live VVTGgRAB at either 24 h before, or 1 h or 3 d after challenge with a 10 to 20 hamster LD_{50} dose of street rabies virus. A second group of 12 hamsters was inoculated with the inactivated preparation according to the same protocol. A control group of 12 hamsters was treated with a conventional rabies virus preparation. As shown in Table 4, both VVTGgRAB preparations protected 50 to 80% of hamsters.[34]

6. Conclusions for the Laboratory Animal Experiments

Experiments with mice and rabbits have demonstrated that introduction of VVTGgRAB by scarification, intradermal or subcutaneous inoculation, oral adminstration, or injection into the footpad all elicit high titers of neutralizing antibodies and resistance to challenge. Further, experiments with hamsters have shown that VVTGgRAB is capable of conferring resistance to rabies if administered immediately after lethal infection. The recombinant VV expressing the rabies virus G-protein (VVTGgRAB) has thus proved to be highly immunogenic and has outperformed conventional rabies vaccines in every comparative protection test in laboratory animals. The question remained, however, whether this recombinant vaccine would be effective in feral or domestic animals and whether it could be developed for large-scale vaccination campaigns.

III. EFFICACY AND INNOCUITY OF THE VACCINIA-RABIES RECOMBINANT VIRUS FOR THE TARGET SPECIES

Since 1985, numerous efficacy and innocuity trials using the VVTGgRAB in captive red foxes have been conducted in France, Belgium, and Canada. The results of efficacy trials in experimental stations were evaluated by the titer of rabies virus (CVS strain)-neutralizing antibodies after vaccination and challenge infection with virulent rabies virus (wild strain).[35–37]

TABLE 5
Efficacy of the VVTGgRAB (Liquid Form) Administered by Several Routes and at Different Doses to Foxes

Number of tested animals	VVTGgRAB Dose (PFU)	VVTGgRAB Route	Rabies seroconversion	Resistance to rabies challenge[a] (months)
2 Adults	1×10^8	Intradermal	2/2	2/2 (1)[b]
2 Adults	1×10^8	Intradermal	2/2	2/2 (3)
2 Adults	1×10^8	Subcutaneous	1/2	2/2 (1)
4 adults	1×10^4	Oral[c]	1/4	2/4 (1)
8 Adults	1×10^6	Oral	4/8	4/8 (1)
4 Adults	1×10^7	Oral	3/4	4/4 (1)
5 Cubs	$1 \times 10^{7.2}$	Oral	5/5	5/5 (1)
4 Cubs	$1 \times 10^{7.2}$	Oral	4/4	4/4 (6)
4 Cubs	$1 \times 10^{7.2}$	Oral	3/4	2/3 (12)
4 Adults	1×10^8	Oral	4/4	4/4 (1)
Scarification of the Oral Mucosa				
8 Adults	1×10^8	Oral	8/8	8/8 (1)
4 Adults	1×10^8	Oral	2/4	4/4 (12)
4 Adults	1×10^8	Oral	3/4	4/4 (18)
6 Adults	$1 \times 10^{9.2}$	Gastric	3/6	3/6 (3)

[a] All unvaccinated control foxes died from rabies challenge.
[b] Time of rabies challenge after vaccination.
[c] Direct instillation into the mouth.

A. EFFICACY OF THE VVTGgRAB IN THE RED FOX

1. Immunizing Potency

All results of the VVTGgRAB efficacy in the target species are reported in Tables 5 and 6. They could be summarized as follows:

1. Twenty-six adult captive foxes were inoculated intradermally, subcutaneously, or orally (by direct installation; with or without scarification of oral mucous membranes) with 10^8 pfu of VVTGgRAB (liquid form).[32] All but one (subcutaneously inoculated) developed high titers of rabies VNA. All foxes resisted rabies challenge performed on day 28 after vaccination. The sole fox exhibiting undetectable levels of rabies-neutralizing antibodies also resisted challenge, suggesting cell-mediated immunity in defence against rabies. Scarification of oral mucous membranes before oral administration did not yield higher titers of neutralizing antibodies. All animals presented only low levels of antibody capable of neutralizing VV.
2. The duration of immunity conferred by the VVTGgRAB (10^8 pfu; oral route) reached a minimum of 18 months.[39]

TABLE 6
Efficacy of Oral Vaccination with VVTGgRAB Lyophilized or Enclosed in a Bait

No. of tested animals	VVTGgRAB Dose	VVTGgRAB Route	Rabies seroconversion	Resistance to rabies challenge[a]
4 Adults	1×10^8	Oral lyophilized form[b]	3/4	3/4 (1 mo)[c]
4 Adults	1×10^8	Oral in chicken head bait	5/5	4/5 (1 mo)
8 Adults	$1 \times 10^{9.2}$	Oral in sponge bait	8/8	8/8 (3 mo)
18 Cubs	1×10^8	Oral in machine-made bait	15/18	16/18 (1 mo)
11 Adults	$2 \times 10^{7.5}$	Oral in machine-made bait	7/11	6/11 (3.5 mo)

[a] All unvaccinated control foxes died from rabies challenge.
[b] Direct instillation into the mouth.
[c] (Time of rabies challenge after vaccination.)

3. Foxes receiving less than 10^8 $TCID_{50}$ of VVTGgRAB showed a clear dose-dependent response, with one out of four, four out of eight, and four out of four animals surviving challenge after oral administration of 10^4, 10^6, and 10^7 $TCID_{50}$ of VVTGgRAB, respectively.[32]
4. The gastric route of administration was less effective.[40]
5. The liquid form of VVTGgRAB presentation seems to be more efficient than the freeze-dried one (when administered by the oral route).[38]
6. A second administration of VVTGgRAB induced an increase of rabies VNA titers (booster effect).
7. The VVTGgRAB ($10^{7.2}$ $TCID_{50}$), administered by the oral route to fox cubs, induces significant levels of rabies virus antibody and protects 92% of them (11/12) from rabies. The duration of immunity conferred reached a minimum of 12 months.[41]

2. Efficacy of a Baiting System

Oral administration is the only route appropriate to the vaccination of wild animals. Accordingly, the vaccine must be presented in a form suitable for ingestion. It has previously been shown that vaccine vehicles (bait + container) could largely affect the contact of the virus with the oral mucosa and, thus, if the baiting system is inappropriate, lower the efficacy of the vaccine. In order to preserve vaccine immunogenicity, both bait and vaccine container have certain properties which are required.[4] As far as the bait is concerned, it should be attractive to the target species (smell, appetency), should not inactivate the virus suspension, should deliver the vaccine in the oral cavity, and should be hydrophic (better contact of the liquid vaccine with the oral cavity). In its turn, the vaccine container (capsule or sachet) should be easily ruptured for delivering the vaccine in the oral cavity, should not be swallowed or rejected intact, and should not inactivate the vaccine. Finally, the ratio bait volume/vaccine liquid volume needs to be as low as possible. The field use of such a vaccine-bait system requires some peculiar properties: the bait must be able to incorporate a biological marker, must be easily manufactured (machine-made) and storable (absence of freezing), and must be resistant to mechanical factors (aerial dropping), physical factors (heat, UV light, rainfall), and biological factors (yeast).

In a first experiment, ''Plasticpak'' capsules containing 10^8 pfu of VVTGgRAB, inserted into chicken heads (one capsule per head into the beak) were distributed to test animals (one per fox). These animals similarly produced high titers of rabies-neutralizing antibodies and most of them resisted severe challenge with rabies virus (Table 6).[38] A good protection against rabies was also obtained when the VVTGgRAB ($10^{9.1}$ pfu) is contained in a sponge bait (Table 6).[40]

The efficacy of VVTGgRAB (10^8 $TCID_{50}$) contained in a new machine-made baiting system has recently been tested (Tables 6 and 7 and Figure 2).[42] In a first trial, 22 captive young foxes were divided into 3 experimental groups of six and a control group of four foxes (see Table 7). Each fox in Groups 1, 2, and 3 were fed one, two, and three vaccine baits, respectively, on successive days. The four unvaccinated control foxes were housed separately. As shown by the incorporation of a tetracycline biomarker into their bones, all the baited foxes ingested at least one bait.

Thirty days after baiting, seroconversion to rabies was observed in 15/18 (83%) of the foxes and seroconversion to vaccinia in 14/18 (77%) foxes. Sixteen of the 18 (89%) baited foxes resisted a rabies challenge 30 d after baiting. One cub was protected against rabies despite the absence of detectable anti-rabies antibodies. These results demonstrate that the bait-sachet system permits a good release of the virus suspension into the mouth and the contact of the virus with the oral cavity.

In a second trial, 11 adult foxes were fed one similar vaccine bait. Seven of them elicited a rabies seroconversion and 6 of them were protected against rabies challenge. These results are less demonstrative than those obtained in the cubs, probably due to the lower bait uptake.[59]

B. INNOCUITY OF THE VVTGgRAB TO THE RED FOX

1. Residual Pathogenicity Assessment

VV (Copenhagen strain) was first tested for innocuity to foxes.[32] Two animals received VV by intradermal injection and two other animals received the VVTGgRAB by the same route; 10^2, 10^4, 10^6, or 10^8 $TCID_{50}$ of virus were injected at triplicate sites and cutaneous reactions monitored for 15 d. Mild, localized inflammation was observed at the sites of injection with 10^6 or 10^8 $TCID_{50}$ of virus, and in one animal at 10^2 and 10^4 $TCID_{50}$. In all cases inflammation regressed spontaneously within 8 d. Cutaneous reaction was significantly greater with the wild-type strain of VV than with the VVTGgRAB. No clinical signs or lesions were recorded in any of the inoculatd foxes during

1, 6, 12, and 18 months after vaccination with the VVTGgRAB. This absence of pathogenicity was observed whatever the dose of inoculation (10^2 to 10^{10} $TCID_{50}$) or route of administration (oral, intramuscular, intraduodenal, subcutaneous, intradermic (scarification), ocular, and intranasal).[32,38–41]

2. Assessment of Excretion and Horizontal Transmission

A vaccine virus used for oral vaccination of wildlife should preferably not be horizontally transmitted to unvaccinated animals. For testing the excretion and potential horizontal transmission of the VVTGgRAB, unvaccinated control animals have been held in close contact with vaccinated ones. Seroconversion has never been observed in these ''contact'' animals. When using this assessment method, no transmission of immunizing amounts of VVTGgRAB occurred in adult and young foxes (except one adult fox bitten by a freshly inoculated animal).[32,38,41]

3. Primary Multiplication Site of the VVTGgRAB

Experiments were designed to determine the multiplication site in foxes of the recombinant virus as compared with that of the parental strain of the virus, by virus isolation, titration, and indirect immunofluorescence.[43] The polymerase chain reaction (PCR) was also used to detect specific viral DNA in several fox organs. Foxes were fed with 10^8 $TCID_{50}$ of either VVTGgRAB or VV and were euthanized 12, 24, 48, or 96 h after inoculation.

a. Virus Isolation

As shown in Table 8, small amounts of virus (10^2 to $10^{4.3}$ $TCID_{50}$/ml) were recovered from the tonsils of five of the seven foxes inoculated with VVTGgRAB, during the first 48 h only. In Group B, VV was detected in the tonsils of the animal after 24 h. Virus was not detected ($<10^{1.5}$ $TCID_{50}$/ml) in any other organ, serum, or feces of the vaccinated foxes, nor in any organ of the control animal.

b. Indirect Immunofluorescence

As shown in Table 8, the indirect immunofluorescence test carried out with anti-VV rabbit polyclonal serum confirmed the presence of the virus in the tonsils of four of the five foxes from which VVTGgRAB had been isolated and in the fox from which VV was isolated. The indirect immunofluorescence test carried out with the anti-rabies glycoprotein monoclonal antibody confirmed the presence of virus in the tonsils of three of the five foxes from which the virus was reisolated. Immunofluorescence was, however, too diffuse to allow precise localization of the virus multiplication in the tonsil.

c. Polymerase Chain Reaction

Determination of the sensitivity of the method — Negative controls consisted of DNA from an uninfected fox. Positive controls were prepared with several dilutions of VVTGgRAB or VV in normal fox DNA; the dilutions

TABLE 7
Bait Uptake, Vaccinia and Rabies-Neutralizing Antibodies, and Protection from Rabies Challenge

Group	Fox no.	Vaccine baiting	Bait uptake[a]	Rabies Antibodies[b]			Vaccinia antibodies[c]			Rabies challenge[d]	Rabies diagnosis	Tetracycline detection
				Day 0	Day 15	Day 30	Day 0	Day 15	Day 30			
A	1	1 Bait on day 0	+	<0.55	+2.55	+3.87	+0.08	+0.43	+0.40	R	–	+
	2		+	<0.55	+11.47	+25.79	+0.07	+0.49	+0.72	R	–	+
	3		–	<0.55	<0.26	<0.55	+0.10	+0.06	+0.04	D (16)	+	+
	4		+	<0.55	+14.29	+6.67	+0.07	+0.40	+0.39	R	–	+
	5		+	<0.55	+0.47	+4.10	+0.05	+0.03	+0.06	R	–	+
	6		+	<0.55	+2.37	+14.96	+0.03	+0.26	+0.22	R	–	+
B	7	1 Bait on day 0	– +	<0.55	>32	>32	+0.16	+0.45	+0.32	R	–	+
	8	1 Bait on day 1	– +	<0.55	>32	>32	+0.07	+0.38	+0.58	R	–	+
	9		+ +	<0.55	+7.39	>32	+0.04	+0.27	+0.35	R	–	+
	10		+ +	<0.55	>32	>32	+0.11	+0.63	+0.81	R	–	+
	11		+ +	<0.55	+25.4	>32	+0.05	+0.56	+0.90	R	–	+
	12		+ +	<0.55	+0.57	+0.93	+0.05	+0.06	+0.04	R	–	+
C	13	1 Bait on day 0	+ – +	<0.55	>32	>32	+0.06	+0.60	+0.68	R	–	+
	14	1 Bait on day 1	– + +	<0.55	>32	>32	+0.05	+0.51	+0.79	R	–	+
	15	1 Bait on day 2	+ + +	<0.55	<0.26	>32	+0.06	+0.55	+0.68	R	–	+
	16		+ + +	<0.55	>32	+30.78	+0.05	+0.38	+0.50	R	–	+
	17		– ? ?	<0.55	<0.26	+1.03	+0.05	+0.02	+0.02	D (18)	+	+
	18		+ + +	<0.55	+11.35	+24.02	+0.02	+0.57	+0.61	R	–	+

D	19	Unvaccinated	—	<0.55	<0.55	<0.55	+0.05	+0.08	+0.04	D (19)	+	–
	20	controls	—	<0.55	<0.55	<0.55	+0.05	+0.04	+0.03	D (16)	+	–
	21		—	<0.55	<0.55	<0.55	+0.04	+0.04	+0.07	D (16)	+	–
	22		—	<0.55	<0.55	<0.55	+0.06	+0.08	+0.05	D (18)	+	–

Note: Foxes were fed wth 1, 2, or 3 machine-made baits containing 10^8 $TCID_{50}$ of VVTGgRAB.

[a] Negative results = vaccine sachet rejected intact.
[b] RFFIT method, results are expressed in iu/ml serum.
[c] ELISA method, results are expressed as the optical density for a 1/160-diluted serum.
[d] R Survived, D died (day of death after challenge).

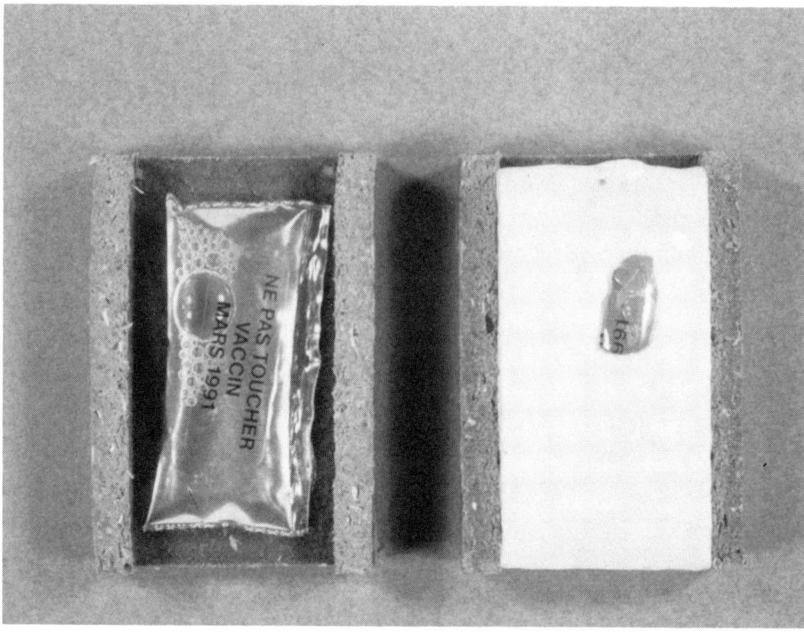

A

FIGURE 2. The vaccine baiting system consists of a bait made of a mixture of plant and animal proteins and fish oil aggregated by a synthetic polymer and containing tetracycline (150 mg/bait) as a biomarker. (A) A sealed plastic sachet containing one dose of 2.5 ml of liquid vaccine ($>10^8$ $TCID_{50}$/dose) is introduced into the bait, which is stitched closed. (B) Each complete bait weighs about 35 g, is stable from $-20°C$ to $+45°C$, and is strong enough to be dropped from an airplane.

ranged from approximately one infectious particle per cell to one infectious particle per 10^6 cells. TK1 and TK2 oligonucleotides were used for the detection of VV and several combinations of oligonucleotide primers (TK1-TK2, G1-G2, TK1-G2, G1-TK2) (Figure 3) were used for the detection of VVTGgRAB. For each virus 10 μl of the amplified product of the 10^{-2} dilution (corresponding to 1 infectious particle/100 cells) was detected by direct gel analysis (Figure 4). For each virus 30 μl of the reaction mixture submitted to two series of 35 PCR cycles were detected by dot blot hybridization, up to and including the 10^{-6} dilution (1 infectious particle/million cells) (Figure 5). No detectable amplification was observed in different samples of normal fox genomic DNA.

Detection of VVTGgRAB or VV in DNA of vaccinated foxes — As shown in Table 9, using TK1 and TK2 as primers and G1 as the probe, VVTGgRAB was detected in the tonsils, the buccal mucosa, and the soft palate of foxes. No virus was detected from any other organ nor from any of the organs of the control fox. After verification of the specificity of the

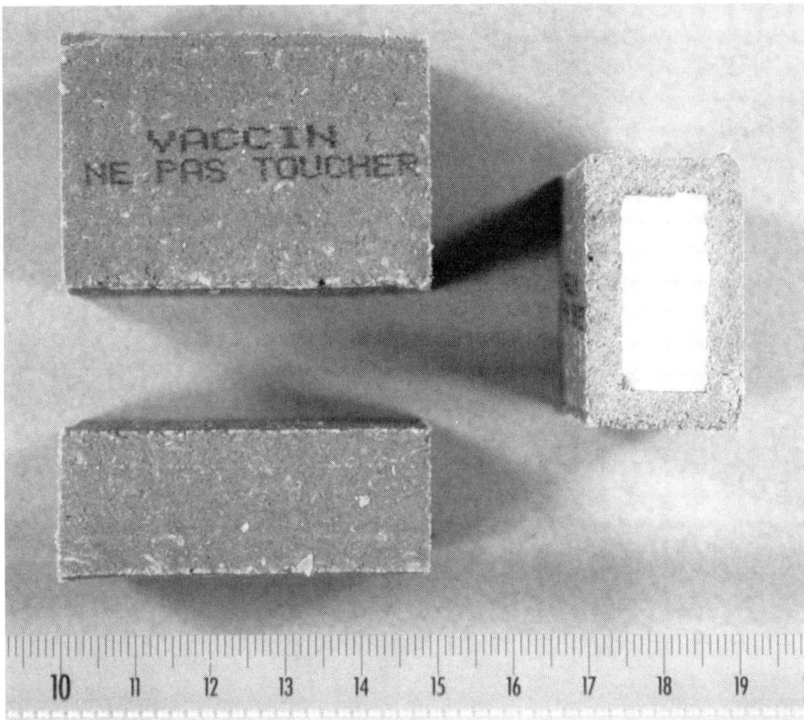

FIGURE 2B.

amplified product in controls, TK1 and TK2 were used as primers and TK1 was used as probe. VV was detected in the tonsils and the buccal mucosa only.

Using different techniques, VVTGgRAB or VV were detected during the first 48 h following vaccination by the oral route, but only in tonsils, buccal mucosa, and soft palate (Figure 6). Similar results have been obtained by others in raccoons, using virus isolation.[44] Results of other unpublished experiments demonstrate that tonsillectomy of the foxes does not affect the protection conferred by vaccination with the VVTGgRAB.[60] Viremia was never observed on days 0, 2, 3, 4, 5, 6, 7, 8, and 14 in inoculated foxes. All these data suggest that orally administered recombinant virus multiplies locally and at a low level. No virus could be detected in salivary glands (parotid and maxillary glands); the risk of transmission by saliva from one animal to another is therefore very low. Furthermore, the fact that VVTGgRAB multiplies only in restricted sites minimizes the potential risk of recombination with other orthopoxviruses. In our experiments, no difference was observed in the replication sites of VVTGgRAB as compared to the parental strain of VV, demonstrating that recombination does not modify the tropism of VV.

TABLE 8
Detection by Viral Isolation and Indirect Immunofluorescence of VVTGgRAB and VV in Tonsils of Foxes Vaccinated by the Oral Route

	Treatment									
	VVTGgRAB (Group A)							VV (Group B)		PBS (Control)
Fox no.	**1**	**2**	**3**	**4**	**6**	**7**	**9**	**5**	**8**	**10**
Time of organ collection (h)	**12**	**12**	**24**	**24**	**48**	**48**	**96**	**24**	**48**	**96**
Virus titer[a]	–	10^{2}	10^{4}	$10^{2.8}$	$10^{4.3}$	$10^{2.8}$	–	$10^{4.3}$	–	–
Dectection of VV antigens by indirect immunofluorescence[b]	–	–	+	+	+	+	–	+	–	–
Detection of rabies glycoprotein by indirect immunofluorescence[b]	–	–	+	–	+	+	–	–	–	–

[a] Virus titers are expressed as $TCID_{50}$/ml.
[b] (+), Viral antigen detected; (–), not detected.

From Thomas, I. et al., *J. Gen. Virol.*, 71, 37, 1990. With permission.

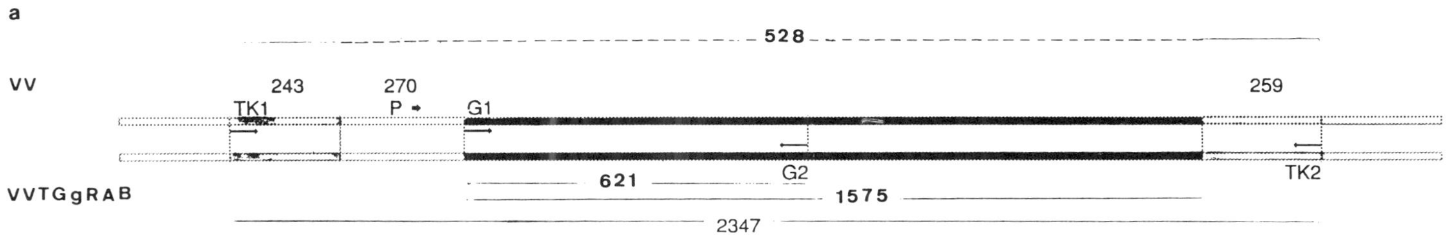

b

TK1	ATG AAC GGC GGA CAT ATT CAG TTG
TK2	TTA TGA GTC GAT CTA ACA CTT TCT
G1	AAA TTC CCT ATT TAC ACG ATA
G2	GCA AGT CTC ACT CCC TTT GGA

FIGURE 3. (a) The cDNA of the rabies glycoprotein gene under the control of a promoter (P) is inserted within the TK gene of the VV genome. Primers TK1 and TK2 are located on the TK gene and primers G1 and G2 are located on the glycoprotein gene. Numbers indicate the size (bp) of the different segments. (b) Sequence of synthetic oligonucleotide primers and complementary oligonucleotide probes. (From Thomas, I. et al., *J. Gen. Virol.*, 71, 37, 1990. With permission.)

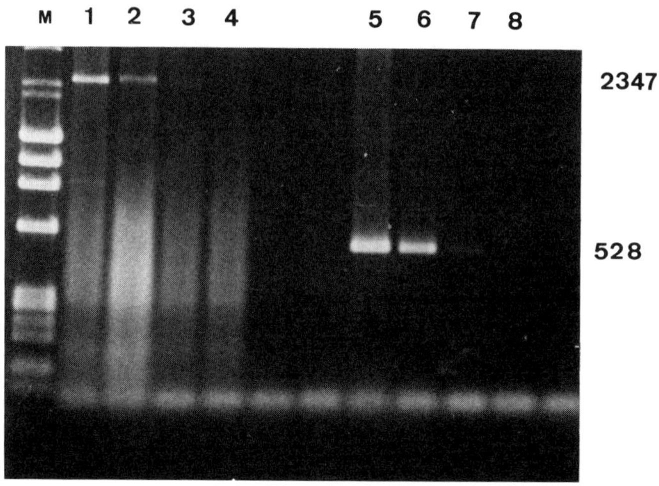

FIGURE 4. Electrophoretic gel analysis of PCR products of different dilutions of VVTGgRAB (lanes 1, 2, 3, 4) or VV (lanes 5, 6, 7, 8) in normal fox DNA. TK1 and TK2 were used as primers. Lane M, size markers (bp) (*Hind*III-digested λ-DNA, *Hae*III-digested φX174 RF DNA); lanes 1 and 5, 1 infectious particle/cell; lanes 2 and 6, 10^{-1} infectious particle/cell; lanes 3 and 7, 10^{-2} infectious particle/cell; lanes 4 and 8, 10^{-3} infectious particle/cell. (From Thomas, I. et al., *J. Gen. Virol.*, 71, 37, 1990. With permission.)

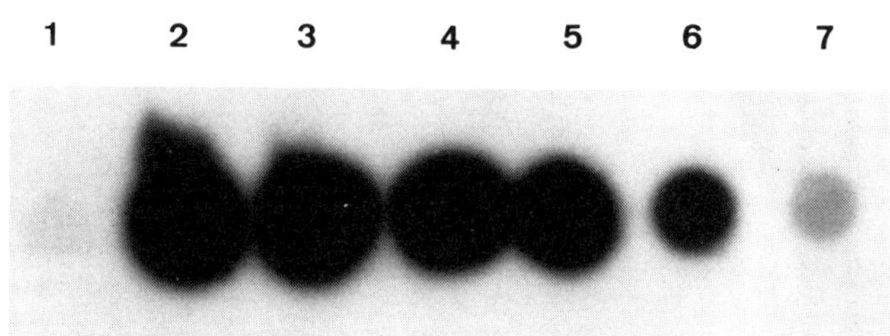

FIGURE 5. Dot blot autoradiograph of PCR products of different dilutions of VVTGgRAB in normal fox DNA. Two series of 35 cycles were performed using TK1 and G2 as primers and G1 as probe. Lane 1, negative control; lane 2, 10^{-1} infectious particle/cell; lane 3, 10^{-2} infectious particle/cell; lane 4, 10^{-3} infectious particle/cell; lane 5, 10^{-4} infectious particle/cell; lane 6, 10^{-5} infectious particle/cell; lane 7, 10^{-6} infectious particle/cell. (From Thomas, I. et al., *J. Gen. Virol.*, 71, 37, 1990. With permission.)

None of these viruses were detected in the brain, suggesting that VVTGgRAB does not multiply in nerve cells. These results were consistent with those of other studies reporting the absence of detectable cytological abnormalities in cerebrospinal fluid from raccoons orally vaccinated with VVTGgRAB.[45]

4. Back Passages in the Red Fox

Foxes were inoculated with virus isolated from fox tonsils 24 h after oral administration (with or without cell culture amplification) to perform back

TABLE 9
Detection by PCR of VVTGgRAB or VV in Orally Vaccinated Foxes.

	Treatment							
	VVTGgRAB (Group A)					VV (Group B)		PBS (Control)
Fox no.	**11**	**12**	**13**	**14**	**15**	**16**	**17**	**18**
Time of organ collection (h)	**24**	**24**	**48**	**48**	**48**	**24**	**48**	**24**
Tonsils	+[a]	+	+	+	+	–[b]	+	–
Buccal mucosa	+	–	+	–	+	+	+	–
Soft palate	+	–	+	–	–	–	–	–
Brain	–	–	–	–	–	–	–	–
Spleen	–	–	–	–	–	–	–	–
Parotid glands	–	–	–	–	–	–	–	–
Maxillary glands	–	–	–	–	–	–	–	–
Liver	–	–	–	–	–	–	–	–
Retropharyngeal lymph nodes	–	–	–	–	–	–	–	–
Submaxillary lymph nodes	–	–	–	–	–	–	–	–
Mesenteric lymph nodes	–	–	–	–	–	–	–	–
Blood	–	–	–	–	–	–	–	–
Feces	–	–	–	–	–	–	–	–

[a] (+), Viral DNA detected.
[b] (–), Viral DNA not detected.

From Thomas, I. et al., *J. Gen. Virol.*, 71, 37, 1990. With permission.

passages.[43] No virus could be isolated in either case after this passage. Neither lesions nor clinical signs were observed in animals vaccinated with these viruses during a 28-d observation period.

5. Interaction Between Rabies Infection and Oral Vaccination

It is also of major importance to preclude some epizootical risks such as the emergence of asymptomatic carriers of rabies virus. This situation could occur in the field by vaccinating naturally infected animals during the incubation period. We have investigated the influence of anti-rabies vaccination on the onset of the disease as well as the delay of death in foxes previously infected with rabies virus.[46] Foxes were divided into 6 experimental groups of 4 animals. On day 0, each fox was experimentally infected with a rabies virus suspension. VVTGgRAB was administered by the oral route to each animal of 3 groups on days 0, 3, or 14, respectively. Foxes of other groups were used as unvaccinated controls or received, on day 0, the parental Copenhagen strain of VV. The length of post-challenge survival and duration of clinical disease were recorded for each animal tested. All foxes, vaccinated or not, died from rabies as confirmed by fluorescent antibody test. Animals vaccinated on days 0 and 3 died after a shorter period of incubation as compared to unvaccinated controls. On the other hand, animals vaccinated

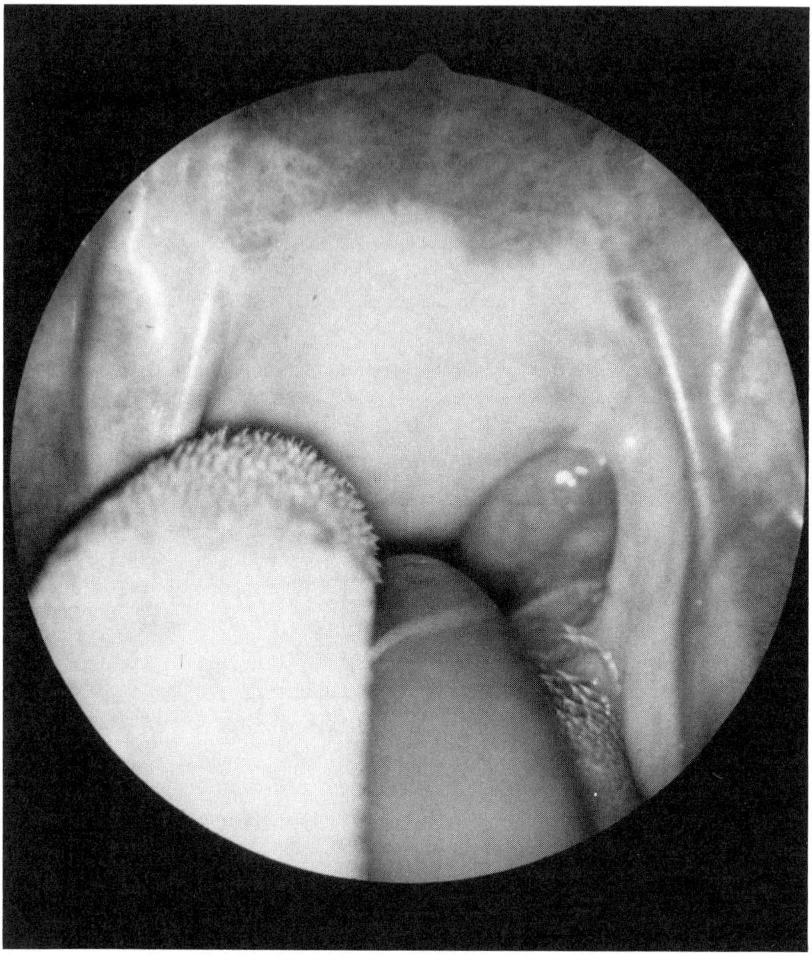

FIGURE 6. Using different techniques, VVTGgRAB or VV were detected during the first 48 hours following vaccination by the oral route, but only in the tonsils, buccal mucosa, and soft palate.

on day 14 post-challenge died later than control animals. Foxes administered with VV died at the same time as unvaccinated controls. These results (reported in Table 10) demonstrate that ''early'' and ''late'' death phenomenons, as consequences of interactions between oral vaccination with VVTGgRAB and rabies infection, can occur in foxes.

Nevertheless, other effects could occur when vaccinating at other time intervals after or before natural infection. In this sense, field data need to be considered: foxes (may) ingest one or several vaccine doses at different times before or after natural contamination; furthermore, the time intervals between successive ingestions can largely vary. Further investigations, including other

TABLE 10
Effect of Oral Inoculation of Foxes with VVTGgRAB (Trial 1) or VV (Trial 2) on the Same Day or Several Days after Rabies Challenge

Treatment	Day of rabies challenge	Day of oral vaccination	Day of death post-challenge	Duration of clinical disease (days)
Trial 1				
4 Foxes VVTGgRAB	0	0	$\bar{x} = 17.25$	$\bar{x} = 2.75$
4 Foxes VVTGgRAB	0	3	$\bar{x} = 17.25$	$\bar{x} = 2.75$
4 Foxes VVTGgRAB	0	14	$\bar{x} = 22.25$	$\bar{x} = 2.75$
4 Unvaccinated foxes	0	—	$\bar{x} = 19.25$	$\bar{x} = 2.75$
Trial 2				
4 Foxes VV	0	0	$\bar{x} = 18.25$	$\bar{x} = 1.00$
4 Unvaccinated foxes	0	—	$\bar{x} = 18.25$	$\bar{x} = 1.25$

time intervals between vaccination and natural infection, need to be performed.

It is also interesting to note that, using the recombinant virus, we could demonstrate that the rabies glycoprotein (the sole rabies virus component of the VVTGgRAB) administered by the oral route can induce the early-death phenomenon when given alone. This effect could not be induced in animals inoculated with the parental strain of VV. It does not mean that the other rabies virus antigens do not participate in the phenomenon, but rather that the glycoprotein may account for the effect.

C. EFFICACY AND INNOCUITY OF VACCINIA-RABIES RECOMBINANT VIRUS FOR NORTH AMERICAN WILD TARGET SPECIES

In North America, the fox is not the predominant target for oral immunization, compared with raccoon (*Procyon lotor*) and skunk (*Mephitis mephitis*), which are the major rabies virus vectors and are refractory to efficient or safe oral immunization using attenuated strains of rabies virus.

1. Efficacy and Innocuity of the VVTGgRAB for the Raccoon

To assess the possibility that vaccinia-rabies recombinant virus might be applied in controlling raccoon rabies, a limited number of raccoons were vaccinated by intramuscular or oral route with $10^{7.8}$ pfu of the virus. For the animals having received the recombinant by intramuscular route (1.0 ml), the administration resulted in a rapid induction of rabies VNA still apparent several months later. For the raccoons having received 1.0 ml of the virus on the oral mucosa, the antibody response was less intensive. However, within 3 weeks of an oral booster inoculation (with or without buccal mucosa scarification), all animals responded with higher levels of antibodies.[47]

Following these encouraging preliminary results, subsequent work was done to further assess the efficacy of the recombinant administered by oral infusion or in baits. Twenty adult raccoons were given 1.0 ml of undiluted recombinant virus titrating 10^8 pfu in a 3-cm polyurethane sponge cube coated with a beef tallow/paraffin wax mixture and containing 100 to 200 mg of tetracycline HCl as a biomarker. Thirteen control animals received a comparable volume (1.0 ml) of cell culture medium in a bait. Baits were placed in each raccoon cage and after 48 h untouched baits were removed. Two animals not having eaten the baits within this period were given 1.0 ml of either 10^8 or 10^6 pfu of the recombinant virus by oral infusion. Blood was collected from all animals at the time of immunization and several times after rabies VNA were determined by rapid fluorescent foci inhibition test (RFFIT).

Vaccinated or control animals were challenged 28 or 205 d after immunization with 0.3 ml of rabies virus MD 5951 ($10^{5.5}$ $MICLD_{50}$) by intramuscular route (cervical musculature). All animals developed rabies VNA following oral infusion or ingestion of bait containing the recombinant virus; and none of the controls developed VNA. Control animals developed clinical signs of rabies virus infection between 10 and 21 d after challenge or died without prior signs. Rabies infection was confirmed in all animals that died. Seventeen of 20 animals immunized with the recombinant survived the challenge and 16 had a secondary VNA response 30 d after challenge (Tables 11 and 12). None of the animals from the preliminary experiment or from the second one have shown any sign of local or general reaction after receiving the recombinant, whatever the route of administration, confirming the innocuity of the recombinant for the raccoon.[48] When administered orally, the recombinant was recovered only from buccal mucosa, tonsils, and submandibular or parotid lymph nodes at necropsy performed during the first 48 h after vaccination (Table 13). No viremia was detected during 14 d of observation after inoculation and no gross or histopathological lesions suggestive of viral infection were found in any tissues collected post-mortem.[44]

2. Horizontal Transmission of the VVTGgRAB in Adult Raccoons

In cage trials with pairs of adult raccoons, 2 of 5 nonimmunized contact animals developed low rabies VNA levels and survived virus challenge, as did the other cagemates immunized with the recombinant. Only the male-female pair combinations displayed any evidence of contact transfer of virus, but the limited number of animals in the study did not allow a definitive interpretation of results.[44]

3. Innocuity of the VVTGgRAB for Pregnant Raccoons and Test for Vertical Transmission to Kits

Two adult female raccoons were immunized with 10^7 pfu (1.0 ml) of

TABLE 11
Induction of Rabies Virus-Neutralizing Antibodies and Protection from Rabies Challenge in Raccoons Following VVTGgRAB Oral Administration — Immediate Immunity Study

Animal	Treatment	Dose	Virus-neutralizing antibody titer units/ml[a] PI/day 0	PC/day 30	PC/day 90	Response to challenge[b]
49	Recombinant	10^8	54.0	464.0	18.0	S
48	in sponge		18.0	163.0	12.0	S
62	bait		18.0	18.0	18.0	S
58			6.0	18.0	18.0	S
59			6.0	18.0	12.0	S
63			2.0	18.0	18.0	S
56			0.6	163.0	18.0	S
57			0.6	18.0	6.0	S
40	Recombinant by	10^6	1.3	0.7	–	D (13)
41	oral infection	10^8	6.0	9.9	36.0	S
1	Controls		<0.3	2.0	–	D (17)
2			<0.3	2.0	–	D (10)
3			<0.3	0.7	–	D (21)
4			<0.3	2.0	–	D (12)
5			<0.3	0.7	–	D (10)
6			<0.3	1.0	–	D (17)

[a] Virus-neutralizing antibody determined after immunization (PI), or after challenge (PC), or on the day of death. All animals were seronegative prior to immunization.
[b] S, Survived; D, died (day of death after challenge).

Adapted from Rupprecht, C. E. et al., *Proc. Natl. Acad. Sci. U.S.A.*, 83, 7947, 1986.

recombinant by intramuscular route within 30 d of parturition. Each gave birth to healthy litters of three to four young. All littermates had rabies VNA at birth comparable to those of the adult females (titers of 1:135), a finding that implies either passive transfer of maternal antibody or active infection *in utero*. Nevertheless, no virus could be isolated from those young.

In another experiment, three suckling raccoons, 3 to 4 weeks old, were replaced with their mother immediately after she received 10^7 pfu (1.0 ml) by mouth. All animals remained healthy, seroconverted within 28 d, and survived rabies virus challenge. The route of transmission in this latter case (e.g., lactation, grooming) was not documented.[44]

To further test the innocuity of the recombinant for pregnant raccoons and the vertical transmission to the offspring, four adult pregnant raccoons having 2 kits each *in utero* (diagnosed by manual palpation and ultrasonography) were vaccinated with $10^{7.8}$ pfu. One female was given the recombinant suspension (1.0 ml) in a sponge bait while the remaining three females were given the same volume of recombinant suspension directly *per os*. All the

TABLE 12
Induction of Rabies-Neutralizing Antibodies and Protection from Rabies Challenge in Raccoons Following VVTGgRAB Oral Administration — 6-Month Duration of Immunity Study

Animal	Treatment	Dose	Virus-neutralizing antibody titer units/ml[a] PI/day16	PI/day 205	PC/day 30	Response to challenge[b]
60	Recombinant	10^8	18.0	3.0	958.6	S
50	in sponge		6.0	0.5	185.2	S
16	bait		3.0	0.1	590.0	S
44			3.0	0.6	47.3	S
61			3.0	0.1	205.1	S
64			3.0	0.3	260.3	S
42			2.0	0.3	7.0	S
43			2.0	0.5	511.2	S
46			2.0	0.1	60.2	D (11)
47			0.6	0.1	18.0	D (16)

[a] Virus-neutralizing antibody determined after immunization (PI), or after challenge (PC), or on the day of death. All animals were seronegative prior to immunization.
[b] S, Survived; D, died (day of death after challenge).

Adapted from Rupprecht, C. E. et al., *Proc. Natl. Acad. Sci. U.S.A.*, 83, 7947, 1986.

females gave birth to healthy kits between 2 to 20 days post-immunization and all females and their offspring remained healthy over the next 3 months. At that time, all animals were bled and vaccinated with a subimmunogenic dose of inactivated rabies vaccine. Three days later the animals were bled again and peripheral blood lymphocytes were collected for determination of cell proliferative indices (CPI). At 90 d following the administration of the recombinant all adult females had rabies VNA titers (GPT: 0.5 IU/ml) as did all the kits. At 3 d pre-, and 2, 8, and 20 d postpartum, the raccoon kits showed GMT equivalent to 0.7, 10.9, 5.6, and 18.8 UI/ml, respectively.

However, only the adult raccoons were sufficiently primed to demonstrate an anamnestic response from the inactivated rabies vaccine. If the reproductive physiology of raccoons is similar to that of other carnivores, it may be expected that passive transfer of immunoglobulins does occur to a significant extent *in utero*, and that rabies VNA may be transferred via colostrum during a relatively narrow temporal window (24 to 48 h post-partum). As there was no evidence of active *in utero* or lactogenic transmission of the recombinant and assessing that newborn raccoons like other carnivores, while partially immunocompetent, possess a comparatively immature immune system for the first 1 to 3 months of life, it seems that kits more likely demonstrated passive immunity only, and no acquired active immunization.

TABLE 13
Distribution of VVTGgRAB in Raccoons Infected by the Oral Route

Tissue	Result on indicated day of immunization						
	1	2	3	5	9	11	14
Brain	−	−	−	−	−	−	−
Buccal mucosa	−	+	−	−	−	−	−
Tongue	−	−	−	−	−	−	−
Lung	−	−	−	−	−	−	−
Stomach	−	−	−	−	−	−	−
Tonsil	+	+	−	−	−	−	−
Esophagus	−	−	−	−	−	−	−
Salivary glands	−	−	−	−	−	−	−
Parotid submandibular lymph nodes	+	+	−	−	−	−	−
Thymus	−	−	−	−	−	−	−
Trachea	−	−	−	−	−	−	−
Duodenum	−	−	−	−	−	−	−
Liver	−	−	−	−	−	−	−
Spleen	−	−	−	−	−	−	−
Kidney	−	−	−	−	−	−	−
Urinary bladder	−	−	−	−	−	−	−
Blood	−	−	−	−	−	−	−
CSF	−	−	−	−	−	−	−

Note: Raccoons having received $10^{7.8}$ plaque-forming units (1.0 ml) orally, tissues were obtained from 1 to 5 animals for viral isolation in BHK 21 cell culture on days indicated; +, recombinant isolated; −, no virus recovered.

Adapted from Rupprecht, C. E. et al., *Rev. Infect. Dis.*, 10(S), 803, 1988.

D. EFFICACY AND INNOCUITY OF THE VVTGgRAB FOR SKUNKS

Striped skunks, which represent another important rabies vector in North America, were vaccinated with the recombinant using the following procedure. Thirty-four skunks, 4 months old, with an equal number of males and females, were divided into five groups and kept separately. Group 1 (8 animals) was given a sponge bait containing 5 ml of the recombinant suspension titrating 2.10^8 pfu/ml. Group 2 (8 animals) was given the vaccine (5.0 ml of the same suspension) into the duodenum using an endoscope. Group 3 (4 animals) received the vaccine (1.0 ml of the same suspension) intramuscularly. Group 4 (6 animals) was given the vaccination by scarification.

Blood samples were taken on day 0, or one day before vaccination, and on days 14, 28, 60, and 90 after vaccination, and the antibody titer was determined by RFFI test. All the skunks and eight non-vaccinated controls (Group 5) were challenged with a suspension of salivary gland from naturally infected skunks ($10^{4.7}$ $MIDLD_{50}$) by the intramuscular route and observed for

90 days. None of the skunks having received the recombinant by any of the described routes have developed local or general reactions, demonstrating the innocuity of the recombinant for this animal species.

At two weeks post-vaccination, only one skunk seroconverted. At four weeks, six of the seven skunks that had eaten the bait had rabies VNA. Neutralizing antibodies were present in the skunks of the intestinal, intramuscular, and scarified groups two weeks after vaccination. Five out of six bait-vaccinated skunks survived the rabies challenge. Four out of eight of the intestinal group, three out of four of the intramuscular group, and five out of six of the scarified group survived, demonstrating the efficacy of the recombinant, while all of the eight animals in the control group developed rabies between the third and the fourth week after challenge[49] (Table 14).

IV. SAFETY OF THE VVTGgRAB FOR LABORATORY, DOMESTIC, AND NONTARGET WILD ANIMAL SPECIES

The use of attenuated rabies virus for oral vaccination of foxes still remains controversial as far as innocuity and stability are concerned, since these virus strains retain pathogenicity in laboratory and wild rodents.[6,9,10] Field trials conducted in Europe have shown that several wildlife nontarget species compete with foxes in consuming "Tübingen" baits consisting of fat and fish meal (World Health Organization Collaborating Centre for Rabies Surveillance and Research, Tübingen, Germany). Bait uptake and tetracycline detection controls performed after vaccination campaigns have proven that mustelids, wild boars, and domestic carnivores could ingest the vaccine bait.[8] A significant proportion of the baits are partially eaten by micromammals. Despite this competition, further uptake by foxes is generally not precluded. However, when baits persist in the field, rodents may eat the whole coating, perforate the plastic capsule, and swallow the vaccine liquid.[8] No case of vaccine-induced rabies has been reported in rodents trapped within vaccination areas.[7] Nevertheless, the method of rabies virus isolation in captured small mammals cannot be considered as an appropriate tool for safety control. This method requires many more specimens per hectare for the detection of rodents with prepatent rabies. Considering the behavior of rabid animals and the action of predators and scavengers, the trapping method does not help to detect patent or dead specimens.

The use of a recombinant vaccinia virus may help to solve some of the problems related to the use of a live, modified rabies virus. Nevertheless, the safety problems of live vaccines (required for oral immunization) must still be investigated when using a recombinant orthopoxvirus. Therefore, it is important to verify the absence of pathogenicity, excretion, and transmission of VVTGgRAB in both target and nontarget animal species.

TABLE 14
Induction of Rabies Virus-Neutralizing Antibodies and Protection from Rabies Challenge in Skunks Following VVTGgRAB Administration — Immediate Immunity Study

		Virus-neutralizing antibody titer I units/ml					
Animal	**Treatment**	**Day 0**	**Day 14**	**Day 28**	**Day 60**	**Day 90**	**Response to challenge[a]**
1	Oral	0	0.41	2.52	2.35	0.85	S
2		0	0	0.17	0.22	0	S
3		0	0	1.43	0.35	0.17	D (19)
4		0	0	2.43	0.65	0.24	S
5		0	0	0	0	0	D (13)[b]
6		0	0	4.61	4.59	1.41	S
7		0	0	0.72	1.20	0.39	S
8[c]		0	0	0	0	0	D (43)
1	Intestinal	0	0	0	0	0	D (26)
2		0	0	0	0	0	D (25)
3		0	3.48	1.35	3.78	0.96	S
4		0	2.13	2.07	1.04	0.28	S
5		0	0	0	0	0	D (25)
6		0	14.4	10.6	3.76	1.70	S
7		0	1.59	0.87	0.80	0.52	S
8		0	2.74	0.85	0.65	0.52	D (27)
1	Intramuscular	0	22.3	12.5	1.20	0.37	D (19)
2		0	21.4	14.4	8.37	3.78	S
3		0	37.4	20.4	2.61	2.87	S
4		0	19.6	15.2	13.8	7.54	S
1	Scarification	0	156	34.3	4.57	2.61	S
2		0	44.5	6.78	1.93	0.35	D (70)
3		0	159	15.7	2.57	1.39	S
4		0	68.5	5.35	2.30	0.70	S
5		0	101	11.1	2.37	0.70	S
6		0	48.4	11.1	2.65	0.70	S

[a] S, Survived; D, died (day of death after challenge).
[b] Death not due to rabies.
[c] Did not eat bait.

Adapted from Tolson, N. D. et al., *Can. J. Vet. Res.*, 51, 363, 1987.

A. INNOCUITY OF THE VVTGgRAB TO LABORATORY AND DOMESTIC ANIMALS

Residual pathogenicity testing — The potential pathogenicity of the VVTGgRAB was evaluated in the following laboratory and domestic animal species:

1. Laboratory mouse inoculated by the oral, intraperitoneal, intradermal, or intramuscular route[33,47]
2. Hamster inoculated by the intramuscular route[34]

3. Rabbit inoculated by the oral, intradermal, or intramuscular route[33,49]
4. Ferret inoculated by the oral route[39,51]
5. Cattle inoculated by the intradermal, subcutaneous, or intramuscular route[38,51]
6. Sheep inoculated by the oral route[50]
7. Pig inoculated by the oral or intramuscular route[39]
8. Dog inoculated by the oral or subcutaneous route[38,51]
9. Cat inoculated by the oral or subcutaneous route[38,51]

Animals were observed daily for a period of 28 d or more post-vaccination. Potential clinical signs and/or pox lesions were recorded. Post-mortem examination was performed on all animals after euthanasia. Except for intradermal inoculation, neither local nor general reaction (clinical sign or lesion) was observed in animals given a dose of 10^8 $TCID_{50}$ or more of the VVTGgRAB. Following intradermal inoculation, and depending upon the animal species, typical pox virus lesions were observed, the size of which varied according to each animal. Furthermore, these lesions are of less importance and duration than those provoked by the parental vaccinia virus.

Assessment of excretion and horizontal transmission — The horizontal transmission of VVTGgRAB between animals has been tested by using control animals maintained in close contact with orally vaccinated ones. Control animals were tested for anti-rabies-neutralizing antibodies in order to check for direct or indirect transmission of the vaccine virus. This transmission test has been done in dogs, cats, cattle, and ferrets.[38,51] Rabies-neutralizing antibodies were never detected in these control, contact animals.

Innocuity to the nude mouse — Because of the difficulty of immunosuppressing animals, particularly wild animals, the safety of the VVTGgRAB was assessed in nude mice. These did not develop any local or general reaction following inoculation into the footpad or by the intraperitoneal route. When intracerebrally administered to mice, the comparison of LD_{50} of the VVTGgRAB with that of the parental Copenhagen strain indicated a reduced capability of the VVTGgRAB to multiply in the brain.[39]

Genetic stability in the laboratory mouse — The genetic stability of the VVTGgRAB was checked by 10 passages in Vero cell cultures, and by 7 passages in mice following footpad and intracerebral inoculation.[39] No evidence of genomic breakdown or of change in biological properties was observed. The inability to recover the virus from the tissues of footpad-inoculated animals after 7 passages, linked to the necessity to perform an amplificative passage on Vero cells between the third and fourth passages by footpad and intracerebral inoculation, is further evidence of the safety of VVTGgRAB.

B. INNOCUITY TO SEVERAL NONTARGET WILD ANIMAL SPECIES

Several nontarget wild species have been chosen for testing because of their opportunist food habits and their presence in the areas where this vaccine may be distributed: daubenton bat (*Myotis daubentoni*), wild boar (*Sus scrofa*), Eurasian badger (*Meles meles*), wood mouse (*Apodemus sylvaticus*), yellow-necked mouse (*Apodemus flavicollis*), bank vole (*Clethrionomys glareolus*), common vole (*Microtus arvalis*), field vole (*Microtus agrestis*), water vole (*Arvicola terrestris*), common buzzard (*Buteo buteo*), kestrel (*Falco tinnunculus*), carrion crow (*Corvus corone*), magpie (*Pica pica*), and jay (*Garrulus glandarius*).[52] Some of them could compete with the natural rabies host, the red fox, in consuming vaccine baits in Europe.

The present study also investigated the efficacy of VVTGgRAB in the immunization and protection of badgers against rabies virus. In Europe, badgers are also considered as a target species for rabies vaccination, since they are often infected. This species is strongly susceptible to the vulpine strain of rabies virus and can excrete large amounts of virus in its saliva. Furthermore, the usual co-occurrence of foxes and badgers in sympatric habitats results in frequent infection of the latter species.

The experimental protocol for each species is detailed in Tables 15 and 16. On day 0, vaccine was administered by direct application into the oral cavity. Unvaccinated animals were either separately housed and used as controls or held in close contact with vaccinated animals. They were given an equal volume of PBS (Phosphate Buffered Saline, pH 7.4). Animals were observed daily for a period of 28 d or more post-vaccination. Potential clinical signs and/or pox lesions were recorded. For rabies and vaccinia serological analysis, blood was collected on day 0 and approximately 1 month (from 28 to 45 days) post-vaccination. Post-mortem examination was performed on all animals after euthanasia.

1. Serological Studies

All the badgers, boars, crows, and magpies included in this study were shown to be serologically negative for rabies VNA on day 0. On day 28 post-vaccination, rabies virus antibody titers exceeding 0.5 I.U./ml appeared in 2 of 12 daubenton bats (16%), 2 of 4 boars (50%), 2 of 6 badgers (33%), 16 of 27 wood mice (59%), 3 of 4 *Apodemus* sp. (75%), 5 of 7 yellow-necked mice (71%), 8 of 13 bank voles (61.5%), 1 of 1 field vole (100%), 2 of 2 common voles (100%), 4 of 5 water voles (80%), none of 5 buzzards (0%), none of 2 kestrels (0%), none of 17 crows (0%), none of 7 magpies (0%), and none of two jays (0%). Rabies serology demonstrates that VVTGgRAB, when orally administered, can elicit seroconversion in all the species of mammals included in this study. The absence of rabies and/or vaccinia seroconversion in birds suggests that the recombinant virus does not multiply in these avian species. On day 0, the sera of all carrion crows had a neutralizing activity on the VV infection of vero cells. This serological reaction is probably nonspecific, since vaccinia and/or rabies-neutralizing antibody titers did not increase in vaccinated crows.

TABLE 15
Experimental Protocol of VVTGgRAB Oral Administration to Several Nontarget Wild Mammals

Species	VVTGgRAB			Number of unvaccinated animals		Observation period (days)
	Number of vaccinated animals	Dose ($TCID_{50}$)	Volume (ml)	Control	Contact[a]	
Daubenton's bat (*Myotis daubentoni*)	12	1×10^{7}	0.05	6	0	19
European badger (*Meles meles*)	6	$1 \times 10^{8.3}$	1	2	2	45
Wild boar (*Sus scrofa*)	4	$1 \times 10^{8.3}$	1	2	1	88
Wood mouse (*Apodemus sylvaticus*)	31	$2–3 \times 10^{6}$	0.03	8	—	28–43
Yellow-necked mouse (*Apodemus flavicollis*)	7	3×10^{6}	0.03	1	—	41
Bank vole (*Clethrionomys glareolus*)	13	2×10^{6}	0.03	5	0	28
Common vole (*Microtus arvalis*)	2	3×10^{6}	0.03	0	0	35
Field vole (*Microtus agrestis*)	1	3×10^{6}	0.03	0	0	35
Water vole (*Arvicola terrestris*)	5	3×10^{6}	0.03	0	0	41

[a] Kept in close contact with inoculated animals.

TABLE 16
Experimental Protocol of VVTGgRAB Oral Administration to Several Nontarget Wild Birds

Species	VVTGgRAB Number of vaccinated animals	VVTGgRAB Dose ($TCID_{50}$)	VVTGgRAB Volume (ml)	Number of unvaccinated controls	Observation period (days)
Common buzzard (*Buteo buteo*)	8	1×10^8	1	0	30–45
Kestrel (*Falco tinnunculus*)	4	1×10^8	1	0	30–45
Carrion crow (*Corvus corone*)	17	1×10^8	1	3	28
Magpie (*Pica pica*)	7	1×10^8	1	3	28
Jay (*Garrulus glandarius*)	2	1×10^8	1	0	28

Tables 17 and 18 indicate vaccinia and rabies serology in badgers and wild boars, respectively. Anti-vaccinia seroconversion was detected in two of four wild boars (50%). As shown in Table 17, two of six badgers (animals 8 and 9) administered VVTGgRAB resisted challenge on day 45 post-vaccination. Badger 9 was protected despite the absence of demonstrable rabies antibodies. The vaccinated badger 10 had produced rabies antibodies, but died accidentally before challenge. When including the unchallenged badger 10, which had produced rabies antibodies, it can be attested that only 50% of vaccinated badgers are protected against rabies infection. The efficacy of VVTGgRAB in the immunization and protection of badgers against rabies is weak when compared to the protection level conferred to other wild target species.

2. Residual Pathogenicity Assessment

Clinical signs and/or pox lesions were not observed in any of the vaccinated animals during the observation period (28 d minimum after vaccination). This period of observation has been chosen as it exceeds the delay of recovery from vaccinia infection in humans.[53] At necropsy, all carcasses were in normal condition and lesions indicative of poxvirus infection were not detected. The absence of mortality, especially in rodents, is a notable advantage of the recombinant virus as compared to the SAD live-modified strains of rabies virus, which retain pathogenicity in small mammals.

3. Assessment of Excretion and Horizontal Transmission

Neither of the unvaccinated wild boars and badgers, kept in contact with vaccinated individuals, elicited any demonstrable rabies VNA. This absence of seroconversion indicates that excretion and horizontal transmission of

TABLE 17
Vaccinia and Rabies Virus-Neutralizing Antibodies and Protection from Rabies Challenge in Badgers Orally Administered with $10^{8.3}$ $TCID_{50}$ of VVTGgRAB

		VNA Titers[a]				
		Rabies		Vaccinia		
Treatment	Animal no.	Day 0[c]	28	Day 0[c]	28	Resistance to challenge[b]
Unvaccinated control	1	<0.5	<0.5	1.05	<0.6	D (34)
	2	<0.5	<0.5	<0.4	0.5	D (20)
Unvaccinated contact	3	<0.5	<0.5	0.7	0.8	D (21)
	4	<0.5	<0.5	<0.6	0.8	D (18)
Vaccinated	5	<0.5	<0.5	<0.6	<0.4	D (27)
	6	<0.5	<0.5	<0.4	<0.5	D (21)
	7	<0.5	<0.5	<0.5	<0.5	D (30)
	8	<0.5	4.84	<0.74	<0.7	S
	9	<0.5	<0.5	0.7	<0.4	S
	10	<0.5	3.8	<0.4	<0.4	ND

[a] Titers of rabies and vaccinia virus-neutralizing antibodies are expressed, respectively, in International Units/ml and in logarithm of the dilution neutralizing 30 $TCID_{50}$ of vaccinia virus.

[b] S, Survived; D, died; ND, accidentally died on day 45; day of death after challenge is in parentheses.

[c] Day 0 = day of vaccination.

TABLE 18
Vaccinia and Rabies Virus-Neutralizing Antibodies in Wild Boars Orally Administered With $10^{8.3}$ $TCID_{50}$ of VVTGgRAB

		VNA Titers[a]				
		Rabies		Vaccinia		
Treatment	Animal no.	Day 0[b]	Day 28	Day 0[b]	Day 28	Day 90
Vaccinated	1	<0.45	4.99	0.58	0.82	1.98
Vaccinated	2	<0.45	<0.45	0.58	<0.35	0.7
Vaccinated	3	<0.45	5.46	0.47	0.82	1.4
Vaccinated	4	<0.45	<0.45	0.82	<0.35	0.47
Unvaccinated control	5	<0.45	<0.45	0.82	<0.35	<0.35
Unvaccinated control	6	<0.45	<0.45	0.7	<0.35	0.58
Unvaccinated contact	7	<0.45	<0.45	0.58	<0.35	<0.35

[a] Titers of rabies and vaccinia virus-neutralizing antibodies are expressed, respectively, in International Units/ml and in logarithm of the dilution neutralizing 30 $TCID_{50}$ of vaccinia virus.

[b] Day 0, day of inoculation.

immunizing amounts of VVTGgRAB vaccine did not occur in those animals. This observation is also an important feature indicating vaccine safety. It precludes the epidemiological risks possibly associated with the use of a recombinant strain of orthopox virus.

The innocuity of this VVTGgRAB vaccine seems to be established in wild animal as well as in domestic and laboratory species. Similar results were obtained in testing several North American nontarget animal species: white-tailed deer (*Odocoileus virginianus*), bobcat (*Lynx rufus*), common opossum (*Didelphis virginiana*), river otter (*Lutra canadensis*), and sea gull (*Larus canus*). All these results confirm the predicted safety linked to this recombinant virus. Inactivation of the thymidine kinase (TK) gene by the insertion of a cDNA sequence coding for rabies glycoprotein affects the remaining pathogenicity of vaccinia virus and markedly attenuates its virulence.[54]

V. FIELD TRIALS WITH THE VACCINIA-RABIES RECOMBINANT VIRUS

A. PRELIMINARY LIMITED FIELD TRIALS

Taking into account all the available experimental data concerning the safety of the VVTGgRAB for target and nontarget species and its efficacy in foxes, initial, limited field trials of fox vaccination were authorized first by the Belgian and then the French Public Health authorities. The sites chosen were military fields, isolated from public and domestic animals, of Marche-en-Famenne in Belgium[55] and Mars la Tour in France; these military fields were located in areas where fox rabies is commonly found (Figure 7). In the Belgian trial, each capsule containing a suspension of 10^8 $TCID_{50}$ of VVTGgRAB was introduced into a chicken head. On October 17 and 18, 1987, a total of 250 vaccine baits were manually delivered over a 6-km^2 area situated in the center of the military field and surrounded by a nonvaccinated buffer zone. By day 15, the bait uptake was 64%. Of 145 rodents trapped in the baited area, only 4 had eaten baits and none showed any lesion that could be related to poxvirus infection.

In France, a machine-made bait was used. One bait was set every 30 m on a grid within 2 areas of the military camp (53 ha). One area (174 baits) was used as control, with baits loaded only with distilled water and a dye (Rhodamine B). Within the second area, 187 baits containing the vaccine were distributed. Two distributions were carried out in France in October, 1988 and May, 1989. By day 15, the bait uptake was 34% within the experimental area and 90% (autumn) and 42% (spring) within the control area. No adverse phenomena (abnormal morbidity or mortality) were observed in wildlife after these VVTGgRAB releases. However, definite proof of a direct contact between the VVTGgRAB and the oral mucosa of any animal was not gained.

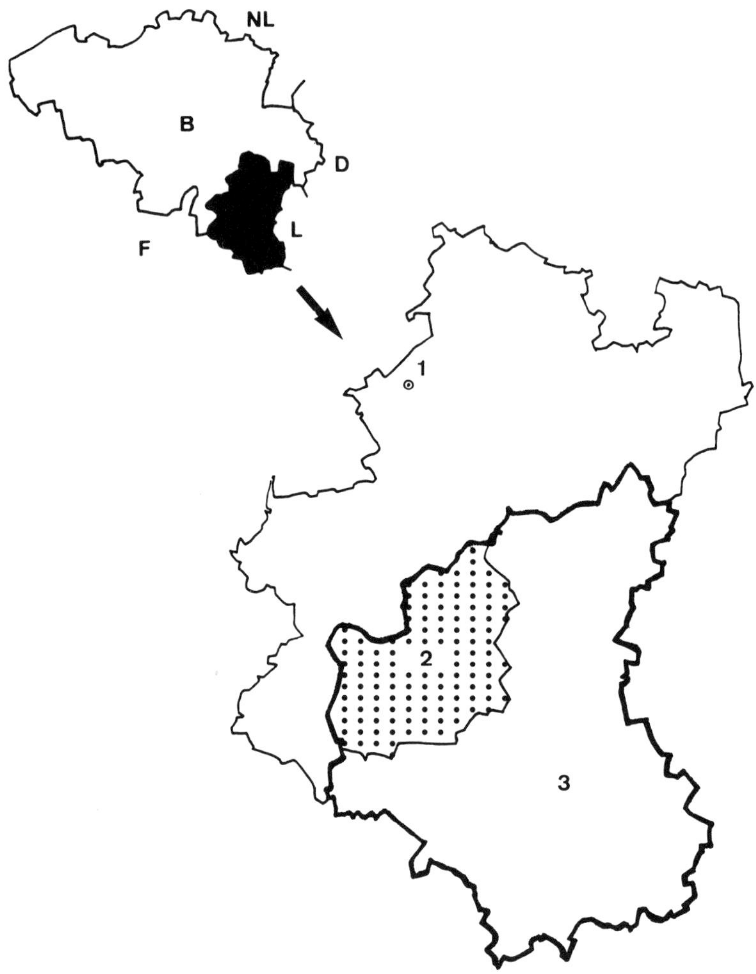

FIGURE 7. Field trials of fox vaccination with the VVTGgRAB in the province of Luxembourg (South Belgium). Baiting areas: (1) military field of Marche-en Famenne (6 km^2, September, 1987); (2) region of Neufchâteau (435 km^2, October, 1988); (3) half part of the province of Luxembourg (2200 km^2, November, 1989, April and October, 1990).

B. LARGE-SCALE FIELD TRIAL

The VVTGgRAB safety having been confirmed, the Belgian Ministry of Public Health on September 21, 1988 gave authorization for an enlarged open-field trial. This latter has been conducted in a 435-km^2 area covering four administrative districts in the Province of Luxembourg (South Belgium) (Figure 7).[56] This vaccination zone was chosen because it has the lowest human population density (42 inhabitants/km^2) in the country, combined with a high incidence of rabies. Furthermore, the region is characterized by various habitats including most of the animal species likely to consume baits.

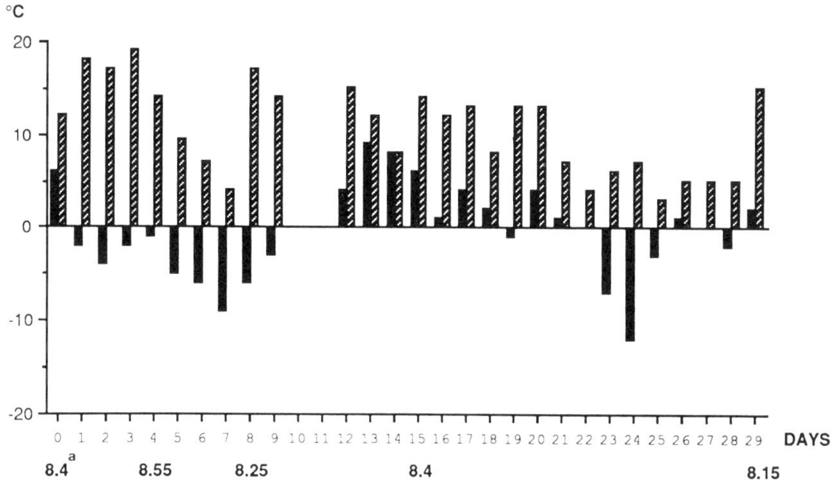

FIGURE 8. Minima and maxima of environmental temperatures recorded after bait distribution (day 0). VVTGgRAB mean titer of two baits removed on days 0, 4, 8, 15, or 29 is given below the axis line.[a] Expressed as the log 10 $TCID_{50}$ ml^{-1}. (From Brochier, B. et al., *Vaccine,* 8, 101, 1990.)

The vaccine used was a suspension of 10^8 $TCID_{50}$ of VVTGgRAB (vol.: 2.2 ml) contained in a plastic sachet. Each vaccine sachet was enclosed in a bait consisting of a mixture of fishmeal and fishoil. Each bait also contained 150 mg tetracycline, used as a long-term biomarker of bait uptake. The vaccine-bait stock was stored at 4°C until the day of distribution in the field. The distribution was performed by 26 forestry rangers and lasted about two weeks (October 24 to November 10, 1988). It was done by hand at a mean density of 15 baits/km^2. After the baiting campaign, both efficacy and innocuity of the method were monitored in the field as well as in the laboratory.

1. Efficacy Assessment

For the bait uptake control, 238 baits were placed and identified in a 20-km^2 zone situated close to the center of the vaccination area. Each bait location was then visited at different times after the distribution. Sixty-five (27%), 128 (54%), 163 (68.5%) and 223 (94%) baits were either consumed or at least removed by animals on days 4, 8, 15, and 30, respectively. These results confirm the efficiency of the baiting system. A significant proportion (56%) of the baits were partially eaten by rodents or coleoptera. Despite this competition, further uptake by foxes was generally not precluded. However, in some cases, rodents ingested the whole coating, perforated the plastic sachet, and swallowed the vaccine liquid.

Despite important variations of environmental temperatures (occurrence of natural freezing-thawing cycles) (Figure 8), the VVTGgRAB titer remained stable after one month in the field. A first significant decrease of the VVTGgRAB titer was observed in the fourth month (Figure 9).

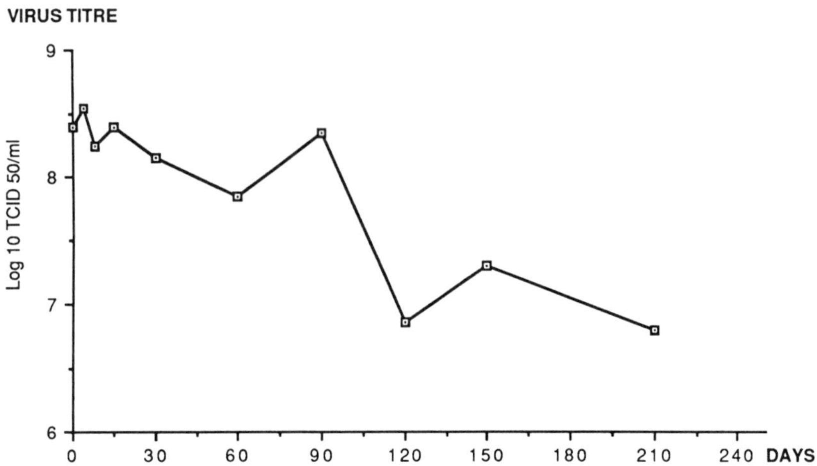

FIGURE 9. Stability of the VVTGgRAB titer under local field conditions (winter 1988–1989 and spring 1989). (From Brochier, B. et al., *Vaccine,* 8, 101, 1990.)

TABLE 19
Bait Uptake by Nontarget Animal Species

Wild boar (*Sus scrofa*)	10[a]	47% TC+[b]	(n = 32)
Dog	DO[c] +10	—	
Stone marten (*Martes foina*)	—	37% TC+	(n = 8)
Polecat (*Mustela putorius*)	10	—	
Cat	DO +10	27% TC+	(n = 11)
Woodmouse (*Apodemus* sp.)	10	12% TC+	(n = 40)
Crow (*Corvus corone*)	—	10% TC+	(n = 10)
Coleoptera	DO	—	

[a] Indirect observation (footprints and feces).
[b] Detection of tetracycline in bones.
[c] Direct observation.

After the vaccination campaign, 222 dead wild animals of 19 species were collected within the vaccination area for post-mortem examination and laboratory testing. After necropsy, the following organs were removed: brain for rabies diagnosis, jaw for tetracycline detection, and blood for the determination of vaccinia and rabies antibodies. Sixteen out of 26 foxes (61%) collected in the vaccination area were found tetracycline positive. This result can be considered as a good score, especially after a first campaign of bait delivery. This biomarking was also helpful for the detection of bait-consuming nontarget species. Among 222 collected specimens, tetracycline was detected in 3/7 stone martens (*Martes foina*), 3/11 domestic or feral cat (*Felis catus*), 5/50 wood mice (*Apodemus* sp.), 15/32 wild boars (*Sus scrofa*) and 1/10 carrion crows (*Crovus corone*) (Table 19). Wild boars, mustelids, and feral cats are thus strong competitors of the foxes for the bait uptake. Serological data were difficult to interpret because of strong discrepancies in the results obtained with RFFIT (macromethod and micromethod) and enzyme-linked immunosorbent assay (ELISA) techniques, using either the rabies virus glycoprotein or the vaccinia virus (Copenhagen strain) as antigens. Blood sample collected from dead animals are of poor quality. When using virus-neutralization serological techniques, such as the RFFIT, nonspecific reactions are often observed. Therefore, ELISA techniques are likely to be a more appropriate tool for serological examinations of blood samples collected in field conditions. Nevertheless, these techniques of detection of antibodies directed against VV and the rabies virus glycoprotein still need to be improved and adapted. Serological controls are thus currently insufficient for evaluating the efficacy of a fox vaccination campaign. This latter would have to be established by the bait uptake rate (using a reliable biomarker) and the evolution of rabies epizootic in the treated area.

The ideal control of efficacy of a vaccination campaign would require the capture of numerous live foxes. Serological results would then be reliable, whatever the technique used. Furthermore, a rabies challenge could then be performed to confirm the protection against rabies in seropositive foxes and to check if some animals are protected even in the absence of detectable rabies VNA.

2. Safety Assessment

No pox-like lesion could be detected in any of the 222 collected animals. Every suspected lesion would have been submitted to three methods of virus identification: virus isolation on cell culture, PCR, and electron microscopy. As far as the safety of this vaccination campaign is concerned, no abnormal morbidity or mortality was noted in wildlife or in domestic animals. Moreover, 12 months of monitoring failed to detect any ecological hazard or problem of public health.

C. LARGE-SCALE FOX VACCINATION CAMPAIGNS

Three fox vaccination campaigns using the VVTGgRAB were carried out in Belgium in November, 1989, April, 1990 and October, 1990. Twenty-five thousand machine-made baits, each containing 10^8 $TCID_{50}$ of VVTGgRAB, were time dropped by helicopters in a 2200-km^2 area in south Belgium (Province of Luxembourg) (Figure 7). The efficacy of the vaccine-bait system used was previously attested in captive foxes as described above (see Section III.A.2 and Figure 2). Baits were dispersed in the field at a mean density of 15 baits/km^2 of vacccinable area. As usual, both efficacy and innocuity of the method were monitored in the field as well as in the laboratory.

In France, several campaigns using the VVTGgRAB have been carried out since 1989. Machine-made baits, each containing 10^8 $TCID_{50}$ were dropped by helicopters over 10,000 km^2. The mean density of bait dropped was 15/km^2 within vaccinable areas.

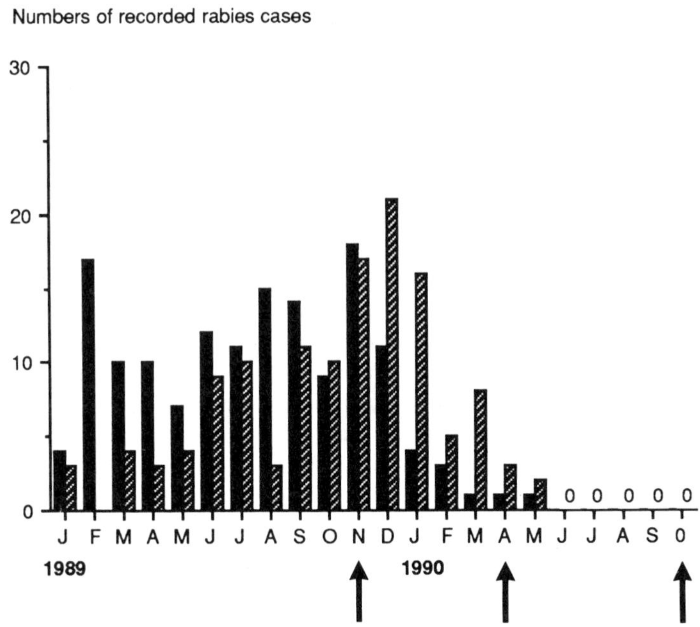

FIGURE 10. Monthly evolution of rabies incidence in a 2200-km^2 area between January 1, 1989 and October 25, 1990. Arrows, campaigns of fox vaccination with the VVTGgRAB; black columns, wild animals; hatched columns, domestic animals.

1. Efficacy and Safety Assessment

Field controls of the bait upake, performed after these releases, have shown that 80 to 96% of the baits were either consumed or at least removed by animals after 30 days. After the first vaccination campaign carried out in Belgium, 23 adult foxes, 33 cats, 9 dogs, 12 wild carnivora, and 24 rodents were collected within the treated area for post-mortem examination, rabies diagnosis, and tetracycline detection. The biomarker has been found in 17 of the 23 (74%) collected foxes, 4/33 cats, 2/9 dogs, and 4/10 stone martens. After the second vaccination campaign, 47 of the 76 (62%) collected foxes (adult and cubs) were tetracycline positive. This result can be considered as an excellent score when it is considered that most cubs confined to the den surroundings, could not have been vaccinated during this spring campaign.

As shown in Figure 10, the rabies incidence has severely decreased in the treated area, despite the high density of fox population observed in 1990. In France, the biomarker has been found in more than 70% of collected foxes, and in several carcasses of miscellaneous other species. The serological survey is still in progress but preliminary results show that more than 80% of foxes seroconverted against rabies.[59] As far as the innocuity is concerned, no pox-like lesion could be detected in any of the collected specimens. (See Reference 61 for more information.)

VI. CONCLUSIONS

Because of its efficacy, innocuity, and heat stability, the VVTGgRAB seems to offer an excellent alternative to the attenuated strains of rabies virus currently used in the field. Furthermore, the bait used is attractive, efficient, and thermostable. Nevertheless, in addition to the efficiency of a vaccine-bait system, the strategy of bait delivery is of major importance for achieving the immunization of 75% of foxes and, subsequently, the disruption of the viral infection cycle. Both spatial and temporal patterns of the bait distribution must be planned according to several natural factors. In this sense, the thermostability of the VVTGgRAB-bait system provides flexibility in planning a campaign.

Baits can be placed by hand or spread from airplane or helicopter. Manual dispersal requires the participation of many distributors; human factors were shown to play a role in the vaccination success or failure. Therefore, the aerial dropping method is currently considered as more effective and should be used during further campaigns.

In conclusion, local eradication of fox rabies could be achieved after three or four vaccination campaigns[7] when the following criteria are met: a potent, safe and thermostable vaccine; an efficient and resistant bait system; a practical and sure method of bait dispersal; and an effective spatial and temporal pattern of bait distribution.[4]

REFERENCES

1. **Baer, G. M.,** Wildlife vaccination, in *The Natural History of Rabies,* Vol. 2, Baer, G. M., Ed., Academic Press, New York, 1975, 261.
2. **Mayr, A., Kraft, H., Jaeger, O., and Haacke, H.,** Orale immunisierung von Fuchsen gegen Tollwut, *Zentralbl. Veterinaermed. Reihe, B.,* 19, 165, 1972.
3. **Steck, F., Wandeler, A., Bichsel, P., Capt, S., and Schneider, L.,** Oral immunisation of foxes against rabies. A field study, *Zentralbl. Veterinaermed.,* 29, 372, 1982.
4. **Wandeler, A. I., Capt, S., Kappeler, A., and Hauser, R.,** Oral immunization of wildlife against rabies: concept and first field experiments, *Rev. Infect. Dis.,* 10 (4), S649, 1988.
5. **Kappeler, A., Wandeler, A., and Capt, S.,** Ten years of rabies control by oral vaccination of foxes in Switzerland, in *Vaccination to Control Rabies in Foxes,* Pastoret, P.-P., Brochier, B., Thomas, I., and Blancou, J., Eds., Office for Official Publications of the European Communities, Brussels-Luxembourg, 1988, 55.
6. **Schneider, L. G. and Cox, J. H.,** Ein Feldversuch zur oralen Immunisierung von Füchsen gegen die Tollwut in der Bundesrepublik Deutschland, *Tieraerztl. Umsch.,* 38, 315, 1983.
7. **Schneider, L. G. and Cox, J. H.,** Eradication of rabies through oral vaccination, in *Vaccination to Control Rabies in Foxes,* Pastoret, P.-P., Brochier, B., Thomas, I., and Blancou, J., Eds., Office for Official Publications of the European Communities, Brussels-Luxembourg, 1988, 22.

8. **Brochier, B., Thomas, I., Iokem, A., Ginter, A., Kalpers, J., Paquot, A., Costy, F., and Pastoret, P.-P.,** A field trial in Belgium to control fox rabies by oral immunization, *Vet. Rec.*, 123, 618, 1988.
9. **Wandeler, A., Bauder, W., Prochaska, S., and Steck, F.,** Small mammals studies in a SAD baiting area, *Comp. Immunol. Microbiol. Infect. Dis.*, 5, 173, 1982.
10. **Leblois, H. and Flamand, A.,** Studies on pathogenicity in mice of rabies virus strains used for oral vaccination of foxes in Europe, in *Vaccination to Control Rabies in Foxes*, Pastoret, P.-P., Brochier, B., Thomas, I., and Blancou, J., Eds., Office for Official Publications of the European Communities, Brussels-Luxembourg, 1988, 101.
11. **Tuffereau, C., Leblois, H., Benejean, J., Coulon, P., Lafaye, E., and Flamand, A.,** Arginine or lysine in position 333 of ERA and CVS glycoprotein is necessary for rabies virulence in adult mice, *Virology*, 172, 206, 1989.
12. **Neurath, A. R., Vernon, S. K., Dobkin, M. B., and Rubin, B. A.,** Characterization of subviral components resulting from treatment of rabies virus with tri(n-butyl) phosphate, *J. Gen. Virol.*, 14, 33, 1972.
13. **Sokol, F., Stancek, D., and Koprowski, H.,** Structural proteins of rabies virus, *J. Virol.*, 7, 241, 1971.
14. **Cox, J. H.,** The structural proteins of rabies virus, *Comp. Immunol. Microbiol. Infect. Dis.*, 5, 21, 1982.
15. **Lepine, P., Atanasiu, P., and Sisman, J.,** Etudes sur le virion rabique, *Symp. Ser. Immunobiol. Stand.*, 1, 31, 1966.
16. **Pinteric, L. and Fenje, P.,** Electron microscopic observation of rabies virus, *Symp. Ser. Immunobiol. Stand.*, 1, 9, 1966.
17. **Dietzschold, B., Cox, J. H., Schneider, L. G., Wiktor, T. J., and Koprowski, H.,** Isolation and purification of a polymeric form of the glycoprotein of rabies virus, *J. Gen. Virol.*, 40, 131, 1978.
18. **Crick, J. and Brown, F.,** Viral subunits for rabies vaccination, *Nature*, 222, 92, 1969.
19. **Wiktor, T. J., György, E., Schlumberger, H. D., Sokol, F., and Koprowski, H.,** Antigenic properties of rabies virus components, *J. Immunol.*, 110, 269, 1973.
20. **Atanasiu, P., Tsiang, H., Perrin, P., Favre, S., and Sisman, J.,** Extraction d'un gène soluble (glycoprotéine) par le triton X100 à partir d'un vaccin antirabique de culture tissulaire de premier explant: résultats d'immunisation et pouvoir protecteur, *Ann. Microbiol.*; 125B, 539, 1974.
21. **Atanasiu, P., Tsiang, H., Perrin, P., and Favre, S.,** Analyse du pouvoir immunogénique et protecteur de la glycoprotéine rabique: comparaison de préparations purifiées par des techniques différentes et résultats, *Ann. Microbiol.*, 127B, 257, 1976.
22. **Cox, J. H., Dietzschold, B., and Schneider, L. G.,** Rabies virus glycoprotein. II. Biological and serological characterization, *Infect. Immun.*, 16, 754, 1977.
23. **Dietzschold, B., Wang, H., Rupprecht, C. E., Celis, E., Tollis, M., Ertl, H., Heber-Katz, E., and Koprowski, H.,** Induction of protective immunity against rabies by immunization with rabies virus ribonucleoprotein, *Proc. Natl. Acad. Sci. U.S.A.*, 84, 9165, 1987.
24. **Anilionis, A., Wunner, W. H., and Curtis, P. J.,** Structure of the glycoprotein gene in rabies virus, *Nature*, 294, 275, 1981.
25. **Lai, C. Y. and Dietzschold, B.,** Amino acid composition and terminal sequence analysis of the rabies virus glycoprotein: identification of the reading frame on the cDNA sequence, *Biochem. Biophys. Res. Commun.*, 103, 536, 1981.
26. **Dietzschold, B., Wiktor, T. J., MacFarlan, R., and Varrichio, A.,** Antigenic structure of rabies virus glycoprotein: ordering and immunological characterization of the large CNBr cleavage fragments, *J. Virol.*, 44, 595, 1982.
27. **Panicali, D. and Paoletti, E.,** Construction of poxviruses as cloning vectors: insertion of the thymidine kinase gene from herpes simplex virus into the DNA of infectious vaccinia virus, *Proc. Natl. Acad. Sci. U.S.A.*, 79, 4927, 1982.

28. **Smith, G. L., Mackett, M., and Moss, B.,** Recombinant vaccinia viruses as new live vaccines, *Biotechnol. Genet. Eng. Rev.*, 2, 383, 1984.
29. **Smith, G. L. and Moss, B.,** Vaccinia virus expression vectors: construction, properties and applications, *Biotechniques,* 2, 5, 306.
30. **Kieny, M.-P, Lathe, R., Drillien, R., Spehner, S., Skory, D., Schmitt, T., Wiktor, T., Koprowski, H., and Lecocq, J.-P.,** Expression of rabies virus glycoprotein from a recombinant vaccinia virus, *Nature,* 312, 163, 1984.
31. **Kieny, M.-P., Desmettre, P., Soulebot, J. P., and Lathe, R.,** Rabies vaccine: traditional and novel approaches, *Prog. Vet. Microbiol. Immunol.*, 3, 73, 1987.
32. **Blancou, J., Kieny, M.-P., Lathe, R., Lecocq, J.-P, Pastoret, P.-P., Soulebot, J. P., and Desmettre, J. P.,** Oral vaccination of the fox against rabies using a live recombinant vaccinia virus, *Nature,* 322, 373, 1986.
33. **Wiktor, T. J., Macfarlan, R. I., Reagan, K., Dietzschold, B., Curtis, P., Wunner, W. H., Kieny, M.-P., Lathe, R., Lecocq, J.-P., Mackett, M., Moss, B., and Koprowski, H.,** Protection from rabies by a vaccinia virus recombinant containing the rabies virus glycoprotein gene, *Proc. Natl. Acad. Sci. U.S.A.*, 81, 7194, 1984.
34. **Wiktor, T. J., Kieny, M.-P., and Lathe, R.,** New generation of rabies vaccine. Vaccinia-rabies glycoprotein recombinant virus, in *Applied Virology Research,* Vol. 1, Kurstak, E., Ed., Plenum Press, New York, 1988, 69.
35. **Smith, J. S., Yager, P. A., and Baer, G. M.,** Epreuve rapide en culture tissulaire pour la détermination de l'anticorps neutralisant, in *La Rage, Techniques de Laboratoire,* 3rd ed., Kaplan, M. M. and Koprowski, H., Eds., Organisation Mondiale de la Santé, Geneva, 1974, 354.
36. **Koprowski, H.,** Epreuve d'inoculation à la souris, in *La Rage, Techniques de Laboratoire,* 3rd ed., Kaplan, M. M. and Koprowski, H., Eds., Organisation Mondiale de la Santé, Geneva, 1974, 88.
37. **Blancou, J., Aubert, M. F. A., Andral, L., and Artois, M.,** Rage expérimentale du Renard roux (*Vulpes vulpes*) L.). I. Sensibilité selon la voie d'infection et la dose infectante, *Rev. Med. Vet.*, 130, 1001, 1979.
38. **Blancou, J., Artois, M., Brochier, B., Thomas, I., Pastoret, P.-P., Desmettre, P., Languet, B., and Kieny, M.-P.,** Innocuité et efficacité du virus recombinant vaccine-rage administré par voie orale chez le renard, le chien et le chat, *Ann. Rech. Vet.*, 20, 195, 1989.
39. **Desmettre, P., Languet, B., Chappuis, G., Brochier, B., Thomas, I., Lecocq, J.-P., Kieny, M.-P., Blancou, J., Aubert, M. F. A., Artois, M., and Pastoret, P.-P.,** Use of a vaccinia rabies recombinant for oral vaccination of rabies vectors, *Vet. Microbiol.*, 23, 227, 1990.
40. **Tolson, N. D., Charlton, K. M., Casey, G. A., Knowles, M. K., Rupprecht, C. E., Lawson, K. F., and Campbell, J. B.,** Immunization of foxes against rabies with a vaccinia recombinant virus expressing the rabies glycoprotein, *Arch. Virol.*, 102, 297, 1988.
41. **Brochier, B., Languet, B., Blancou, J., Kieny, M.-P., Lecocq, J.-P., Costy, F., Desmettre, P., and Pastoret, P.-P.,** Use of a recombinant vaccinia-rabies virus for oral vaccination of fox cubs (*Vulpes vulpes,* L.) against rabies, *Vet. Microbiol.*, 18, 103, 1988.
42. **Brochier, B. M., Languet, B., Artois, M., Zanker, S., Guittre, C., Blancou, J., Chappuis, G., Desmettre, P., and Pastoret, P.-P.,** Efficacy of a baiting system for fox vaccination against rabies with vaccinia-rabies recombinant virus, *Vet. Rec.*, 127, 165, 1990.
43. **Thomas, I., Brochier, B., Languet, B., Blancou, J., Peharpre, D., Kieny, M.-P., Desmettre, P., Chappuis, G., and Pastoret, P.-P.,** Primary multiplication site of the vaccinia-rabies glycoprotein recombinant virus administered to foxes by the oral route, *J. Gen. Virol.*, 71, 37, 1990.

44. **Rupprecht, C. E., Hamir, A. N., Johnston, D. H., and Koprowski, H.,** Efficacy of a vaccinia-rabies glycoprotein recombinant virus vaccine in raccoons *(Procyon lotor), Rev. Infect. Dis.,* 10(S), 803, 1988.
45. **Hanlon, C. A., Ziemer, E. L., Hamir, A. N., and Rupprecht, C. E.,** Cerebrospinal fluid analysis of rabid and vaccinia-rabies glycoprotein recombinant, orally vaccinated raccoons *(Procyon lotor), Am. J. Vet. Res.,* 50, 364, 1989.
46. **Brochier, B., Blancou, J., Aubert, M. F. A., Kieny, M.-P, Desmettre, P., and Pastoret, P.-P.,** Interaction between rabies infection and oral administration of vaccinia-rabies recombinant virus to foxes *(Vulpes vulpes), J. Gen. Virol.,* 70, 1601, 1989.
47. **Rupprecht, C. E., Wiktor, T. J., Johnston, D. H., Hamir, A. N., Dietzschold, B., Wunner, W. H., Glyckman, L. T., and Koprowski, H.,** Oral immunization and protection of raccoons (*Procyon lotor*) with a vaccinia-rabies glycoprotein recombinant virus vaccine, *Proc. Natl. Acad. Sci. U.S.A.,* 83, 7947, 1986.
48. **Tolson, N. D., Charlton, K. M., Stewart, R. B., Campbell, J. B., and Wiktor, T. J.,** Immune response in skunks to a vaccinia virus recombinant expressing the rabies virus glycoprotein, *Can. J. Vet. Res.,* 51, 363, 1987.
49. **Wiktor, T. J., Macfarlan, B., Dietzchold, B., Rupprecht, C., and Wunner, W. H.,** Immunogenic properties of vaccinia recombinant virus expressing the rabies glycoprotein, *Ann. Virol. (Inst. Pasteur),* 136E, 405, 1985.
50. **Soria Baltazar, R., Blancou, J., and Artois, M.,** Résultats de l'administration par voie orale au mouton de deux vaccins contenant un virus de la rage modifié (SAD B19) ou un recombinant du virus de la vaccine et de la rage (187 XP), *Ann. Med. Vet.,* 131, 481, 1987.
51. **Brochier, B., Languet, B., Blancou, J., Thomas, I., Kieny, M.-P., Lecocq, J.-P., Desmettre, P., and Pastoret, P.-P.,** Innocuité du virus recombinant vaccine-rage chez quelques espèces non-cibles, in *Vaccination to Control Rabies in Foxes,* Pastoret, P.-P., Brochier, B., Thomas, I., and Blancou, J., Eds., Office for Official Publications of the European Communities, Brussels-Luxembourg, 1988, 118.
52. **Brochier, B., Languet, B., Blancou, J., Thomas, I., Kieny, M.-P., Costy, F., Desmettre, P., and Pastoret, P.-P.,** Use of recombinant vaccinia-rabies virus for oral vaccination of wildlife against rabies: innocuity to several non-target bait consuming species, *J. Wildl. Dis.,* 25, 540, 1989.
53. **Fenner, F.,** Poxviruses, in *Virology,* Fields, B. N., Knipe, D. M., Chanock, R. M., Melwick, J. L., Roizman, B., and Shope, R. E., Eds., Raven Press, New York, 1985, 661.
54. **Buller, R. M. L., Smith, G. L., Cremer, K., Notkins, A. L., and Moss, B.,** Decreased virulence of recombinant vaccinia virus expression vectors is associated with a thymidine kinase-negative phenotype, *Nature,* 317, 813, 1985.
55. **Pastoret, P.-P., Brochier, B., Languet, B., Thomas, I., Paquot, A., Bauduin, B., Kieny, M.-P., Lecocq, J.-P., De Bruyn, J., Costy, F., Antoine, H., and Desmettre, P.,** First field trial of fox vaccination against rabies using a vaccinia-rabies recombinant virus, *Vet. Rec.,* 123, 481, 1988.
56. **Brochier, B., Thomas, I., Bauduin, B., Leveau, T., Pastoret, P.-P., Languet, B., Chappuis, G., Desmettre, P., Blancou, J., and Artois, M.,** Use of a rabies-vaccinia recombinant virus for the oral vaccination of foxes against rabies, *Vaccine,* 8, 101, 1990.
57. **Sureau, P.,** personal communication.
58. **Dietzschold, B.,** personal communication.
59. **Artois, M.,** unpublished results.
60. **Thomas, I. et al.,** unpublished results.
61. **Brochier, B., Kieny, M.-P., Costy, F., Coffens, P., Bauduin, B., Lecocq, J.-P., Languet, B., Chappuis, G., Desmettre, P., Afiademomyo, K., Libois, R., and Pastoret, P.-P.,** Large-scale eradication of rabies using recombinant vaccinia-rabies vaccine, *Nature,* 354, 520, 1991.

Chapter 6

IMMUNOGENICITY AND ANTIGEN PRESENTATION

Marion E. Andrew, Barbara E. H. Coupar, and David B. Boyle

TABLE OF CONTENTS

I. INTRODUCTION

The technology to construct recombinant vaccinia viruses has opened up a new avenue to the production of safe and efficacious vaccines. The first antigens expressed from recombinant viruses were viral glycoproteins, such as rabies glycoprotein[1] or influenza hemagglutinin,[2] that were expressed on the surface of infected cells or hepatitis B surface antigen,[3] which formed particles indistinguishable from those formed in the sera of patients infected with hepatitis B virus. These antigens proved to be protective when the recombinant vaccinia viruses were used to immunize laboratory animals. However, it was realized quickly that recombinant viruses were useful tools for biological experiments, as a range of proteins with biological functions could be expressed and characterized in cells infected with recombinant vaccinia viruses. Recombinant vaccinia viruses allowed identification of the individual antigens that stimulate T or B cells and protective immunity. Also, they have been powerful tools in defining some of the requirements for antigen presentation and immunogenicity and it is those studies which form the topic of this chapter.

Vaccinia virus has a wide host range and recombinant viruses have been tested in a range of animal species, including mice, rabbits, monkeys,[4] cattle,[5] sheep,[6] raccoons,[7] and skunks.[8] This chapter will largely cover studies done in mice and rabbits and care should be taken in extrapolating results to other species, particularly those sections which deal with the route of inoculation and co-expression of cytokines.

II. VACCINIA VIRUS GENOTYPE AND PHENOTYPE

A. VACCINIA VIRUS STRAINS AND MUTANTS

Most recombinant vaccinia viruses have been constructed using the WR strain of virus; however, there have been recombinants made using the NYBH,[9]

Lister,[10] Copenhagen,[1] and LIVP[11] strains or attenuated mutants from the Lister[10] or WR[12] strains. In all cases, an immune response was elicited to the expressed foreign gene, but few direct comparative studies of immunogenicity of a single antigen expressed by different vaccinia virus strains have been done. In one study,[9] recombinant viruses which expressed the gp340 gene from Epstein-Barr virus were constructed using either the WR or NYBH strains of vaccinia virus. Cottontop tamarins were inoculated either intradermally or by skin scarification and challenged with a tumorigenic dose of Epstein-Barr virus. Protection against the development of tumors was observed in the animals inoculated with the WR recombinants, but not in animals inoculated with recombinants which were constructed using the NYBH strain. Antibody titers to vaccinia virus were higher in the animals that were inoculated with the WR recombinants than in the NYBH-inoculated animals. The WR strain was selected originally for mouse neurovirulence and may grow to higher titers than NYBH in the cottontop tamarins, as it does in mice.[13]

Recombinant vaccinia viruses which express rabies virus glycoprotein G have been constructed using Copenhagen[1] and NYBH[14] strains. Although direct comparisons have not been reported in the literature, it was concluded[14] that the recombinant viruses constructed with the Copenhagen strain were more immunogenic than the recombinant viruses constructed using the NYBH. However, all of the recombinants protected mice against a lethal challenge with rabies virus.

Shida et al.[10] used Lister strain and LC16mO, a strain selected from Lister for lower neurovirulence, to construct recombinant viruses which expressed the *env* gene from human T-cell leukemia virus. The *env* gene was inserted into the vaccinia virus hemagglutinin (HA) gene and both recombinant viruses were attenuated, compared to their parent strains, which resulted in lower anti-vaccinia antibody titers in animals vaccinated with the recombinant viruses. However, both the vaccinia-specific and human T-cell leukemia virus-specific antibody responses were higher in the rabbits which received the recombinant constructed with LC16mO than in the rabbits which received the Lister-based recombinant virus.

B. THYMIDINE KINASE PHENOTYPE

Many recombinant vaccinia viruses have been constructed by insertion of a foreign gene into the thymidine kinase (TK) gene of vaccinia virus,[15] resulting in recombinant viruses which are phenotypically TK negative (TK^-). Although TK^- viruses replicate well in continuous cell lines, their replication *in vivo* is reduced,[16,17] perhaps due to infection of cells with low or no endogenous TK activity. Although Buller et al.[16] reported that they detected no differences in vaccinia virus-specific neutralizing antibody titers in mice immunized with TK^- recombinant viruses, compared to mice infected with parental TK^+ virus, we[17] have detected differences in immune responses to both vaccinia virus antigens and to a foreign gene product expressed from TK^+ and TK^- recombinant viruses. In our experience, antibody titers to vaccinia virus correlated with the TK phenotype of the immunizing viruses; that is, titers were highest in mice which were inoculated with VV-WR (TK^+), lowest in mice inoculated with VV-HA (TK^-) and intermediate in mice inoculated with VV-HA-TK (a virus made phenotypically TK^+ by the insertion of the herpes simplex virus [HSV] TK gene into the HindIII F region and expressing approximately 10% of the TK activity expressed by VV-WR[18]). Also, antibody titers to HA were higher in mice inoculated with VV-HA-TK, than in mice inoculated with VV-HA. The discrepancy in results between References 16 and 17 may be due to differences in route of inoculation used (intravenous and footpad[17] vs. intradermal scarification at the base of the tail[16]). Buller et al. reported that there were no differences in the titer of TK^- and TK^+ viruses at the site of inoculation, whereas in the footpad of mice, VV-WR replicated to higher titers than VV-HA-TK and titers of VV-HA were lower than those of VV-HA-TK. We have shown also that the TK phenotype has a similar effect on the stimulation of primary vaccinia-specific cytotoxic T-lymphocyte (CTL) responses,[17] but that there was no effect on the level of CTL memory.

As indicated above, the effect of TK phenotype on immunogenicity of recombinant viruses correlated with the degree of replication *in vivo*. As well as differences in virus replication in the footpad,[17] we have also observed differences in virus titer in the ovary[13] and larger lesions were observed in rabbits inoculated intradermally with TK^+ virus, compared to TK^- virus. On the whole, it would appear that the immunogenicity of an antigen expressed by a recombinant vaccinia virus is dependent upon the extent of virus replication. Olmsted et al.[19] observed that, in five monkey species, the size of the lesion at the site of vaccination correlated positively with the titer of antibody to the co-expressed respiratory syncytial virus (RSV) antigen. Consequently, attenuation of vaccinia virus recombinants may compromise their efficacy as vaccines.

III. HOST/VIRUS INTERACTIONS

A. ROUTE OF INOCULATION

During the smallpox eradication campaign, vaccination was performed by dermal scarification, often using a bifurcated needle. However, with experimental animals and recombinant vaccinia viruses, a variety of inoculation routes have been used. Mice have been inoculated intravenously or intraperitoneally and occasionally intranasally or intracerebrally, as well as by more peripheral routes of inoculation in the footpad or scarification at the base of the tail. The rabies-vaccinia recombinant viruses have been used for oral vaccination in wildlife species, such as raccoons,[7] foxes,[20] and skunks.[8] This route of inoculation has produced high serum antibody titers and protection against challenge with rabies virus.

The LD_{50} of vaccinia virus is dependent upon the route of inoculation used, being lower following intracerebral or intranasal routes than following other routes of inoculation. Immunogenicity is affected also by the route of inoculation. When intravenous, intraperitoneal, intranasal, footpad, and tail scarification inoculation were compared, antibody responses to both vaccinia virus antigens and to influenza HA were highest when mice were inoculated intravenously and lowest when they were inoculated intraperitoneally.[17] The other routes of inoculation gave intermediate titers.

When the same routes of inoculation were used to inoculate mice with VV-HA and primary, vaccinia virus-specific CTL assayed 6 d later, both intravenous and intraperitoneal inoculation resulted in high levels of CTL activity, particularly at a higher inoculation dose (10^7 plaque forming units [pfu]). When 10^4 pfu was inoculated, there was always a low, but detectable CTL response following intravenous inoculation, but after intraperitoneal inoculation with 10^4 pfu, some mice failed to develop detectable levels of CTL activity. The other routes of inoculation gave low CTL responses to 10^7 pfu and no response to 10^4 pfu. When secondary CTL responses were assayed by *in vitro* restimulation of spleen cells, all mice that received 10^7 pfu of virus, by any route, developed a high secondary response. In mice that received 10^4 pfu, only those that were inoculated intravenously had consistently high levels of secondary CTL. Mice that were inoculated intraperitoneally with 10^4 pfu virus sometimes had a high response, whereas mice inoculated by other routes always had low responses.

The route of inoculation also affects the generation of mucosal antibody responses. Small et al.[21] immunized mice either intranasally or intradermally with a recombinant virus expressing the influenza HA and then challenged the mice with influenza virus. Intransally immunized mice were protected from both pneumonia (lower respiratory tract infection) and nasal infection (upper respiratory tract), whereas the intradermally immunized mice were protected only from pneumonia. Protection of the upper respiratory tract by intranasal immunization correlated with the generation of specific mucosal IgA, whereas intradermal immunization generated serum IgG responses and no mucosal IgA response.

In a similar study, Murphy et al.[22] examined the effect of route of inoculation in overcoming the suppressive effects of passively acquired antibody. The antibody response to RSV antigens following intradermal inoculation with a recombinant vaccinia-RSV virus was suppressed in cotton rats that had previously received antibody specific for RSV. However, when the recombinant virus was administered intranasally, the suppression was not apparent.

B. IMMUNIZATION OF ANIMALS WITH EXISTING IMMUNITY

1. Active Immunity

The widespread use of vaccinia virus during the Smallpox Eradication Programme has provided a body of data on the risks associated with its use. That experience provides background data on which the potential users of recombinant vaccinia virus vaccines can draw. However, it has meant that there is a proportion of the population with immunity to vaccinia virus. As vaccination ceased over a decade ago, the seropositive proportion of the human population is declining, as are the antibody titers in those who were vaccinated. However, if a recombinant vaccinia virus is used as a vaccine, then the levels of immunity to vaccinia virus within the community will increase. Although recombinant vaccinia viruses which express multiple vaccine antigens can be constructed,[23,24] revaccination with a different recombinant vaccinia virus may be required at some point in time. For these reasons, it is important to know how existing immunity to vaccinia virus will affect the immunogenicity of a novel antigen expressed by a recombinant vaccinia virus vaccine.

In our experiments, only some mice which were immunized with vaccinia virus and reimmunized 4 weeks later with a recombinant vaccinia virus expressing influenza HA, had detectable titers of antibody specific for HA.[25] However, titers were lower than in mice that did not have prior exposure to vaccinia virus. The group of mice with anti-vaccinia titers before immunization with the recombinant virus, VV-HA, also had higher mortality and morbidity upon challenge with influenza virus. The titer of HA-specific antibody and the level of protection correlated inversely with the anti-vaccinia titer at the time of immunization with VV-HA. This indicates that as immunity to the viral vector wanes, the possibility of inducing protective immunity with a new recombinant virus will increase.

This is supported by a report of an accidental inoculation of a laboratory worker with a recombinant vaccinia virus expressing vesicular stomatitis virus G antigen.[26] The person had been vaccinated 30 years prior to the exposure and after inoculation with the recombinant virus, developed antibody titers to vesicular stomatitis virus that were indistinguishable from those in naive mice inoculated with the recombinant virus.

Similar results have been reported by Rooney et al.[27] using recombinant viruses that express glycoprotein D from HSV-1. They found that existing immunity to a recombinant vaccinia virus expressing influenza HA, decreased the levels of neutralizing antibody and protective immunity to HSV-1 which were stimulated by subsequent immunization with a recombinant vaccinia virus expressing glycoprotein D. However, as has been found in other systems, secondary inoculation with the recombinant HSV-vaccinia virus resulted in a significant boost to the HSV-specific antibody response. This may indicate that the level of viral replication required for the stimulation of a secondary response is lower than is required to stimulate a primary response.

Although there is no evidence that foreign gene products expressed from recombinant vaccinia viruses are incorporated into the virion itself, it is possible that existing immunity to the pathogen from which the foreign antigen is derived will decrease the immunogenicity of a recombinant vaccinia virus. This question was elegantly addressed by Flexner et al.,[28] who inoculated mice, previously infected with either influenza or HSV, with a recombinant

TABLE 1
Effect of Existing Immunity on VV-HA Antibody and Virus Titers

Primary immunization	Antibody titer[a] Vaccinia	Antibody titer[a] Influenza	Virus[b] titer (pfu)
VV-WR	12570	<10	<2.00
Influenza virus	466	320	2.63 ± 0.59
NIL	466	53	2.24 ± 0.62

[a] 4 Weeks after the indicated primary immunization, mice were reinoculated with 10^7 pfu VV-HA intravenously and bled 2 weeks later. Sera were assayed for vaccinia virus-specific antibodies by ELISA and for influenza virus-specific antibodies by hemagglutination inhibition.

[b] 3 Weeks after the indicated primary immunization, mice were reinoculated with 10^7 pfu VV-HA into a hind footpad and the foot removed 2 days later for assay of virus.

virus expressing antigens from both herpes and influenza viruses and showed that the protective immunity to herpes in the influenza plus recombinant virus-inoculated mice (and vice versa) was not decreased. We have similar results (Table 1) that show that existing influenza immunity does not decrease the ability of VV-HA to replicate mice, nor does it affect the level of primary antibody to vaccinia virus.

2. Passive Immunity

Passive transfer of immune serum leads to different results. When Johnson et al.[29] transferred serum from influenza-immune mice and inoculated the recipient mice with a recombinant virus expressing influenza HA, the antibody response to HA was depressed. Similarly, Murphy et al.[30] demonstrated that passive transfer of RSV-immune serum to cotton rats suppressed RSV-specific antibody titers following inoculation of the recipient animals with recombinant vaccinia virus which expressed either RSV F or G antigens. These results suggest that the immune regulation following passive transfer of immune serum differs considerably from that in an animal that has recovered from infection. The complexity of immune recognition of vaccinia virus-expressed antigens in immune animals is illustrated in a study in which rabies-infected foxes were inoculated with a vaccinia-rabies recombinant virus on days 0, 3, and 14 after infection with the rabies virus.[31] The foxes that received the recombinant vaccinia virus on days 0 or 3 died from rabies after a shorter incubation period than unvaccinated controls, whereas the foxes that received the recombinant vaccinia virus on day 14 died later than control animals.

3. Clearance of Recombinant Virus by Activated CTL

Infection with influenza virus some weeks before inoculation with an influenza HA-expressing recombinant virus does not reduce replication of the recombinant virus (Table 1). In contrast, mice inoculated with influenza virus and challenged intracerebrally four days later with a lethal dose of recombinant vaccinia virus expressing either HA or nucleoprotein (NP) from influenza virus, were protected against death from vaccinia virus.[32] The protection appears to be mediated by T-cells as it was abrogated by treatment of the mice with a monoclonal antibody to Thy-1. Anti-influenza antibodies appear not to be involved, as mice were protected against VV-NP and NP is not displayed on the surface of infected cells and is not a target of virus-neutralizing antibodies. Protection was apparent only when the mice were immunized with influenza virus a few days prior to inoculation with the recombinant vaccinia virus, perhaps indicating a requirement for activated CTL. Replication of VV-HA and VV-NP in the ovaries of mice is also controlled by infection of the mice a few days earlier with influenza virus. Because of the high titers to which wild-type and recombinant vaccinia viruses replicate in murine ovaries, T-cell responses to antigens of interest can be rapidly detected by infection of mice with the virus from which the antigen was derived, intravenous or intraperitoneal challenge with the appropriate vaccinia virus recombinant, and then assay of the level of vaccinia virus replication in the ovaries.

IV. VACCINIA VIRUS PROMOTERS

A. LEVEL OF EXPRESSION AND ANTIBODY TITERS

Vaccinia virus recombinants are most commonly constructed using the vaccinia virus promoter, P7.5, derived from a 7.5-kDa protein gene which maps to the inverted terminal repeats, and which has both early and late functions.[33] However, a number of other promoters, including ones from other orthopoxviruses and from nonorthopoxviruses, are available for the expression of foreign genes from recombinant vaccinia viruses. Using recombinant vaccinia viruses that expressed influenza HA under the control of P7.5 or PF (an early promoter from the HindIII F region) or PL11 (which regulates the late expression of an 11-kDa gene from the HindIII F region), we demonstrated[34] that the level of influenza HA expressed under the control of PF was significantly lower than that expressed under the control of the other two promoters. The antibody responses to HA stimulated by each of the 3 recombinant viruses were not dramatically different, despite the fact that there was approximately 3-fold less HA expressed under the control of PF.

Other workers have demonstrated that the level of expression of an antigen from a recombinant vaccinia virus is affected by the promoter used to control expression of the antigen.[14,35] Esposito et al.[14] reported that rabies virus G expression was 6- to 10-fold higher when under the control of PL11, compared to P7.5, but that antibody levels and protection were not affected by the lower levels of G expressed under the control of P7.5. Taken together with the data on the effect of virus dose and TK phenotype on antibody levels,[17] these data suggest that the extent of viral replication is more important to the immunogenicity of a recombinant vaccinia virus-expressed antigen than the level of antigen expressed in an infected cell. However, it is possible that very much higher levels of expression, e.g., 100- to 1000-fold or greater, may overcome the influences of route of inoculation or TK^- phenotype. High-expressing promoters such as those described by Davison and Moss[36] may achieve such effects.

B. T-CELL RECOGNITION OF LATE-EXPRESSED ANTIGENS

Whereas neither the level of antigen, nor the temporal regulation, of antigen expression (i.e., whether an early or late promoter is used), appears to affect antibody titers, CTL activity was affected by the temporal regulation of antigen expression by vaccinia virus promoters. When our three recombinant vaccinia viruses which expressed influenza HA under the control of PF, P7.5, or PL11 were used to infect target cells for recognition by influenza virus-specific CTL, late-expressed HA was not recognized.[34] Similarly, late-expressed HA could not prime CTL, nor could it restimulate influenza-immune splenocytes in CBA/H mice. In BALB/c mice, all 3 recombinant viruses could prime mice for a CTL response to HA, but late-expressed HA failed to restimulate influenza virus-immunized mice. These results suggest that, although HA expressed late in vaccinia virus replication can be detected on the surface of the infected cell by antibody and can stimulate an antibody response, it may be unable to form peptides that associate with class I major histocompatibility complex (MHC) antigens for CTL recognition. As vaccinia virus inhibits host protein synthesis, a protein which is required for the processing of antigen and/or production of functional MHC/peptide complex may be decreased or absent late in virus infection. Likely candidates include proteases for peptide formation, proteins required for trafficking of peptides, or class I MHC antigens themselves. Alternatively, vaccinia virus-encoded protease inhibitors[37,38] may block antigen-processing events required for MHC/peptide association late in infection.

In further work in collaboration with Townsend et al.,[39] using cloned CTL specific for NP epitopes, it was shown that the defect can be selective; that is, CTL specific for amino acids 50 to 63 could recognize late-expressed NP, whereas CTL specific for amino acids 365 to 379 could not recognize late-expressed NP. The same paper reports that the defect in late presentation could be overcome by increasing the rate of degradation of the antigen. This was achieved by using an HA gene that had its signal sequence deleted, by expressing a COOH-terminal fragment of NP or by fusing the NP gene to a coding sequence for ubiquitin.

The amount of functional MHC/peptide complex which reaches the cell surface will depend on the intracellular concentrations of both the peptide and class I MHC. As a vaccinia virus-expressed class I MHC antigen ($H\text{-}2K^d$) can present antigens that are expressed from the same recombinant virus,[40] viruses that co-express both influenza antigens and H-2K^d late in virus infection should test whether co-expression of class I MHC will overcome the defect. Use of early and late promoters for the expression of foreign antigens has provided a technique for partial elucidation of the mechanism of processing of antigens for recognition by T-cells. Further examples of the use of recombinant vaccinia viruses in the study of antigen processing are described in Sections V and VI.

In another example of deficient T-cell recognition of late-expressed antigen, Wachsman et al.[41] compared late- and early-expressed HSV glycoprotein D. Antibody levels to early and late glycoprotein D expressed under the control of early or late promoters were indistinguishable, but delayed-type hypersensitivity and T-cell proliferation to late antigen were reduced. More significantly, the protective efficacy of the recombinant vaccinia virus which expressed glycoprotein D under the control of the late promoter, PL11, was dramatically reduced, compared to the recombinant virus which expressed glycoprotein D under the control of the early plus late promoter, P7.5.

These defects in recognition of antigens expressed under the control of late vaccinia virus promoters indicate that late promoters are unsuitable for use in recombinant vaccinia viruses which are to be used as vaccines, unless a T-cell response is not required. Also, viruses constructed for the purpose of examining T-cell epitopes would be better constructed with candidate antigens expressed under the control of a promoter which has early function.

P7.5 has both early and late functions and has been widely used in the construction of recombinant viruses. Although studies with cytosine arabinoside, which prevents DNA synthesis and therefore late promoter function, show that approximately 50% of antigen under the control of P7.5 is expressed early in virus infection,[15] the presence of a poxvirus early transcription termination signal in the foreign gene can prevent early protein expression. Earl et al.[42] have shown that the poxvirus early termination sequence, TTTTTNT (T_5NT), in the human immunodeficiency virus *env* gene prevented early expression of *env* under the control of P7.5. When site-directed mutagenesis was used to change the T_5NT sequence (TTT to TTC, both of which code for phenylalanine), early expression of *env* was restored. These studies suggest that any failure to detect T-cell responses to the product of a recombinant vaccinia virus-encoded gene which contains T_5NT, may be due to lack of early-expressed antigen, rather than lack of inherent immunogenicity of the antigen.

In the study of T_5NT effects on *env* expression by Earl et al.,[42] T-cell responses were not reported, but, unlike the earlier examples,[34,41] antibody levels were increased when the termination signal was removed. It may be that recognition of late-expressed antigen by the helper subset of T-cells is affected also in some cases and may explain why the antibody responses to some late-expressed antigens are depressed. Some of our unpublished results show that recognition of late antigens by CD4$^+$ (presumptively class II MHC-

TABLE 2
T-Cell Responses to Late- and Early-Expressed Antigens

Antigen	Promoter	Recognition by CTL: *In vivo*[a]	*In vitro*[b]	On targets[c]	Proliferation[d]
HA[e]	Early	+	+	+	+
	Late	SD	−	−	−
HAFMDV[f]	Early	+	+	+	+
	Late	+	−	−	−
L⁻HA[g]	Early	+	−	+	+/−
	Late	+	+/−	+/−	−
NP[h]	Early	+	+	+	+
	Late	+/−	+/−	Epitope dependent	+

SD, Strain dependent; BALB/c +, CBA/H −.

[a] Ability of the indicated antigen to prime mice for *in vitro* stimulation of CTL with influenza virus.
[b] Ability of the indicated antigen to stimulate CTL *in vitro* from the spleens of influenza virus-primed mice.
[c] Ability of the indicated antigen to sensitize target cells for lysis by influenza virus-specific CTL.
[d] Ability of the indicated antigen to stimulate proliferation of influenza virus-specific $CD4^+$ T-cells.
[e] Native influenza virus HA; anchored on the cell surface.
[f] Denatured HA; remains intracellular, presumably remaining in the endoplasmic reticulum.
[g] Signal sequence-deleted HA; remains in the cytoplasm.
[h] Native influenza virus NP.

restricted) T-cells was, in some cases, depressed or absent. The pattern of recognition of a panel of antigens by class II-restricted T-cells differed from the pattern of recognition by class I-restricted T-cells of the same antigens (Table 2). This is consistent with the idea that the density of functional MHC/peptide complex on the surface of an infected cell is highly dependent upon the antigen itself and the stability and location within the cell of the peptides which are derived from the antigen.

Table 2 also illustrates that the defect in recognition of late-expressed antigens varies according to the form of antigen (i.e., native or denatured) and with the assay system. Epitope and mouse strain-dependent responsiveness has been demonstrated for only some antigens. However, a more comprehensive analysis may reveal mouse strain and epitope-dependent response for other, perhaps all, antigens.

V. EXPRESSION OF MHC ANTIGENS

A. ALLORECOGNITION

To date there have been no reports of *in vivo* or *in vitro* primary CTL responses to viral antigens expressed from recombinant vaccinia viruses. This is likely to be due to the low precursor frequency of CTL to a single viral antigen (e.g., 1:10,000 NP-specific CTL in influenza virus-primed mice[43]). However, the precursor frequency for alloreactive CTL, that is, CTL that recognize and lyse cells bearing foreign MHC antigen, is much higher (e.g., 1 in 1438 for H-2K^d-specific CTL in unprimed C57BL/6 mice[44]). The nature of allorecognition by T-cells has been a controversial area of cellular immunology over the years. Even since the identification and cloning of the T-cell receptor, it is not clear how the same T-cells can recognize a peptide from a foreign protein in association with self class I MHC antigen or a foreign MHC antigen, most likely in association with a self peptide.

We constructed recombinant vaccinia viruses which expressed the murine class I MHC antigen, H-2K^d, either alone or co-expressed with influenza virus antigens, NP or HA. Despite the higher frequency of alloreactive CTL normally generated by *in vitro* stimulation of lymphocytes with allogeneic spleen cells, inoculation of CBA/H (H-2^k) mice with a recombinant virus expressing H-2K^d generated very low levels of recognition of H-2K^d target cells, even after *in vitro* expansion of the CTL for 3 days in the presence of the T-cell growth factor, interleukin 2 (IL-2).[45] Possible reasons for the low stimulation of H-2K^d-specific CTL by vaccinia virus-expressed H-2K^d include *in vivo* suppressive mechanisms and low levels of antigen expression by the cells infected with the recombinant vaccinia virus. As the vaccinia virus-expressed H-2K^d also stimulated only low levels of CTL *in vitro* and was poorly recognized as an alloantigen on the surface of target cells infected with the recombinant virus, *in vivo* suppression is an unlikely reason. When the amount of H-2K^d on the surface of cells infected with the recombinant virus was compared to the level of expression on P815 mastocytoma cells, it was found that the vaccinia-expressed antigen was 10-fold lower.

A possible explanation for the difference in levels of alloreactivity stimulated by endogenous vs. vaccinia virus-expressed H-2K^d may lie in the affinities of the CTL which were selected. If cells within the spleen population express relatively high levels of class I MHC on their surface, they will be recognized by a few precursors with high-affinity receptors for that antigen plus many more precursors T-cells with receptors having a much lower affinity. When the total population is assayed using target cells that express high levels of MHC, high- and low-affinity CTL will recognize the antigen; however, when target cells expressing lower levels of MHC (e.g., recombinant vaccinia virus-infected target cells) are used, only the CTL which bear high-affinity receptors will recognize the target cells. Similarly, recombinant vaccinia virus-infected cells, either *in vivo* or *in vitro*, will stimulate only the high-affinity CTL.

B. MHC-RESTRICTED RECOGNITION OF ANTIGENS

Using the same recombinant vaccinia virus which was used for the work on allorecognition, restriction of vaccinia virus antigens by the vaccinia virus-expressed $H\text{-}2K^d$ was tested. When the recombinant virus was used either to stimulate $H\text{-}2^d$ vaccinia virus-immune splenocytes or to infect target cells, the vaccinia virus antigens were recognized in association with the vaccinia virus-expressed $H\text{-}2K^d$ as effectively as in association with $H\text{-}2K^d$ expressed from a cellular genome. In this case, it would appear that the low level of expression of $H\text{-}2K^d$ did not affect its recognition as part of a peptide/MHC complex. This would suggest that the mean affinity of T-cell receptors that recognize peptide/self MHC complex is higher than those that recognize alloantigen. One could postulate that during normal virus infection, virus-derived peptides compete with self peptides, so that the density of any one peptide/MHC complex on the surface of a cell would be low and only CTL with relatively high-affinity receptors would be stimulated. One would also need to postulate that synthesis of $H\text{-}2K^d$ and vaccinia antigens from the same recombinant genome allows some preferential association of those antigens, to the exclusion, to at least some extent, of self peptides. That this may be so is supported by results with recombinant vaccinia viruses that co-express influenza NP and $H\text{-}2K^d$.[40]

Target cells infected with two recombinant vaccinia viruses, one expressing NP and the other $H\text{-}2K^d$, were not lysed by $H\text{-}2^d$-restricted CTL specific for NP, whereas target cells infected by a recombinant virus which expressed both NP and $H\text{-}2K^d$ were lysed. That is, there was better recognition of the antigen when the MHC antigen was expressed from the same genome. Close proximity of the synthesis of NP and $H\text{-}2K^d$ may be required, though the mechanism of peptide association with MHC is unknown and it is difficult to envisage how a cytoplasmic and endoplasmic reticulum-enclosed protein can associate.

These experiments indirectly support the notion that the defect in recognition of late-expressed antigens is due to vaccinia virus-induced shut down of host MHC antigen synthesis. Peptides appear to preferentially associate with newly synthesized MHC antigens[46] and the level of both MHC and peptide within the appropriate intracelluar compartment or locale may be critical. This hypothesis is supported further by the data in Table 3, which show that recognition of late-expressed $H\text{-}2K^d$ (under the control of PL11) by either vaccinia virus-specific BALB/c ($H\text{-}2K^d$) CTL or alloreactive CTL is lower than recognition of $H\text{-}2K^d$ under the control of P7.5. This occurs despite the fact that PL11 is a strong promoter, resulting in 6- to 10-fold greater expression levels of a foreign gene product.[14]

When similar experiments to the NP plus $H\text{-}2K^d$ experiments were performed using HA, rather than NP, it was found that HA could be recognized in association with vaccinia virus-expressed $H\text{-}2K^d$, whether the $H\text{-}2K^d$ was expressed from the same virus or a different virus to the HA.[40] HA and $H\text{-}2K^d$ are both cotranslationally inserted into the endoplasmic reticulum and travel via the Golgi to the cell surface, whereas NP remains in the cytosol. How this difference in cellular location affects the MHC/peptide interaction

TABLE 3
Recognition of a Late-Expressed MHC Antigen

CTL	% Specific lysis of L929 cells infected with: VV-KD/6[a]	VV-KD/8B[b]	VV-KD/11A[b]	VV-KD/11B[b]	VV-TK[c]
BALB anti-vaccinia (1:1)	60	31	10	21	7
CBA anti-BALB (30:1)	45	21	22	23	25

[a] Recombinant vaccinia virus expressing H-2K^d under the control of P7.5 (early + late).
[b] Recombinant vaccinia virus expressing H-2K^d under the control of PL11 (late).
[c] Nonrecombinant vaccinia virus.

is unknown. This result, together with the variation in T-cell recognition of different late-expressed viral antigens[39] (Table 2) is compatible with the idea that the level of peptide/MHC complex attained during infection will vary from protein to protein and may be dependent, not only upon the level of synthesis of the native protein, but also on the rate of degradation to produce the peptide, the stability of the peptide, the affinity of the peptide for the available MHC antigens, and, possibly the cellular location of the peptides.

VI. EXPRESSION OF RECOMBINANT DNA ENGINEERED GENES

A. DEFINITION OF ANTIGENIC EPITOPES

It was once believed that MHC class I-restricted CTL recognized whole proteins which were expressed at the cell surface; however, it is now known that CTL recognize peptides which are noncovalently associated with class I MHC antigens on the surface of the cell. Once it was realized that tertiary protein structure was not required for T-cell epitopes, then it was possible to express either truncated[47–49] or chimeric[50] genes from recombinant vaccinia viruses to define the epitopes which are recognized in association with a particular MHC allele. Recombinant vaccinia virus expression of truncated genes of influenza virus NP was used in this way to identify T-cell epitopes in NP.[39]

As B-cell eptitopes are more likely to be conformational, truncated genes are less likely to define B-cell epitopes. However, in some cases this technique can provide useful information. Olmsted et al.[51] expressed the N-terminal 230 amino acids of human RSV G protein (298 amino acids) from a recombinant vaccinia virus and detected G-specific antibodies and protection in cotton rats immunized with the recombinant virus. Highly conformational epitopes in viral antigens that are represented by more than one serotype may be mapped by the construction of chimeric genes. This technique was used by Keil and Wagner,[52] who constructed a series of chimeric genes between the vesicular stomatitis virus G-proteins of the Indiana and New Jersey strains. The chimeric proteins were expressed by recombinant vaccinia virus and then assayed by Western blot for recognition by a panel of monoclonal antibodies, enabling definition of the epitopes which were recognized by each of the monoclonal antibodies.

B. IMMUNOGENICITY AND ANTIGEN PRESENTATION OF VARIANT ANTIGENS

Recombinant vaccinia viruses have already been used to investigate a variety of questions in the area of immunogenicity and antigen presentation. They offer the advantage that they can be used both *in vivo* and *in vitro* and the technology to insert genes which have been altered in some way into a recombinant vaccinia virus is relatively quick. Following are examples of some of the ways that recombinant vaccinia viruses have been used to answer such questions.

1. Single-Amino Acid Changes

One of the earliest recombinant vaccinia viruses constructed expressed the G-glycoprotein from rabies virus. The first recombinant virus constructed by Wiktor et al.[1] produced a form of G-protein which was not transported to the surface of the infected cell and was not recognized by monoclonal antibodies that bind only to native antigen. This recombinant virus failed to protect animals against subsequent challenge with rabies virus. Analysis of the N-terminal amino acids revealed a leucine at position 8, rather than proline. When a new recombinant virus was constructed using a gene that coded for proline at position 8, the G-protein was transported to the cell surface, was highly immunogenic, and the recombinant virus stimulated good protective immunity. This is a good example of the profound changes in protein processing and immunogenicity which can result from a single amino acid change.

Single-amino acid changes can have profound affects on the host range and pathogenicity of influenza viruses. Growth of influenza viruses in embryonated chicken eggs selects for variants, as does growth in tissue culture cells. Rota et al.[53] expressed the HA from two antigenically different subpopulations of B/England/222/82 from recombinant vaccinia viruses. The influenza virus subpopulations had been grown either in eggs or in canine kidney cells and there was a single amino acid difference between their HAs. Experiments with the recombinant viruses demonstrated that, although the HA from the tissue culture-grown virus was less immunogenic in mice, both recombinant viruses provided protection against a lethal challenge with influenza virus. In this case, using recombinant vaccinia viruses avoided the problem that is often found with RNA viruses, which have a high mutation rate; that is, there is often a minor population of variants within any pool of virus, which may complicate results.

2. Peptide Expression

Peptide vaccines usually require conjugation of the peptide to a protein carrier and the use of an adjuvant to obtain good immune responses. Lew et al.[54] compared the ability of vaccinia virus to deliver a malaria peptide antigen with delivery of the peptide in several adjuvant systems. Although the titers of antibody in response to the peptide-carrier antigen in either complete Freund's or muramyl dipeptide adjuvant were higher than those to the vaccinia-expressed antigen, the recombinant-expressed antigen stimulated higher titers than peptide-carrier with alum adjuvant and also stimulated antibodies with an affinity that was approximately the same as the affinity of antibodies stimulated by peptide-carrier in muramyl dipeptide.

3. Truncated and Chimeric HA

As indicated in previous sections, recombinant vaccinia viruses have been used in a variety of ways to investigate the processing of antigens for recognition by T-cells. Following are two examples of altered HA genes that have been expressed in recombinant vaccinia viruses to investigate the requirements for the association of antigens with class I MHC antigens. In the first,[55] the signal sequence that is responsible for the insertion of influenza virus HA co-translationally into the endoplasmic reticulum was removed. The signal-deleted HA was not detected on the surface of the infected cell and its position on a polyacrylamide gel was consistent with an unglycosylated form, suggesting that it had remained within the cytoplasm. Despite these changes, the signal-deleted HA was as efficiently recognized as native HA by CTL which had been raised against influenza virus. This experiment indicated that not only were intracellular proteins recognized by CTL, but also that the location of the protein within the cell was not critical for recognition.

In the second example, we expressed from vaccinia virus, an HA gene which had the coding sequence for a foot-and-mouth disease virus VP1 antigenic peptide inserted within the HA-coding sequence.[56] When the chimeric HA-VP1 protein was expressed, no HA in native conformation could be detected. However, using an antiserum which recognizes denatured HA, a product was detected within infected cells. Not surprisingly, animals inoculated with the recombinant virus did not produce antibody which bound to whole influenza virus in an ELISA assay or which inhibited hemagglutination. However, HA-specific CTL were able to lyse target cells infected with the recombinant virus, indicating that native conformation is not required for recognition of antigen by CTL. Interestingly, although the foot-and-mouth disease virus VP1 peptide is highly immunogenic, especially in guinea pigs, we were unable to detect antibodies specific for foot-and-mouth disease virus when guinea pigs were inoculated with the recombinant virus (unpublished observations). It is possible that the VP1 epitope was not accessible within the chimeric HA or that proteolytic processing of the chimeric protein cleaved the epitope.

C. CELL SURFACE-EXPRESSED ANTIGENS

In terms of antibody response and protective immunity, the most successful antigens expressed from recombinant vaccinia viruses have been viral glycoproteins which are anchored in the cell surface membrane of the infected cell (e.g., rabies G, influenza HA, rinderpest virus HN and F). There have been few reports of parasite antigens successfully expressed from recombinant vaccinia viruses. Langford et al.[57] expressed a malarial antigen from a recombinant vaccinia virus. The antigen, a repetitive sequence from the S antigen, was secreted from the recombinant virus-infected cell and was poorly immunogenic in rabbits and mice. However, when the transmembrane domain from an immunoglobulin gene was engineered onto the S antigen gene to produce a cell surface-anchored version of the malaria antigen, and the altered gene expressed from a recombinant vaccinia virus, it stimulated high titers of antibodies in infected animals.

We have expressed several rotavirus antigens from recombinant vaccinia viruses. VP7 is the serotype specific antigen from the outer coat of rotavirus to which most virus-neutralizing antibodies are directed. VP7 is normally cotranslationally inserted into the endoplasmic reticulum, where it remains in both rotavirus and recombinant vaccinia virus-infected cells. Antibody responses were obtained in seronegative rabbits inoculated with VP7-expressing recombinant virus,[58] however in seronegative mice antibody responses were either low or undetectable. In an attempt to increase the immunogenicity of the vaccinia virus-expressed VP7, the gene was engineered by modification of the signal sequence and the addition of the membrane anchor region of the influenza HA and then inserted into vaccinia virus. The new cell surface-expressed VP7 stimulated higher titers of antibody in rabbits and readily detectable titers in mice.[59] The increased immunogenicity was not due merely to the extracellular location of the anchored VP7, as the immunogenicity of a secreted form of VP7 was not significantly increased, compared to the native intracellular VP7. The cell surface-expressed VP7 also stimulated T-cells, in a proliferation assay, more efficiently than the native intracellular VP7.[60]

Recently, we constructed recombinant vaccinia viruses which express either the native intracellular rotavirus VP6 or a cell surface-anchored VP6 and have recorded higher antibody responses in animals inoculated with the cell surface-anchored VP6, compared to the intracellular VP6.[60] T-cell responses to the cell surface VP6 have not yet been tested.

Similar results have been obtained with the bovine leukemia virus envelope protein expressed in vaccinia virus.[61] The envelope glycoprotein is processed into gp30, which remains membrane anchored, and gp51, which is noncovalently attached to gp30 on the outside of the infected cell. When the *env* gene (gp30 + gp51) was expressed in vaccinia virus, antibodies to gp51 were readily detected in animals inoculated with the recombinant virus. However, when a truncated form of the *env* gene expressing gp51, but not gp30, was inserted into vaccinia virus, gp51 remained intracellular and there was no antibody response to gp51. These results suggest that cell surface location of antigens may be a mechanism by which the immunogenicity of antigens can be increased and examples exist where the failure of recombinant vaccinia virus-expressed antigen to reach the surface of the infected cell has decreased the antibody response.[62,63] However, expression of an antigen on the surface of an infected cell does not guarantee stimulation of antibody. Expression of the gp70 of feline leukemia virus from a recombinant vaccinia virus resulted in an accumulation of gp70 on the surface of infected cells, but no antibody response was detected in cats or mice.[64]

The mechanism by which cell surface-anchored antigens increase antibody responses is not known. It may operate via increased helper T-cells or directly on B-cells, or may increase both B- and T-cell stimulation. There is no *a priori* reason why cell surface-anchored antigens should be more readily recognized by either B- or T-cells. Any direct stimulation of B cells may be due to a clustering of the antigen on the surface of the infected cell. This would facilitate multipoint binding of antigen to the B-cell, thus more easily allowing the cell to get over the threshold number of B-cell receptors that need to bind antigen to trigger activation of the B-cell. Similarly, for class II-restricted helper T-cell activation, clustering of antigen within a patch on the surface of an infected cell would facilitate uptake of multiple copies of the antigen, either by a phagocytic cell or a B-cell, which would then more efficiently act as an antigen-presenting cell with a sufficiently high density of MHC/peptide on its surface to trigger a helper T-cell.

VII. LYMPHOKINES

A. EFFECTS IN IMMUNOCOMPETENT ANIMALS

1. IL-2 and Antibody- and Cell-Mediated Immune Responses

Lymphokines are produced by T-cells and other cells of the immune system and act via specific receptors to stimulate differentiation and/or proliferation in their target cells. As most of their effects defined by *in vitro* assays appear to enhance the immune response, expression of lymphokines by viral vectors might be expected to increase the immune responses to antigens co-expressed by the virus.

IL-2 was defined first as a T-cell growth factor that was required for the culture and cloning of CTL. Receptors for IL-2 have since been documented on the surface of B-cells and IL-2 has been recognized as a potent inducer of natural killer (NK) and lymphokine-activated killer cells. Expression of IL-2 by a recombinant vaccinia virus did not increase antibody responses when animals were inoculated with the recombinant virus; in fact, antibody titers to either vaccinia antigens or to co-expressed influenza virus HA were lower than in mice inoculated with recombinant virus not expressing IL-2 (Table 4). These lower antibody responses may be due to the more rapid

TABLE 4
Antibody Titers in Outbred Nude and CBA/H Mice

Immunizing virus[a]	Nude mice		CBA/H	
	HAI	Anti-VV	HAI	Anti-VV
VV-HA-TK	<10–10[b]	951	320	3118
VV-HA-IL-2	<10	283	160	600

[a] Mice were inoculated with 10^6 pfu virus intravenously and bled 14 d later. Sera were tested for antibody specific for HA by hemagglutination inhibition (HAI) and for vaccinia virus-specific antibody by ELISA.

[b] Range of antibody titers in a group of 4 mice. Data for the remaining groups are presented as the geometric means of 4 antibody titers.

clearance of IL-2-expressing vaccinia virus from mice, compared to the clearance of virus that does not express IL-2.[13] CTL are thought to play a large role in the recovery of mice from vaccinia virus infection. However, the difference in clearance of IL-2 expressing vaccinia viruses, vs. control virus, from murine ovaries or lungs was observed by day 2 after infection, before it would be expected that virus-specific CTL could be expanded and recruited to the site of infection.

NK activity was higher in mice inoculated with IL-2-expressing vaccinia virus, than in mice which were inoculated with non-IL-2-expressing vaccinia virus.[65] The elevation in NK activity was apparent by day 1 after infection and lasted for approximately 4 d. In beige mice, which are deficient in NK cells, IL-2-expressing vaccinia virus was not cleared as quickly as it was from CBA/H mice, which have a normal complement of NK cells. However, in beige mice the IL-2-expressing virus still was cleared more quickly than non-IL-2-expressing vaccinia virus. As the CTL and antibody responses in beige mice were not elevated in response to IL-2-expressing vaccinia virus, this suggests that the IL-2-mediated virus clearance was acting partially via NK cells and partially via another, unidentified, mechanism.

2. IL-2 and Effects on Ir Gene Control

Ir gene defects in immune responsiveness result in the failure of some antigens to stimulate antibody or CTL responses in association with particular MHC alleles. It has been reported that administration of IL-2 together with antigen can overcome some Ir gene defects.[66] In the case of the malaria Resa antigen, no antibodies are stimulated in BALB/c mice and the Ir gene defect has been identified as a lack of helper T-cell stimulation for Resa-specific B-cells. When BALB/c mice were inoculated with a recombinant vaccinia virus that expressed both the Resa antigen and IL-2, the vaccinia virus-expressed IL-2 did not overcome the failure to stimulate helper T-cells and there was no stimulation of an antibody response.[67]

In the B10.A(2R) (H-2K^kD^b) strain of mouse, the CTL response to vaccinia virus antigens is restricted to the K^k allele and there is no response in association with D^b. There appears to be an *in vivo* suppression which is preventing the activation of D^b-restricted, vaccinia virus-specific CTL, as D^b-restricted precursor CTL are present and can be activated after transfer into irradiated mice or by stimulation *in vitro*. When B10.A(2R) mice were inoculated with an IL-2-expressing vaccinia virus, the vaccinia-expressed IL-2 did not overcome the Ir gene nonresponsiveness.[68]

3. Other Lymphokines

Other published results with recombinant vaccinia virus-expressed lymphokines include enhanced expression of MHC antigens *in vitro* by γ-interferon[69] and the failure of granulocyte-macrophage colony-stimulating-factor to resolve a murine Leishmania major infection.[70] In unpublished results, we have noted that vaccinia virus-expressed interleukin 4 (IL-4) did not affect titers of antibody, but lowered CTL activity and that the lower CTL activity did not correlate with more rapid clearance of virus. We also observed that there was considerable mortality in female mice inoculated with recombinant viruses that express IL-4, whereas there was no mortality or morbidity in mice that received the same dose of non-IL-4-expressing vaccinia virus. As this mortality occurred only in female mice, it probably was associated with the high level of virus replication that occurred in the ovaries of the mice.[13] However, the IL-4-expressing viruses did not replicate to higher titers than the control viruses, so mortality is likely to be due to a direct or indirect toxic effect of the IL-4.

Thus, although vaccinia virus-expressed IL-2 does increase nonspecific lytic activity, it does not appear to offer any advantage in the vaccination of normal animals to achieve an enhanced immune response; nor have there been reports of other lymphokines offering increases in vaccine efficacy.

B. EFFECTS IN IMMUNODEFICIENT MICE

1. IL-2 in Nude Mice

Nude mice have a genetic defect that prevents the development of the thymus (as well as other effects, including hairlessness) and therefore prevents the maturation of functional T-cells. Nude mice readily succumb to viral infections which are cleared from heterozygous littermates. Nude mice inoculated with vaccinia virus develop generalized vaccinia infection, which is lethal. However, nude mice inoculated with a recombinant vaccinia virus which expressed IL-2, cleared the virus within 9 to 12 d.[71,72] Antibody titers in the nude mice inoculated with the IL-2-expressing virus were lower than in the mice that received the control virus, as was observed in normal CBA/H mice (Table 4). Again, this probably reflects the more rapid clearance

TABLE 5
Effect of VV-HA-IL-2 on Clearance of VV-HA-TK

Virus in opposite foot	VV-HA-TK[a] mean titer[b]	Standard deviation (SD)
VV-HA-IL-2[c]	4.33	1.72
NIL	4.03	1.78

Virus[d]	No. mice	No. cleared	Mean titer[b]	SD
VV-HA-IL-2 + VV-HA-TK	13	2	2.99	1.27
VV-HA-TK	14	1	4.49	1.55
VV-HA-IL-2	18	10	2.45	1.10

[a] Nude mice were inoculated with 10^6 pfu VV-HA-TK into a hind footpad and the foot removed for virus assay 15 d later.
[b] Log_{10} plaque-forming units.
[c] 10^7 pfu VV-HA-IL-2 or no virus into the opposite hind footpad.
[d] Nude mice were inoculated with 10^7 pfu VV-HA-IL-2, 10^6 pfu VV-HA-TK, or a mixture of the two viruses, into a hind footpad and the foot removed 15 d later for virus assay.

of the IL-2-expressing virus. Although there have been reports of high doses of IL-2 stimulating functional CTL in nude mice, we detected no CTL activity in either infected or recovered nude mice.[73] However, NK activity was increased in mice that were inoculated with IL-2-expressing vaccinia virus[65] and may have accounted for the ability of the nude mice to clear the infection. Fifty percent of nude mice which have recovered from infection with a recombinant virus expressing IL-2 plus influenza virus HA, survive a challenge with a lethal dose of either vaccinia virus or influenza virus.[73]

We were not able to detect IL-2 in the serum of nude mice that had been inoculated with IL-2-expressing vaccinia virus and experiments with mixed infection of nude mice indicated that the vaccina virus-expressed IL-2 acted over very short distances (Table 5). When IL-2-expressing virus was inoculated into one hind footpad and control (non-IL-2-expressing) virus into the opposite footpad, the IL-2 had no effect on the clearance of the control virus. When a mixture of the two viruses was inoculated into the same footpad, the virus titers were reduced, compared to the titers in the footpads which were inoculated with only VV-HA-TK. The mean titer in the group that received the mixed inoculum was not significantly different from the mean titer in the group inoculated only with VV-HA-IL-2. However, although there was no statistical difference between the mean virus titers in the two groups of mice (that received VV-HA-IL-2 + VV-HA-TK or VV-HA-IL-2 alone), the proportion of mice that completely cleared virus was lower in the group that received VV-HA-TK + VV-HA-IL-2.

2. IL-2 in Sublethally Irradiated Mice

Sublethal irradiation of mice provides another model of immunodeficiency in which to test the effects of vaccinia virus-expressed IL-2. Sublethally irradiated mice, like nude mice, have difficulty in clearing vaccinia virus. Co-expression of IL-2 by the virus offered some, but not complete, protection from mortality from recombinant vaccinia virus.[74] No antibody or CTL responses were detected in mice that had recovered from infection with an IL-2-expressing vaccinia virus, but NK activity was elevated, particularly in the first few days after infection. As in the nude mice, there was some protection against a lethal influenza or vaccinia virus challenge in sublethally irradiated mice which had recovered from infection with an IL-2-expressing recombinant vaccinia virus.[74] Sublethally irradiated mice could survive a lower dose of non-IL-2-expressing vaccinia virus (10^5 pfu rather than 10^7 pfu). Therefore, it was possible to test whether mice that had recovered from 10^5 pfu VV-HA-IL-2 had better protective anamnestic immune responses than mice that had recovered from 10^5 pfu VV-HA-TK. When immunized mice were challenged with a lethal dose of either vaccinia or influenza virus, mice that had been immunized with either VV-HA-TK or VV-HA-IL-2 were equally protected, indicating that the IL-2 had not increased the protective immunogenicity of the viral antigens.[74]

These results with nude and sublethally irradiated mice demonstrate that although IL-2 does not increase the immunogenicity of an antigen expressed from a live viral vector, it may protect immunodeficient recipients of the vaccine from complications due to the viral vector.

3. Other Lymphokines

Immunodeficient (nude and sublethally irradiated) mice survive infection with recombinant vaccinia virus expressing γ-interferon and in normal mice γ-interferon-expressing virus is cleared from ovaries with a similar time course to IL-2-expressing virus.[75] However, vaccinia virus-expressed γ-interferon did not elevate levels of lysis by NK cells.[75] As in normal mice, IL-4 appeared to be toxic in immunodeficient mice; that is, both nude mice and sublethally irradiated mice died more quickly after inoculation of IL-4-expressing virus than after inoculation of a control virus.[76]

Recombinant viruses that express more than one lymphokine have been constructed in our laboratory and are currently being assessed for *in vivo* activity. Initial results suggest that co-expression of IL-2 decreased the toxic effects of IL-4, however, this is likely to be due to the more rapid clearance of the virus and therefore a lower dose of IL-4 delivered to the animal. Additive and synergistic effects of lymphokines have been reported in many *in vitro* assays for T- and B-cell function. Recombinant vaccinia viruses that express multiple lymphokines should aid in the extension of these studies into *in vivo* models. Although individual lymphokines have not yet been shown to be effective in increasing the immunogenicity of vaccine antigens expressed from recombinant viruses, it is possible that expression of multiple lymphokines may increase immunogenicity, particularly of poor immunogens.

VIII. CONCLUSIONS

Since the development of techniques to construct recombinant vaccinia viruses, their construction and use have spread to many research laboratories, worldwide. Undoubtedly, their major application has been in the identification of protective antigens and the construction of candidate vaccines. However, along the path to vaccine development, there have been revelations about how antigens can be optimally presented to the immune system. Some of these insights will be applicable to other vaccine strategies, as well as to recombinant vaccinia virus vaccines.

However, recombinant vaccinia viruses also have been used in basic research into immune recognition and regulation. Their relative ease and rapidity of construction, their ability to replicate *in vitro* and *in vivo* and their wide host range have made them powerful, cost-effective laboratory tools. From the knowledge derived from these experiments, we can acknowledge that immunity to an infectious agent is a delicately balanced response. The effectiveness of an immune response appears to be dependent upon the concentration or density of antigen (or antigen/MHC) which, in turn, is affected by the attenuation of the virus, the level of existing immunity, the level of MHC antigens, the cellular location of antigens, and their rate of degradation and the rate of clearance of the virus by early immune responses, including lymphokine-mediated mechanisms.

REFERENCES

1. **Wiktor, T. J., Macfarlan, R. K., Reagan, K. J., Dietzschold, B., Curtis, P. J., Wunner, W. H., Kieny, M.-P., Lathe, R., Lecocq, J.-P., Mackett, M., Moss, B., and Koprowski, H.,** Protection from rabies by a vaccinia virus recombinant containing the rabies virus glycoprotein gene, *Proc. Natl. Acad. Sci. U.S.A.*, 81, 7194, 1984.
2. **Bennink, J. R., Yewdell, J. W., Smith, G. L., Moller, C., and Moss, B.,** Recombinant vaccinia virus primes and stimulates influenza haemagglutinin-specific cytotoxic T cells, *Nature*, 311, 578, 1984.
3. **Smith, G. L., Mackett, M., and Moss, B.,** Infectious vaccinia virus recombinants that express hepatitis B virus surface antigen, *Nature*, 302, 490, 1983.
4. **Spriggs, M. K., Collins, P. L., Tierney, E., London, W. T., and Murphy, B. R.,** Immunization with vaccinia virus recombinants that express the surface glycoproteins of human parainfluenza virus type 3 (PIV3) protects patas monkeys against PIV3 infection, *J. Virol.*, 62, 1293, 1988.
5. **Wedman, E. E., Smith, A. W., and Oliver, R. E.,** Immunogenicity, pathogenicity, and transmissibility of a recombinant vaccinia virus in calves, *Am. J. Vet. Res.*, 49, 2018, 1988.
6. **Boyle, D. B., Coupar, B. E. H., Parsonson, I. M., Bagust, T. J., and Both, G. W.,** Responses of cattle, sheep and poultry to a recombinant vaccinia virus expressing a swine influenza haemagglutinin, *Res. Vet. Sci.*, 41, 40, 1986.

7. **Rupprecht, C. E., Wiktor, T. J., Johnston, D. H., Hamir, A. N., Dietzschold, B., Wunner, W. H., Glickman, L. T., and Koprowski, H.,** Oral immunization and protection of raccoons (*Procyon lotor*) with a vaccinia-rabies glycoprotein recombinant virus vaccine, *Proc. Natl. Acad. Sci. U.S.A.*, 83, 7947, 1986.
8. **Tolson, N. D., Charlton, K. M., Stewart, R. B., Campbell, J. B., and Wiktor, T. J.,** Immune response in skunks to a vaccinia virus recombinant expressing the rabies virus glycoprotein, *Can. J. Vet. Res.*, 51, 363, 1987.
9. **Morgan, A. J., Mackett, M., Finerty, S., Arrand, J. R., Scullion, F. T., and Epstein, M. A.,** Recombinant vaccinia virus expressing Epstein-Barr virus glycoprotein gp340 protects cottontop tamarins against EB virus-induced malignant lymphomas, *J. Med. Virol.*, 25, 189, 1988.
10. **Shida, H., Hinuma, Y., Hatanaka, M., Morita, M., Kidokoro, M., Suzuki, K., Maruyama, T., Takahashi-Nishimaki, F., Sugimoto, M., Kitamura, R., Miyazawa, T., and Hayami, M.,** Effects and virulences of recombinant vaccinia viruses derived from attenuated strains that express the human T-cell leukemia virus type I envelope gene, *J. Virol.*, 62, 4474, 1988.
11. **Chelyapov, N. V., Antonova, T. P., Yanova, N. N., and Chernos, V. I.,** Antigenic properties of vaccinia virus and of the virus recombinant strains expressing heterologous genes, *Acta Virol.*, 32, 409, 1988.
12. **Rodriguez, D., Rodriguez, J.-R., Rodriguez, J. F., Trauber, D., and Esteban, M.,** Highly attenuated vaccinia virus mutants for the generation of safe recombinant viruses, *Proc. Natl. Acad. Sci. U.S.A.*, 86, 1287, 1989.
13. **Karupiah, G., Coupar, B., Ramshaw, I., Boyle, D., Blanden, R., and Andrew, M.,** Vaccinia virus-mediated damage of murine ovaries and protection by virus-expressed interleukin 2, *Immunol. Cell Biol.*, 68, 325, 1991.
14. **Esposito, J., Brechling, K., Baer, G., and Moss, B.** Vaccinia virus recombinants expressing rabies virus glycoprotein protect against rabies, *Virus Genes*, 1, 7, 1987.
15. **Mackett, M., Smith, G. L., and Moss, B.,** A general method for the production and selection of infectious vaccinia virus recombinants expressing foreign genes, *J. Virol.*, 49, 857, 1984.
16. **Buller, R. M. L., Smith, G. L., Cremer, K., Notkins, A. L., and Moss, B.,** Decreased virulence of recombinant vaccinia virus expression vectors is associated with a thymidine kinase-negative phenotype, *Nature*, 317, 813, 1985.
17. **Andrew, M. E., Coupar, B. E. H., and Boyle, D. B.,** Humoral and cell-mediated immune responses to recombinant vaccinia viruses in mice, *Immunol. Cell Biol.*, 67, 331, 1989.
18. **Coupar, B. E. H., Boyle, D. B., and Both, G. W.,** Effect of in vitro mutations in a vaccinia virus early promoter region monitored by herpes simplex virus thymidine kinase expression in recombinant vaccinia virus, *J. Gen. Virol.*, 68, 2299, 1987.
19. **Olmsted, R. A., Buller, R. M. L., Collins, P. L., London, W. T., Beeler, J. A., Prince, G. A., Chanock, R. M., and Murphy, B. R.,** Evaluation in non-human primates of the safety, immunogenicity and efficacy of recombinant vaccinia viruses expressing the F or G glycoprotein of respiratory syncytial virus, *Vaccine*, 6, 519, 1988.
20. **Brochier, B. M., Languet, B., Blancou, J., Kieny, M.-P., Lecocq, J.-P., Costy, F., Desmettre, P., and Pastoret, P.-P.,** Use of recombinant vaccinia-rabies virus for oral vaccination of fox cubs (Vulpes vulpes, L) against rabies, *Vet. Microbiol.*, 18, 103, 1988.
21. **Small, P. A., Smith, G. L., and Moss, B.** Intranasal vaccination with a recombinant vaccinia virus containing influenza hemagglutinin prevents both influenza virus pneumonia and nasal infection: intradermal vaccination prevents only viral pneumonia, in *Vaccines 85*, Lerner, R. A., Chanock, R. M., and Brown, F., Eds., Cold Spring Harbor Laboratory, New York, 1985, 175.

22. **Murphy, B. R., Collins, P. L., Lawrence, L., Zubak, J., Chanock, R. M., and Prince, G. A.,** Immunosuppression of the antibody response to respiratory syncytial virus (RSV) by pre-existing serum antibodies: partial prevention by topical infection of the respiratory tract with vaccinia virus-RSV recombinants, *J. Gen. Virol.*, 70, 2185, 1989.
23. **Perkus, M. E., Piccini, A., Lipinskas, B. R., and Paoletti, E.,** Recombinant vaccinia virus: immunization against multiple pathogens, *Science*, 229, 981, 1985.
24. **Coupar, B. E. H., Andrew, M. E., and Boyle, D. B.,** A general method for the construction of recombinant vaccinia virus expressing multiple foreign genes, *Gene*, 68, 1, 1988.
25. **Andrew, M. E.,** Protective efficacy of a recombinant vaccinia virus in vaccinia-immune mice, *Immunol. Cell Biol.*, 67, 339, 1989.
26. **Jones, L., Ristow, S., Yilma, T., and Moss, B.,** Accidental human vaccination with vaccinia virus expressing nuceloprotein gene, *Nature*, 319, 543, 1986.
27. **Rooney, J. F., Wohlenberg, C., Cremer, K. J., Moss, B., and Notkins, A. L.,** Immunization with a vaccinia virus recombinant expressing herpes simplex virus type 1 glycoprotein D: long-term protection and effect of revaccination, *J. Virol.*, 62, 1530, 1988.
28. **Flexner, C., Murphy, B. R., Rooney, J. F., Wohlenberg, C., Yuferov, V., Notkins, A. L., and Moss, B.,** Successful vaccination with a polyvalent live vector despite existing immunity to an expressed antigen, *Nature*, 335, 259, 1988.
29. **Johnson, M. P., Meitin, C. A., Bender, B. S., and Small, P. A.,** Passive immune serum inhibits antibody response to recombinant vaccinia virus, in *Vaccines 88*, Ginsberg, H., Brown, F., Lerner, R. A., and Chanock, R. M., Eds., Cold Spring Harbor Laboratory, New York, 1988, 189.
30. **Murphy, B. R., Olmsted, R. A., Collins, P. L., Chanock, R. M., and Prince, G. A.,** Passive transfer of respiratory syncytial virus (RSV) antiserum suppresses the immune response to the RSV fusion (F) and large (G) glycoproteins expressed by recombinant vaccinia viruses, *J. Virol.*, 62, 3907, 1988.
31. **Brochier, B. M., Blancou, J., Aubert, M. F. A., Kieny, M.-P., Desmettre, P., and Pastoret, P.-P.,** Interaction between rabies infection and oral administration of vaccinia-rabies recombinant virus to foxes (Vulpes vulpes), *J. Gen. Virol.*, 70, 1601, 1989.
32. **Doherty, P. C., Allan, W., Boyle, D. B., Coupar, B. E. H., and Andrew, M. E.,** Recombinant vaccinia viruses and the development of immunization strategies using influenza virus, *J. Infect. Dis.*, 159, 1119, 1989.
33. **Venkatesan, S., Baroudy, B. M., and Moss, B.,** Distinctive nucelotide sequences adjacent to multiple initiation and termination sites of an early vaccina virus gene, *Cell*, 25, 805, 1981.
34. **Coupar, B. E. H., Andrew, M. E., Both, G. W., and Boyle, D. B.,** Temporal regulation of influenza hemagglutinin expression in vaccinia virus recombinants and effects on the immune response, *Eur. J. Immunol.*, 16, 1479, 1986.
35. **Sharif, A., Mars, M., Dru, A., and Beaud, G.,** Expression of herpes simplex virus thymidine kinase, mediated by vaccinia virus early promoters, *Res. Virol.*, 140, 103, 1989.
36. **Davison, A. J. and Moss, B.,** Structure of vaccinia virus early promoters, *J. Mol. Biol.*, 210, 749, 1989.
37. **Kotwal, G. J. and Moss, B.,** Vaccinia virus encodes two proteins that are structurally related to members of the plasma serine protease inhibitor superfamily, *J. Virol.*, 63, 600, 1989.
38. **Smith, G. L., Howard, S. T., and Chan, Y. S.,** Vaccinia virus encodes a family of genes with homology to serine proteinase inhibitors, *J. Gen. Virol.*, 70, 2333, 1989.
39. **Townsend, A., Bastin, J., Gould, K., Brownlee, G., Andrew, M., Coupar, B., Boyle, D., Chan, S., and Smith, G.,** Defective presentation to class I-restricted cytotoxic T lymphocytes in vaccinia-infected cells is overcome by enhanced degradation of antigen, *J. Exp. Med.*, 168, 1211, 1988.

40. **Andrew, M. E., Coupar, B. E. H., Boyle, D. B., and Blanden, R. V.,** Recognition by major histocompatibility complex class I-restricted cytolytic T lymphocytes of cells expressing vaccinia-encoded viral and class I proteins, *Eur. J. Immunol.*, 17, 1515, 1987.
41. **Wachsman, M., Aurelian, L., Smith, C. C., Perkus, M. E., and Paoletti, E.,** Regulation of expression of herpes simplex virus (HSV) glycoprotein D in vaccinia recombinants affects their ability to protect from cutaneous HSV-2 disease, *J. Infect. Dis.*, 159, 625, 1989.
42. **Earl, P. L., Hugin, A. W., and Moss, B.,** Removal of cryptic poxvirus transcription termination signals from the human immunodeficiency virus type 1 envelope gene enhances expression and immunogenicity of a recombinant vaccinia virus, *J. Virol.*, 64, 2448, 1990.
43. **Pala, P., Townsend, A. R. M., and Askonas, B. A.,** Viral recognition by influenza A virus cross-reactive cytotoxic T (Tc) cells: the proportion of Tc cells that recognize nucleoprotein varies between individual mice, *Eur. J. Immunol.*, 16, 193, 1986.
44. **MacDonald, H. R., Cerottini, J.-C., Ryser, J.-E., Maryanski, J. L., Taswell, C., Widmer, M. B., and Brunner, K. T.,** Quantitation and cloning of cytolytic T lymphocytes and their precursors, *Immunol. Rev.*, 51, 93, 1980.
45. **Coupar, B. E. H., Andrew, M. E., Boyle, B. D., and Blanden, R. V.,** Immune responses to H-2K^d antigen expressed by recombinant vaccinia virus, *Proc. Natl. Acad. Sci. U.S.A.*, 83, 7879, 1986.
46. **Cox, J. H., Yewdell, J. W., Eisenlohr, L. C., and Bennink, J. R.,** Antigen presentation requires transport of MHC class I molecules from the endoplasmic reticulum, *Science*, 247, 715, 1990.
47. **Klavinskis, L. S., Whitton, J. L., and Oldstone, M. B. A.,** Molecularly engineered vaccine which expresses an immunodominant T-cell epitope induces cytotoxic T lymphocytes that confer protection from lethal virus infection, *J. Virol.*, 63, 4311, 1989.
48. **Reay, P. A., Jones, I. M., and Brownlee, G. G.,** A region of the influenza A/NT/60/68 PB2 protein containing an antigenic determinant recognized by murine H-2D^d restricted cytotoxic T lymphocytes, *Virology*, 167, 261, 1988.
49. **Sweetser, M. T., Braciale, V. L., and Braciale, T. J.,** Class I MHC-restricted recognition of cells expressing a gene encoding a 41 amino acid product of the influenza hemagglutinin, *J. Immunol.*, 141, 3324, 1988.
50. **Gould, K. G., Scotney, H., Townsend, A. R., Bastin, J., and Brownlee, G. G.,** Mouse H-2^k-restricted cytotoxic T cells recognize antigenic determinants in both the HA1 and HA2 subunits of the influenza A/PR/8/34 hemagglutinin, *J. Exp. Med.*, 166, 693, 1987.
51. **Olmsted, R. A., Murphy, B. R., Lawrence, L. A., Elango, N., Moss, B., and Collins, P. L.,** Processing, surface expression, and immunogenicity of carboxy-terminally truncated mutants of G protein of human respiratory syncytial virus, *J. Virol.*, 63, 411, 1989.
52. **Keil, W. and Wagner, R. R.,** Epitope mapping by deletion mutants and chimeras of two vesicular stomatitis virus glycoprotein genes expressed by a vaccinia virus vector, *Virology*, 170, 392, 1989.
53. **Rota, P. A., Shaw, M. W., and Kendal, A. P.,** Comparison of the immune response to variant influenza type B hemagglutinins expressed in vaccinia virus, *Virology*, 161, 269, 1987.
54. **Lew, A. M., Anders, R. F., Edwards, S. J., and Langford, C. J.,** Comparison of antibody avidity and titre elicited by peptide as a protein conjugate or as expressed in vaccinia, *Immunology*, 65, 311, 1988.
55. **Townsend, A. R. M., Bastin, J., Gould, K., and Brownlee, G. G.,** Cytotoxic T lymphocytes recognize influenza haemagglutinin that lacks a signal sequence, *Nature*, 324, 575, 1986.
56. **Jennings, P. A., Boyle, D. B., and Andrew, M. E.,** Class I MHC-restricted cytotoxic T cells efficiently recognize haemagglutinin that is defective in protein folding and cell surface expressions, *Mol. Immunol.*, 25, 1371, 1988.

57. **Langford, C. J., Edwards, S. J., Smith, G. L., Mitchell, G. F., Moss, B., Kemp, D. J., and Anders, R. F.,** Anchoring a secreted plasmodial antigen on the surface of recombinant vaccinia virus-infected cells increases its immunogenicity, *Mol. Cell. Biol.*, 6, 3191, 1986.
58. **Andrew, M. E., Boyle, D. B., Coupar, B. E. H., Whitfeld, P. L., Both, G. W., and Bellamy, A. R.,** Vaccinia virus recombinants expressing the SA11 rotavirus VP7 glycoprotein gene induce serotype-specific neutralizing antibodies, *J. Virol.*, 61, 1054, 1987.
59. **Andrew, M. E., Boyle, D. B., Whitfeld, P. L., Lockett, L. J., Anthony, I. D., Bellamy, A. R., and Both, G. W.,** The immunogenicity of VP7, a rotavirus antigen resident in the endoplasmic reticulum is enhanced by cell-surface expression, *J. Virol.*, 64, 4776, 1990.
60. **Andrew, M. E., Boyle, D. B., Coupar, B. E. H., Bellamy, A. R., Reddy, D., and Both, G. W.,** Improved immunogenicity of cell surface-anchored rotavirus antigens, in *Vaccines 91*, Brown, F., Chanock, R., Ginsberg, H., and Lerner, R., Eds., Cold Spring Harbor Laboratory, New York, 1991, 271.
61. **Kumar, S., Andrew, M. E., Boyle, D. B., Brandon, R. B., Lavin, M. F., and Daniel, R. C. W.,** Expression of bovine leukaemia virus envelope gene by recombinant vaccinia virus, *Virus Res.*, 17, 131, 1990.
62. **Deubel, V., Kinney, R. M., Esposito, J. J., Cropp, C. B., Vorndam, A. V., Monath, T. P., and Trent, D. W.,** Dengue 2 virus envelope protein expressed by a recombinant vaccinia virus fails to protect monkeys against dengue, *J. Gen. Virol.*, 69, 1921, 1988.
63. **Seki, M., Sashiyama, H., Hayami, M., and Shida, H.,** Intracellular processing and immunogenicity of the envelope proteins of human T-cell leukemia virus type 1 that are expressed from recombinant vaccinia viruses, *Virus Genes*, 3, 235, 1990.
64. **Gilbert, J. H., Pedersen, N. C., and Nunberg, J. H.,** Feline leukemia virus envelope protein expression encoded by a recombinant vaccinia virus: apparent lack of immunogenicity in vaccinated animals, *Virus Res.*, 7, 49, 1987.
65. **Karupiah, G., Coupar, B. E. H., Andrew, M. E., Boyle, D. B., Phillips, S. M., Mullbacher, A., Blanden, R. V., and Ramshaw, I. A.,** Elevated natural killer cell responses in mice infected with recombinant vaccinia virus encoding murine IL-2, *J. Immunol.*, 144, 290, 1990.
66. **Kawamura, H., Rosenberg, S. A., and Berzofsky, J. A.,** Immunization with antigen and interleukin 2 in vivo overcomes Ir gene low responsiveness, *J. Exp. Med.*, 162, 381, 1985.
67. **Langford, C.,** personal communication, 1987.
68. **Mullbacher, A., Ramshaw, I. A., and Coupar, B. E. H.,** Vaccinia-interleukin 2 recombinant virus or exogenous interleukin 2 does not alter the magnitude or immune response gene defects of the cytotoxic T-cell response to vaccinia virus in vivo, *Scand. J. Immunol.*, 29, 1, 1989.
69. **Kohonen-Corish, M. R. J., Blanden, R. V., and King, N. J. C.,** Induction of cell surface expression of HLA antigens by human IFN-γ encoded by recombinant vaccinia virus, *J. Immunol.*, 143, 623, 1989.
70. **Corcoran, L. M., Metcalf, D., Edwards, S. J., and Handman, E.,** GM-CSF produced by recombinant vaccinia virus or in GM-CSF transgenic mice has no effect in vivo on murine cutaneous leishmaniasis, *J. Parasitol.*, 74, 763, 1988
71. **Ramshaw, I. A., Andrew, M. E., Phillips, S. M., Boyle, D. B., and Coupar, B. E. H.,** Recovery of immunodeficient mice from a vaccinia virus/IL-2 recombinant infection, *Nature*, 329, 545, 1987.
72. **Flexner, C., Hugin, A., and Moss, B.,** Prevention of vaccinia virus infection in immunodeficient mice by vector-directed IL-2 expression, *Nature*, 330, 259, 1987.
73. **Andrew, M. E.,** unpublished data, 1987.

74. **Andrew, M. E., Karupiah, G., Boyle, D. B., Blanden, R. V., Mullbacher, A., Ramshaw, I. A., and Coupar, B. E. H.,** Effects of vaccinia virus-expressed interleukin 2 on the immune system of sublethally irradiated mice, *Microb. Pathogen.*, 10, 363, 1991.
75. **Kohonen-Corish, M. R. J., King, N. J. C., Woodhams, C., and Ramshaw, I. A.,** Immunodeficient mice recover from infection with vaccinia virus expressing interferon-γ, *Eur. J. Immunol.*, 20, 157, 1990.
76. **Andrew, M. E. and Coupar, B. E. H.,** unpublished data, 1990.

Chapter 7

SAFETY AND ATTENUATION OF VACCINIA VIRUS

R. Mark L. Buller and Gregory J. Palumbo

TABLE OF CONTENTS

I. INTRODUCTION

A major obstacle in the use of vaccinia virus as a ''universal'' vaccine vector for human lies in the rather high frequency of both minor and major vaccinia virus-associated complications following immunization. This review examines the safety record of the vaccinia virus strains utilized in the smallpox eradication program and the recent approaches taken to reduce further the frequency of vaccinia virus-associated complications.

II. SAFETY OF VACCINIA VIRUS

Concerns about the safe use of live recombinant vaccinia virus vaccines stem from the experience with various vaccinia virus strains during the Smallpox Eradication Programme. Abnormally high mortality associated with primary immunization with vaccinia virus was documented in the U.K. in the early 1900s, although vaccinia virus was not officially recognized as the cause.[1] With time, however, it was generally recognized that immunization with vaccinia virus was not without risks. In the definitive history of smallpox and its eradication, Fenner and colleagues[2] identified two major groups of vaccinia virus complications, abnormal skin eruptions and disorders affecting the central nervous system; a brief description based on their work is presented here.

A. VACCINIA VIRUS-ASSOCIATED COMPLICATIONS FOLLOWING IMMUNIZATION

1. Abnormal Skin Eruptions

Accidental infections — The most common, and probably the most under-reported, complication of immunization with vaccinia virus was the transfer of virus from the primary vaccination site to elsewhere on the vaccinee's body, especially the eyelids, vulva, and perineum, or to another individual (Table 1).

Generalized vaccinia — Infection of the human skin by vaccinia virus can result in a systemic infection with an associated transient viremia.[3–6] On rare occasions, individuals with no obvious prior pathological conditions show, by 6 to 9 d post-infection (p.i.), the appearance of a generalized rash which is distinct from the "centrifugal" distribution of the lesions of smallpox. With time, the individuals usually recovery completely.

Eczema vaccinatum — On exposure to vaccinia virus, individuals with a history of eczema can develop lesions distal from the site of inoculation but in areas of the skin which were, or had previously been, eczematous. These regions become intensely inflamed and, on occasion, the infection can spread to normal tissue. The illness is severe and is associated with mortality (0.07% of cases[2]), especially in infants who have large areas of skin involved.

Erythema multiforme — Erythema multiforme is an acute inflammatory systemic disease which is thought to be secondary to infections with, but not limited to, measles, mumps, herpes simplex, and vaccinia viruses. With vaccinia virus, the rash develops with the onset of constitutional symptoms and evolves rapidly to the vesicular stage. The individual usually recovers completely.

Progressive vaccinia (vaccinia necrosum) — Vaccinia virus immunizations of persons suffering from any of a wide array of immune dysfunctions can result in the failure of the primary lesion to heal. Secondary lesions sometimes appear, and all lesions become progressively larger, with death being the expected outcome (in 36% of cases within 2 to 5 months of onset of disease[2]) with lack of treatment.

2. Central Nervous System Disorders

Central nervous system (CNS) disorders are the most serious of the vaccinia virus-associated complications and, unfortunately, a predisposition cannot be predicted. These disorders can be associated with incomplete recovery of the patient and a high mortality rate. Subjects who presented with postvaccinial encephalitis had, in reality, one of at least three distinct pathological disorders: coincidental encephalitis, encephalopathy, or encephalitis. The distinction between these latter two pathologies was made by Vries[7] and is supported by differences in the age of the patient, symptomatology, and the incubation period of the illness.

Coincidental or concomitant encephalitis — As pointed out by Fenner et al.,[2] in any population there will be a certain frequency of encephalitis, the result of any number of agents. For instance, a survey in New Jersey in 1965 and in Florida in 1968 showed that, in a population of "normal" children, 2.86 and 2.29 cases of encephalitis, respectively, were detected per million children per 28-day period.[8] Therefore, without additional

TABLE 1
Complications of Smallpox Vaccination in Ten States of the U.S. for the Year 1968

Vaccination status and age (years)	Accidental infection	Generalized vaccinia	Eczema vaccinatum	Erythema multiforme	Progressive vaccinia	Post-vaccinial encephalitis	Other	Total
Primary vaccination[a]								
<1	507.0	394.4	14.1	436.6	—	42.3	154.9	1,549.3
1—4	577.3	233.4	44.2	157.7	3.2	9.5	236.6	1,261.8
5—19	371.2	139.7	34.9	87.3	—	8.7	214.0	855.9
≥20	606.1	212.1	30.3	30.3	—	—	636.4	1,515.2
Total	529.2	241.5	38.5	164.6	1.5	12.3	266.2	1,253.8
Revaccination[a]								
<1	—	—	—	—	—	—	—	—
1—4	198.1	—	—	72.7	—	—	18.2	200.0
5—19	47.7	9.9	2.0	2.0	—	—	23.9	85.5
≥20	25.0	9.1	4.5	9.1	6.8	4.5	54.5	113.6
Total	42.1	9.0	3.0	10.0	3.0	2.0	39.1	108.2

Note: Complications expressed as cases per million vaccinations.

[a] NYCBH strain of vaccinia virus was used.

From Lane, J. M. et al., *J. Infect. Dis.*, 122, 303, 1970; cited in Fenner, F. et al., *Smallpox and Its Eradication*, World Health Organization, Geneva, 1988.

confirmatory tests, it is reasonable to expect that one to three cases of encephalitis per million vaccinia virus immunizations may be coincidental.

Encephalopathy — Encephalopathy occurred in infants under 2 years of age approximately 6 to 10 d (mean, 8.6 ± 2.3) after immunization with vaccinia virus and was characterized often by convulsions, hemiplegia, and aphasia.[9,10] On recovery from disease, the subject was often left with cerebral impairment. When death occurred, it was on average 10.5 ± 3.0 days after immunization.

Encephalitis — This complication occurred in subjects over 2 years of age approximately 11 to 15 d (mean, 12.3 ± 2.1 days) after immunization with vaccinia virus. The onset of the disease was marked by fever, vomiting, and other symptoms of cerebral involvement. The case fatality rate was approximately 35%, with the day of death on the average 16.3 ± 3.0 days after immunization.[9]

B. FREQUENCY OF COMPLICATIONS

In the U.S., the frequency of vaccinia virus-associated complications (New York City Board of Health; NYCBH strain) was thoroughly examined in 1968 through a national survey[11] and in a ten-state survey.[12] Higher frequencies of vaccinia virus complications were noted in the ten-state study than the national survey, presumably due to a more intensive effort to uncover all cases of complications. For this reason, the ten-state survey results are presented here (Table 1). Data from the national survey are presented in Chapter 1, Table 4.

The majority of vaccinia virus complications reported occurred after primary immunization and not revaccination, except in the case of progressive vaccinia, which was not unexpected. From Table 1, it is clear that there is a relatively high probability of a complication from primary immunizations with vaccinia virus (1253.8 cases per million immunizations or 0.13% for all ages) and a frequency of 0.0015% (15.4 cases per million immunizations) for life-threatening complications involving the CNS in children 4 years of age and younger. What is also clear from this table is that the frequency of post-vaccinial encephalitis is significantly higher in individuals under the age of one; for this reason, immunization with vaccinia virus was recommended for the second year of life. The frequency of vaccinia virus-associated complications observed with the NYCBH strain was lower than with most other strains employed during the smallpox eradication program.

C. CONTRAINDICATIONS TO IMMUNIZATION WITH VACCINIA VIRUS

In areas in which smallpox was not endemic, four conditions were generally accepted as contraindicators for immunization with vaccinia virus: immune disorders, eczema, pregnancy, and disorders of the central nervous system.[2] These criteria still appear reasonable for any future immunization program involving humans and vaccinia virus recombinants.

TABLE 2
The Effect of Vaccinia Virus Strain on the Frequency of Death Due to Post-Vaccinial Encephalitis in Infants in the Netherlands and Austria

Country	Period	Vaccinia strain	Post-vaccinial encephalitis deaths per million immunizations
Netherlands[a]	1959–1962	Copenhagen	19.4
	1963–1970	Lister	6.4
Austria[b]	1959–1966	Bern	14.0
	1967–1970	Lister	0

[a] Data from Reference 13.
[b] Data from Reference 14.

D. VACCINIA VIRUS STRAIN VARIATION IN PATHOGENICITY

The complication frequencies from vaccinia virus immunization varied among countries, due in part to differences in the diagnostic criteria and reporting requirements, but also due to the use of different strains of vaccinia virus in the respective immunization programs. The importance of vaccinia virus strain on the frequency of complications was underscored when the frequency of deaths from post-vaccinial encephalitis was observed to decline with the replacement of the Copenhagen strain in the Netherlands[13] and the Bern strain in Austria[14] with the Lister strain of vaccinia virus (Table 2). Additional epidemiological investigations by Marennikova[15] showed higher levels of reported post-vaccinial complications with the Tashkent strain (18 per million immunizations) than the B-51 (10 per million immunizations) or EM-63 strains (7 per million immunizations). These findings were supported by clinical trials in which a number of vaccinia virus strains were compared, and the Lister and NYCBH strains were found to be less reactogenic than the Copenhagen and Bern Strains.[16,17] Based on a number of parameters,[2] the relative ''virulence'' of various vaccinia virus strains is presented in Table 3. The NYCBH strain of vaccinia virus is one of the least virulent, while retaining satisfactory immunogenicity.

III. VACCINIA VIRUS ''VIRULENCE'' GENES

The correlation of certain virus strains with higher frequencies of vaccinia virus-associated complications suggested that ''virulence'' had a genetic basis and, therefore, could be modified. Only a limited number of ''virulence'' genes have been identified, some of which are shown in Figure 1 (see also Table 2, Chapter 3 for additional insertion sites, some of which affect virulence). Of these genes, some may be essential for virus replication

TABLE 3
Relative Virulence of Vaccinia Virus Strains Employed in the Smallpox Eradication Programme

Vaccinia virus strain[a]	Relative virulence
Tashkent	High
Copenhagen	High
Ikeda	High
Bern	Moderate
B-5	Moderate
Lister	Moderate
EM-63	Low
NYCBH	Low

[a] For additional information on strain pedigrees and virulence, see Fenner et al.[2]

in nongrowing cells; others may simply optimize the replication process or be needed during the infection of certain cell types (host range) or for protection from host defense mechanisms *in vivo*.

The virus-encoded genes which are specific for virus replication in the CNS or spread from the primary site of inoculation in the skin to internal organs will be some of the most useful targets for attenuation of vaccinia virus. Other virus genes which are important in virus replication in the skin may be less suitable targets for inactivation, as their loss may reduce recombinant virus replication (following intradermal [i.d.] immunization) to a point where there would be insufficient antigen to trigger a protective immune response. Of the known virulence genes, only the thymidine kinase (TK) and hemagglutinin (HA) genes have been evaluated in a number of animal models for their contribution to vaccinia virus pathogenesis.

A. HOST RANGE GENES

There have been at least three (tissue culture) host range genes identified in poxviruses (Table 4). Of interest is the observation that, in certain situations, the loss of one host range gene can be compensated for by the presence of another.[18] Although there are currently no published reports that any of the host range genes identified in orthopoxviruses are important for virus replication in tissues within the animal, it is reasonable to expect that this will be the case.

Human host range gene *ORF K1L (hr-1)* — The first host range gene to be identified (*hr-1; Copenhagen-ORF K1L; WR-K0L;* Figure 1) encodes a 29-kDa polypeptide that is expressed early in infection and permits vaccinia virus replication in human cells.[19] A vaccinia virus deletion mutant lacking this gene replicates in chick embryo fibroblasts and BHK21 cells, but not in

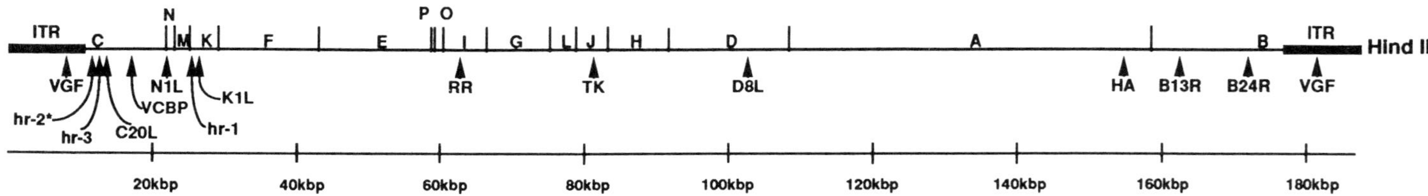

FIGURE 1. Map position of genes nonessential for vaccinia virus replication *in vitro*. The genome depicted is for the WR strain of vaccinia virus.[55,127] VGF, Vaccinia growth factor (ORF C6R); *hr-2*, ORF C10L - C13L, not active in vaccinia; *hr-3*, ORF C18L; ORF C20L; VCBP, vaccinia complement-binding protein (C21L); N1L, homology to tumor necrosis factor; *hr-1*, ORF KOL; K1L, homology to SERPIN; RR, ribonucleotide reductase (large subunit) (ORF I4L); TK, thymidine kinase (ORF J2R); D8L, transmembrane protein; HA, hemagglutinin (ORF A56R); B13R, homology to SERPIN; B24R, homology to SERPIN.

TABLE 4
Poxvirus Tissue Culture Host-Range Genes

		Growth of vaccinia virus in cell lines derived from:				
Gene deletion	**Gene insertion**	**Monkey (Vero)**	**Human (MRC-5)**	**Pig (LLC-PK1)**	**Rabbit (RK-13)**	**Hamster (CHO)**
None	None	Yes	Yes	Yes	Yes	No
hr-1	None	Yes	Yes	Yes	No	No
hr-3	None	Yes	Yes	Yes	Yes	No
hr-1 + *hr-3*	None	Yes	No	No	No	No
hr-1 + *hr-3*	*hr-1*	Yes	Yes	Yes	Yes	No
hr-1 + *hr-3*	*hr-2*	Yes	Yes	Yes	Yes	Yes
hr-1 + *hr-3*	*hr-3*	Yes	Yes	Yes	No	No

Summarized from Perkus, M. E. et al., *Virology,* 179, 276, 1990.

RK13 or human cell lines MRC-5, KB, or Hep2.[19–22] In human cells, only early genes are transiently expressed, followed by rapid inhibition of both host and virus protein synthesis. A related gene has been found in other orthopoxviruses and in fowlpox.[23] This 29-kDa polypeptide has homology to ankyrin, with the presence of two ankyrin-like repeats.[24] Ankyrins are thought to control interactions between integral membrane proteins and cytoskeletal elements. Interestingly, the severely attenuated MVA strain of vaccinia virus[25] (which was derived from the dermovaccinia strain ankara) has three major deletions relative to the WR strain, one of which affects the 29-kDa human host range protein.[26] It is not known, however, if the attenuated phenotype of the MVA strain is directly or indirectly the result of the loss of this gene.

Chinese hamster ovary host range gene (*hr-2*) — All wild-type vaccinia virus strains that have been studied do not productively infect Chinese hamster ovary (CHO) cells,[27-29] although a second orthopoxvirus, cowpox virus, does. Vaccinia virus infection of CHO cells appears restricted to early gene expression, followed by an extremely rapid inhibition of both virus and host protein synthesis, which was completed within 2 h p.i.,even at a relatively low multiplicity of infection (m.o.i.) of 5 plaque forming units (pfu). The infection, although abortive, results in cell death.[22a] The recombination of the *hr-2* gene from cowpox virus (which encodes a 77-kDa polypeptide) into vaccinia virus allows a productive replication cycle in CHO cells. Like *hr-1*, this gene also has been shown to contain an ankyrin-like repeat.[24] Further studies have shown that other vaccinia virus strains possess sequences homologous to the cowpox gene, although the ORF is not intact, and with the WR strain, it is broken into four short ORFs: C10L, C11L, C12L, and C13L.[22a,30] With the Copenhagen strain of vaccinia virus, DNA sequences homologous to the *hr-2* gene are lacking.[31]

Human host range gene *ORF C7L (hr-3)* — Recently, a third poxvirus host range gene (*hr-3;* Copenhagen-ORF C7L; WR — ORF C18L) has been

TABLE 5
Relative Virulence of Selected Vaccinia Virus Strain WR Mutants by i.c. Route in the Mouse

Gene inactivated[a]	Mouse strain	LD_{50} of mutant virus relative to wild type[b]	Ref.
TK	BALB/cByJ	40000	41
VGF	BALB/cByJ	20000	38
ORF N1L	BALB/cAn/NCr	13000	82
HA	DDY	4600	99
HA	BALB/cByJ	3500	87
VCBP	BALB/cAn/NCr	57	130
ORF C20L	BALB/cByJ	7.5—50	87
RR	BALB/cByJ	17	42
None	NA[c]	1	

[a] In most cases, the gene was inactivated by insertion of a DNA sequence alone or in combination with the deletion of sequences. VGF, Vaccinia growth factor; TK, thymidine kinase; RR, large subunit of ribonucleotide reductase; VCBP, vaccinia complement binding protein; HA, hemagglutinin.
[b] Value was calculated as LD_{50} mutant $\div$ LD_{50} wild type.
[c] NA, Not applicable.

identified. This gene can substitute for *hr-1* in tested cell lines derived from humans and pigs, but not rabbits.[18]

B. GENES THAT STIMULATE MACROMOLECULAR SYNTHESIS

1. Epidermal Growth Factor Homolog

Vaccinia virus expresses a glycosylated, secretory protein of 22 to 26 kDa, named vaccinia growth factor (VGF), which has significant amino acid homology with epidermal growth factor (EGF) and transforming growth factor-α.[32–34] A series of studies found that VGF competed with EGF for binding to EGF membrane receptors, induced phosphorylation of tyrosine residues in the EGF receptor, and was mitogenic for certain cells.[35–37]

The contribution of VGF to virus pathogenesis was evaluated by comparing the replication and spread of a VGF minus mutant with wild-type virus. *In vivo,* the VGF minus mutant replicated to lower levels in brains of mice following intracranial (i.c.) inoculations; also, there was a concomitant increase in the virus dose required for lethality of 50% of the inoculated mice (LD_{50})[38] (Table 5). The VGF minus mutant was also shown to replicate less well than wild-type virus when inoculated by the i.d. route into rabbit skin and onto the chorioallantoic membrane (CAM) of the 12-day-old chick embryo.[38,39] These studies, taken together, are consistent with VGF inducing localized hyperplasia around the focus of infection, which likely provides additional, metabolically active cells for virus replication.

2. Thymidine Kinase

The thymidine kinase (TK) gene is expressed early during infection but is not strictly required for replication in tissue culture cells. Presumably, this enzyme is important in nondividing cells which lack sufficient or accessible concentrations of TTP for virus DNA replication. The TK gene is important for a productive infection of animals, because TK minus mutants are severely attenuated by the i.c. and i.d. routes of infection of the mouse and rabbit, respectively.[40,41] A large number of the vaccinia virus recombinants which have been evaluated as vaccine candidates are TK minus as a result of recombination of the foreign gene into the TK gene (see Tables 7 and 8).

3. Ribonucleotide Reductase

Ribonucleotide reductase (RR) is a two-subunit enzyme encoded by vaccinia virus that catalyzes the reduction of ribonucleotides to their corresponding deoxyribonucleotides.[42] The virus probably encodes its own RR in order to decrease its dependence on the cellular RR, which is expressed in a cell cycle-dependent manner. A vaccinia virus mutant lacking an intact RR large-subunit gene appeared to replicate as efficiently as wild-type virus after intradermal infection of the rabbit, but had a 10-fold higher LD_{50} after i.c. inoculation of the mouse[42] (Table 5).

C. GENES THAT INHIBIT HOST DEFENSE MECHANISMS

The animal host has a number of nonspecific and specific (immune) defense mechanisms which limit poxvirus replication. First, the relatively large size of the virus aids in its clearance by phagocytic cells of the reticuloendothelial system.[43] Second, the early modifications of the plasma membrane of the host cell with virus-specified polypeptides provides the means for antibody-dependent cell-mediated cytotoxicity or cytotoxic T-lymphocyte killing of infected cells prior to the production of progeny virus.[44–49] Third, the apparent inability of the poxvirus genome to stably exist in a host cell precludes the maintenance of the virus in the animal in a ''cryptic'' state. In spite of these features, poxviruses are relatively successful animal pathogens, as evidenced by the maintenance of variola viruses in humans until the recent smallpox eradication program. Perhaps this success is due in part to a growing list of virus-encoded functions that either have been shown to interact with a host defense mechanism, or the circumstantial evidence is strong enough to suspect that this may be the case. It is not easy to predict how the inactivation of these types of genes will affect virus pathogenesis and foreign protein immunogenicity of vaccinia virus recombinants.

1. Serine Protease Inhibitors

A family of proteins, which show homology at the amino acid level, contains some members which inhibit serine proteases, and they have been termed SERPINs (serine protease inhibitor).[50] Some SERPINs exert control over a variety of critical events associated with connective tissue turnover, coagulation, fibrinolysis, complement activation, and inflammatory reactions.[50] Recently, three distinct genes with amino acid homology to SERPINs were found in vaccinia virus and are probably present in other poxviruses: 38 K-like[51] (also named SPI-2, VVB13R), ORF K1[52] (also named VVK2L, SPI-3), and SPI-1[53] (also: VVB24R). These three genes are transcribed early during infection[55] and may have a cytoplasmic site of action, except ORF K1, which has a potential signal sequence at the amino terminus and a hydrophobic sequence near the carboxy terminus suggestive of an extracellular or membrane-associated location.[51,54,55]

Of these proteins, the cowpox 38-kDa protein has been investigated most thoroughly. A number of studies strongly suggest that this protein inhibits the production of a chemoattractant, possibly a leukotriene-like molecule.[56,56a] A cowpox virus mutant lacking the 38-kDa protein was very restricted in its replication and spread in the ectoderm (one of the three embryonic germ layers) of the CAM of the chick embryo, although it was still able to seed the spleen and liver at a frequency only slightly lower than wild-type virus.[128]

2. Vaccinia Virus Complement Binding Protein

The deduced amino acid sequence of a major secretory protein of 35-kDa has been shown to have structural similarity to a family of eukaryotic proteins which contains at least six complement control proteins. The 35-kDa protein has the highest degree of sequence similarity to the protein that binds the C4b fragment of complement and has been termed vaccinia virus complement binding protein (VCBP).[57] The protein has been shown to block the classical complement pathway and to bind to C4b.[58] It is synthesized continually throughout infection and has been shown to be expressed by cowpox virus (CPV) and probably ectromelia virus. If VCBP effectively blocks the antibody-mediated classical complement cascade *in vivo,* then this may explain the relative unimportance of antibody in recovery from a primary poxvirus infection. A mutant lacking an intact gene is attenuated in the mouse LD_{50} assay (Table 5). Interestingly, the glycoprotein C of herpes simplex viruses 1 and 2 has been shown to have receptor binding activity for complement C3b.[59,60]

3. Anti-Interferon Properties of Poxviruses

Currently, there are no specific gene products that have demonstrated anti-interferon properties; however, there are a number of reports that interferon-mediated effects can be disrupted during poxvirus infections in certain cells. For example, studies have suggested that vaccinia virus can inhibit the interferon-induced protein kinase pathway and that this inhibition is the result of an early gene product.[61–64] A second interferon pathway, which involves the activation of latent endoribonuclease L, thereby leading to an increased degradation of viral mRNAs,[65–68] may be inhibited by virus-induced ATPase and phosphatase activities.[69–71] Additional support for the existence of virus genes that modify the antiviral effects of interferon comes from the studies of Youngner and colleagues,[72] who showed that prior infection of RK-13 cells with vaccinia virus was able to partially reverse the ''interferon''-mediated inhibition of vesicular stomatitis virus replication.

4. Immunosuppressive Properties of Vaccinia Virus

Although the immunosuppressive effects of myxoma and a derivative virus, malignant rabbit fibroma virus, are documented,[73,74] there are no data to date to suggest that vaccinia virus encodes gene products which block the generation of an immune response.

D. OPEN READING FRAMES NOT ESSENTIAL FOR GROWTH

There are a large number of open reading frames (ORFs) which are nonessential for replication of poxviruses in tissue culture.[30–41,75–80] In certain cases, the identified ORFs have yielded homologies with proteins of known functions, although the significance of these findings to the biology of the virus remains obscure.

ORF C20L — This ORF encodes a 38-kDa polypeptide of unknown function which is expressed early during infection.[81] ORF C20L has been used as a site for the introduction of foreign genes into the genome (see Tables 7 and 8).

ORF N1L — The ORF N1L encodes a 12-kDa polypeptide which is expressed early and late during vaccinia virus infection, and is secreted from the infected cells. The protein has a limited similarity in sequence with both tumor necrosis factor and an adenovirus protein which has been demonstrated to prevent tumor necrosis factor-mediated lysis of adenovirus cells. A mutant virus lacking an intact ORF N1L is severely attenuated in the mouse LD_{50} assay[82] (Table 5).

ORF D8L — The ORF D8L encodes a protein with a predicted molecular mass of 35.4 kDa which is expressed late during infection and appears to be a component of the virion membrane. The structure of the protein has two interesting features. First, it has a stretch of 19 hydrophobic amino acids flanked by hydrophilic amino acids, reminiscent in structure of the transmembrane anchor sequences in several transmembrane proteins. Second, the protein exhibits a high degree of homology with a mammalian enzyme, carbonic anhydrase[83] (optimized score of 441 out of a possible total of 1541; release 18; National Biomedical Research Foundation protein database). Carbonic anhydrase is important in regulating the concentration of dissolved CO_2 in tissue and plasma by catalyzing the hydration of CO_2 to H_2CO_3. The contribution of this gene to virus pathogenesis is unknown.

Hemagglutinin — The vaccinia virus hemagglutinin (HA; Copenhagen — A56R) is a glycoprotein which is synthesized late during the infection cycle and is detected at the external surface of the plasma membrane after the commencement of virus assembly.[84] The function of this gene is unknown,[85] and it has been used for the insertion of foreign genes[85–87] (see

Table 7). Recombinant viruses lacking a functional HA gene are attenuated in the mouse LD_{50} assay (Table 5).

IV. ATTENUATION OF VACCINIA VIRUS AND RECOMBINANT VIRUSES

Interest in the attenuation of vaccinia virus has spanned many decades. The earliest studies showed that attenuation of vaccinia virus pathogenesis could be achieved through sequential passage of the virus either in tissue explants or cell lines. More recent studies have shown that attenuation can be achieved by the inactivation of single genes. Most recently, virus recombinants have been constructed in which lymphokine genes have been recombined into the virus in an attempt to enhance the immunogenicity of the foreign protein and/or attenuate the virus.

A. SERIAL PASSAGE

Over the years, a number of the vaccinia virus strains were further attenuated either through design, as was the case with the MVA,[25,88] the DIs,[89] and the LC16 strains[90] or as a consequence of multiple passages through various tissue/organ culture systems, as with the CVI and CVII strains.[91–93] Only the LC16 strains of vaccinia virus were considered sufficiently immunogenic for continued consideration in immunization programs.

The LC16 prototype strain was produced from the Lister strain by 36 serial passages through rabbit kidney cells at 30°C and the selection of the most temperature-sensitive (at 40°C) clone from a group of 25. A further six passages of this clone on primary rabbit kidney cells and the selection of a medium-sized pock (2 to 3 mm) on the CAM of the chick embryo resulted in the LC16m0 strain. The further passage of LC16m0 and the selection of a small pock (0.5 to 1 mm) from the CAM of the chick embryo yielded the LC16m8 strain.[94] The only genetic difference between the Lister parent and the LC16m8 strain (as detected by restriction endonuclease analysis) was the appearance of a new *Xho*I site in the *Hin*dIII D fragment of the LC16m8 genome.[95,96]

The LC16 strains were shown to be less pathogenic than the CV-I, EM-63, Lister, or NYCBH strains following intracerebral inoculations of monkeys[90,97] (Table 6). The LC16m8 strain had no reported severe immunization-associated complications in a trial with 10,000 children 5 years of age and younger, and showed a take rate and immunogenicity comparable with the traditional vaccinia virus strains, but with a smaller average induration.[94]

B. INACTIVATION OF SPECIFIC "VIRULENCE" GENES

The effects on virus pathogenesis of the recombination of various foreign genes into the TK, HA, or C20L ORFs of the WR, Lister, Wyeth (NYCBH), or LC16 strains of vaccinia virus are summarized in Table 7.

TABLE 6
Mortality of Cynomolgus Monkeys Following Intracerebral Inoculation with Various Vaccinia Virus Strains

Vaccinia virus strain	LD_{50}
CVI	$\leq 6.3 \times 10^5$ pfu
Ikeda	$\leq 6.3 \times 10^5$ pfu
NYCBH	2.9×10^6 pfu
Lister	4.4×10^6 pfu
EM-63	1.4×10^7 pfu
DIs	$\geq 6.3 \times 10^8$ pfu
LC16	$\geq 6.3 \times 10^8$ TCID

From Morita, M. et al., *Arch. Virol.*, 53, 197, 1977.

The TK gene has been one of the major target genes for construction of vaccinia virus recombinants. What is clear from studies with the very virulent WR strain of virus is that inactivation of the TK gene usually results in a significant effect on the pathogenicity of the virus in all animal models (and by various routes) in which the virus grows well (mouse, rabbit, nonhuman primates, etc.; Table 7). When less virulent strains of vaccinia virus are used, such as the Wyeth (NYCBH) strain, the effect of a TK minus phenotype on pathogenicity is less. Similarly, inactivation of the HA gene in the virulent WR strain resulted in a dramatic increase in the LD_{50} by the i.c. or i.p. routes of infection in the mouse, whereas with the Lister strain, there was a slight decrease in the LD_{50}[99] (Table 7, entries 14, 16, 17, i.c.; entry 18, 20, i.p.)

A comparison of neurovirulence in the rabbit of the LC16, the Lister, and the WR strains, and TK minus derivatives of each parent virus (which expressed the heptitis B surface antigen [HBs Ag]) indicated that the LC16 parents and their recombinants were attenuated to a similar extent and were the most attenuated of the viruses tested. The WR viruses were the most virulent, while the Lister viruses were intermediate between the two.[98] Inactivation of the TK gene did not result in a significant increase in attenuation for all strains tested by this route (Table 7, entries 9 to 11). No comparisons of the immunogenicity of the HBs Ag were made among the various recombinants.

Hemagglutinin minus derivatives of LC16m0, Lister, and WR viruses, which expressed the human T-cell leukemia virus type I (HTLV-I) envelope gene, were evaluated for immunogenicity and protection in rabbits. The LC16m0 recombinant elicited a high titer of anti-HTLV-I antibody in rabbits and better ''protected'' them against HTLV-I infection than a comparable Lister recombinant virus.[99] The WR strain recombinant, however, yielded the highest level of anti-HTLV-I *env* antibody in the rabbit and also ''protected'' against HTLV-I infection.[86] These results suggest that *env* antigen is more

TABLE 7
The Effect on Vaccinia Virus Pathogenesis of Insertional Inactivation of a Virus Gene by a Foreign Gene

Vaccinia virus strain[a]	Host	Route of infection[b]	Foreign gene inserted[c]	Virus gene inactivated[d]	Effect on vaccinia virus pathogenesis[e]	Ref.
Wyeth	Mouse, BALB/cByJ	IC	HBV sAg	TK	C	41
WR	Mouse, BALB/cByJ	IC	HSV-1 Glycoprotein D or influenza virus HA	TK	B	41
WR	Mouse, BALB/cByJ	IP	Influenza HA or HSV-1 glycoprotein D	TK	G	41
WR	Mouse, CBA/H	FP	Influenza HA	TK	G	124
WR	Mouse, BALB/c	IN	RSV G	TK	A	125
Wyeth	Guinea pig	ID	Lassa fever nucleotprotein	TK	H	126
Wyeth	Rabbit	ID	HBV sAg	TK	F	40
WR	Rabbit	ID	HBV sAg	TK	D	40, 87
WR	Rabbit	IC	HBV sAg	TK	H	98
Lister	Rabbit	IC	HBV sAg	TK	H	98
LC16m8	Rabbit	IC	HBV sAg	TK	H	98
WR	Rhesus, patas	ID	RSV F or G	TK	E	122
WR	Chimp	ID	HBV sAg	TK	E	118
Lister	Mouse, DDY	IC	HTLV-I pro *env*	HA	H	99

LC16m0	Mouse, DDY	IC	HTLV-I pro *env*	HA	H	99
WR	Mouse, DDY	IC	HTLV-I pro *env*	HA	B	99
WR	Mouse, BALB/cByJ	IC	LacZ	HA	B	87
Lister	Mouse, DDY	IP	HTLV-I pro *env*	HA	H	99
LC16m0	Mouse, DDY	IP	HTLV-I pro *env*	HA	H	99
WR	Mouse, DDY	IP	HTLV-I pro *env*	HA	B	99
Lister	Rabbit	ID	HTLV-I pro *env*	HA	E	99
LC16m0	Rabbit	ID	HTLV-I pro *env*	HA	E	99
WR	Rabbit	ID	HTLV-I pro *env*	HA	E	99
WR	Rabbit	ID	LacZ	HA	D	87
WR	Mouse, BALB/cByJ	IC	Influenza HA	ORF C20L	C	87

[a] WR derived from NYCBH by mouse brain passage; Wyeth derived directly from NYCBH; LC16 strains derived from Lister strain.

[b] Route of virus inoculation: IC, intracranial; IP, intraperitoneal; IN, intranasal; FP, footpad; ID, intradermal.

[c] HBV sAg, Hepatitis B virus surface antigen; HSV-1, herpes simplex virus 1; HA, hemagglutinin; RSV G, respiratory syncytial virus G protein; HTLV-I, human T-cell leukemia virus type I; LacZ, β-galactosidase gene.

[d] The virus gene into which the foreign gene was recombined: TK, thymidine kinase; HA, hemagglutinin.

[e] Comment key: A, increased attenuation of virus, decreased morbidity, less mortality; B, greater than 1000-fold increase in virus dose to reach LD_{50}; C, 10- to 100-fold increase in virus dose to reach LD_{50}; D, decreased size of lesions, and/or decreased ulceration of the lesions, and/or greater than 100-fold increase in virus dose to reach erythema $dose_{50}$ (ED_{50}); E, decreased size of lesions, and/or decreased ulceration of the lesions, and/or between 50- and 100-fold increase in virus dose to reach ED_{50}; F, decreased size of lesions, and/or decreased ulceration of the lesions, and/or less than 50-fold increase in virus dose to reach ED_{50}; G, decreased virus yields recovered from tissues of infected animals; H, no attenuation of the virus, similar pathogenicity as wild-type virus.

immunogenic in the WR strain of vaccinia virus than in LC16m0 or Lister strains. It was further noted that the loss of the hemagglutinin gene through insertion of the HTLV-I *env* gene resulted in a similar magnitude of attenuation for the Lister, WR, and LC16m0 virus strains, as indicated by a decrease in the erythema diameter $dose_{50}$ following i.d. injection of rabbits (Table 7, entries 21 to 23). Consistent with previous studies[98] was the low-level neurovirulence in the rabbit model of the LC16m0 and the derivative virus.[99] In summary, the LC16 series of recombinant viruses grow well in the skin, are immunogenic for the foreign antigen, and show very low neurovirulence in rabbit and mouse models. Although the immunogenicity of the foreign gene product is usually greater in recombinants using the WR strain vs. other strains such as the LC16m0,[86,99] the WR strain pedigree makes it unacceptable for human vaccine applications.

C. EXPRESSION OF LYMPHOKINES

In an attempt to further attenuate vaccinia virus (and perhaps increase the immune response to foreign antigens), several laboratories recently undertook a series of investigations in which an immunomodulator was incorporated into the vaccine vector itself. Some recombinants which expressed murine (mu) or human (hu) interleukin (IL)-2,[87,100–102] tumor necrosis factor (TNF)-α,[103] and murine interferon (IFN)-γ[104] (Table 8) were more attenuated *in vivo* than control virus, whereas others which expressed lymphokines IL-1, IL-3, IL-4, IL-5, or IL-6 had similar virulence to control virus.[103,105,108]

The lymphokine genes have been inserted into several different locations within the vaccinia virus genome: the TK gene in *Hin*dIII-J, the hemagglutinin gene in *Hin*dIII-A, the 38-kDa gene in *Hin*dIII-C, and a site in *Hin*dIII-F (Table 8). In order to quantify the effect of the lymphokine gene on the host immune response, an additional reporter gene such as influenza hemagglutinin (HA) or nuclear protein (NP) was also recombined into the vaccinia virus genome (Table 8).

Several consistent observations have been made in studies with recombinant vaccinia viruses expressing IL-2 (VV-IL2). First, the VV-IL2 recombinant virus is greatly attenuated in comparison to wild-type or control recombinant viruses in susceptible animals as determined by skin lesion size and ulceration (rabbits, monkeys, and mice), morbidity, and mean survival time (in both euthymic normal and athymic nude mice).[87,100–102,104,106] Consistent with these observations was a lower yield of recombinant VV-IL2 virus recovered from tissues, including the primary site of inoculation and secondary sites for virus replication including lung, brain, spleen, liver, and ovaries of infected athymic nude and X-irradiated mice and, to a lesser extent, normal mice.[87,100,104,106,108] Second, the level of attenuation of the virus would appear to be partially dependent on the route of virus inoculation.[87] Third, the precise location in the virus genome of the IL-2 gene influences the level of expression of IL-2, although there was no detectable effect on the level of virus attenuation.[87] Together, these results suggest that the expression of

TABLE 8
Effect of Lymphokine Genes on the Immunogenicity and Pathogenicity of Vaccinia Virus Recombinant Viruses[a]

Lymphokine gene		Reporter gene				Effect on infection		
Product	Site[b]	Product	Site[b]	Host	Route[c]	Pathogenicity[d]	Immune response[e]	Ref.
hu-IL2	TK or HA or ORF C20L	—	—	NZW Rabbit	ID	AC	N	87
hu-IL2	TK or HA or ORF C20L	—	—	BALB/cByJ, or BALB/c nude, or B10.A (5R) mice	IC, IP, IV, or ID	ABCD	ACD	87
	ORF C20L	FLU-HA or NP	TK					
hu-IL2	ORF C20L	—	—	Patas, rhesus, and squirrel monkeys	ID	CD	C	102
hu-IL2	ORF C20L	FLU-HA	TK					
hu-IL2	ORFF?[g]	gp160 (HIV-1)	?	Pig-tailed macaque	ID, SC	C	C	101
hu-IL2[f]	TK	—	—	Chimpanzee	ID	N	E	115
mu-IL2	ORFF?[g]	FLU-HA	TK	CBA/H, Swiss nude, CBA (bg/bg) X-irrad mice	SC, IV, IP, or IN	ACD	ACEFG	100, 105, 106, 108, 109
mu-IL2	TK	—	—	C57BL/6, CBA/H, B10.A, B10.A(2R), or C3H.H-2^0 mice	IV	AC	EH	116
mu-IL4	ORFF?[g]	FLU-HA	TK	CBA/H, Swiss nude, and X-irrad mice	IV or IN	F	BCK	108
mu-IL5	?	HA	?	Mice?	IN	N	J	103

TABLE 8 (continued)
Effect of Lymphokine Genes on the Immunogenicity and Pathogenicity of Vaccinia Virus Recombinant Viruses[a]

Lymphokine gene		Reporter gene				Effect on infection		
Product	Site[b]	Product	Site[b]	Host	Route[c]	Pathogenicity[d]	Immune response[e]	Ref.
bo-IFN-γ	TK	—	—	VERO, 143 (TK-), BT cells		N	N	103
mu-IFN-γ	ORFF?[g]	IL-2	TK	BALB/c, CBA/H,	IV	ABCD	BCIL	104
mu-IFN-γ	ORFF?[g]	—	—	Swiss nude, or				105
hu-IFN-γ	ORFF?[g]	—	—	X-irrad mice				
mu-TNF-α	ORFF?[g]	HA?	TK?	CBA/H, Swiss nude, CBA/H (nu/nu) mice		ABD	BCE	105

[a] All recombinant viruses were WR strain of vaccinia virus except Reference 115.
[b] Sites of insertion of lymphokine gene or reporter gene into virus genome.
[c] Route of viral inoculation: ID, intradermal; IC, intracranial; IP, intraperitoneal; SC, subcutaneous; IV, intravenous; and IN, intranasal.
[d] Comment key: A, increased attenuation of virus, decreased morbidity; B, increased survival of inoculated animals; C, decreased number of lesions, and/or decreased size of lesion, and/or decreased ulceration of lesions, and/or accelerated healing of lesions; D, decreased virus yields recovered from tissues of infected animals; E, antiviral activity demonstrated *in vitro*, i.e., decreased level of virus replication; F, no attenuation of the virus, similar pathogenicity as wild-type virus; N, not determined.
[e] Comment key: A, no change in protection following challenge with a virulent pathogen, and no change in immunogenicity; B, decreased immunogenicity; C, level of specific antibody to reporter gene unchanged or decreased; D, level of specific antibody to FLU-HA or FLU-NP was increased in B10.A(5R) mice immunized with VV-IL2; E, no significant increase in levels and acceleration of T-cell proliferative, and/or cytolytic, and/or memory responses; F, increase in cytolytic activity from NK cells and a cell lacking NK or T-cell surface markers; G, increase in production of IFN-γ; H, immune response in mice with IR defect remains unchanged; I, increase in MHC class I and class II antigens; J, increase in IgA-producing B-cells; K, decrease in CTL activity; L, no increase in NK cell cytolytic activity; N, not determined.
[f] Copenhagen strain of vaccinia virus. VV-IL2 given at the same time as VV-gp 160 (LAV-1_{BRU}).
[g] Contains HSV-TK. Precise location within *Hin*dIII-F ORF was not given.

IL-2 in some way limits the ability of the virus to replicate and spread in the infected host. Recent studies suggest that IL-2 acts in mice at the local site of infection by stimulating an increase in natural killer (NK) cell activity, including an increase in cytolytic activity and the production of the lymphokine IFN-γ.[106,109] The data further suggest that IFN-γ produced by NK cells, as well as some non-NK cells, is more important for limiting virus replication *in vivo* than NK cytolytic activity.[106] One effect of IFN-γ secretion may be to increase the MHC class I concentration on cells surrounding a focus of infection.[107]

In individuals who have cell-mediated immune deficiency syndromes, severe vaccinia virus infections occur which can often result in death.[110–114] One intent of adding an immunomodulator such as IL-2 was to enhance the immune system at the site of virus infection and, specifically, to increase the concentration of a lymphokine which could be lacking in an individual with a cellular immune deficiency, thereby amplifying the immune response. The results to date suggest that recombinant vaccinia viruses that express IL-2 and another immunogenic protein fail to increase the level of specific antibody to that antigen in normal mice without an IR gene deficiency,[87,100,107] or in monkeys.[101,102] The data indicate that the amount of specific antibody induced by VV-IL2 recombinant viruses is lower in most species examined, probably reflecting a decrease in virus replication and a lower level of immunogen.[87,100,104] Additional studies have also demonstrated that there are no significant increases in the levels or acceleration of T cell proliferative, cytolytic, or memory responses following immunization with a VV-IL2 recombinant virus in mice or chimpanzees.[109,115,116]

The inclusion of other lymphokines has also given mixed results on virus pathogenicity and immunogenicity. A recombinant virus expressing IL-4 is no more attenuated than control virus *in vivo,* where there is a significant decrease in immunogenicity as measured by lower levels of antibody to FLU-HA and significantly lower cytotoxic T-lymphocyte (CTL) responses.[108] In contrast, it would appear that VV-IL5 recombinant virus can increase the number of IgA-secreting cells locally in the lung following intranasal inoculation.[129] It will be important to determine if the IL-5 recombinant virus also increases the number of IgA memory cells and whether expression of this lymphokine results in better protection than the control virus on challenge. In sum, experimental results to date suggest that the expression of lymphokines (except IL-5) by vaccinia virus has failed to increase the immune response of the immunized individual above control levels, although these recombinants protect mice equally well as control virus to challenge with the appropriate pathogen.

A potential complication of the incorporation of immune modulators in recombinant vaccinia viruses has also been demonstrated recently, but in only one animal model system. In immunodeficient nude mice treated with mAb to IFN-γ, there is an increase in VV-HA-IL2 virus recovered from internal organs which correlated with an increase in systemic levels of IL-2 and, most importantly, the side effects associated with higher systemic levels of IL-2, e.g., vascular leak syndrome.[106] This latter observation suggests that there is a crucial balance that must be maintained between virus replication and immunogenicity vs. systemic toxic side effects which can occur from high doses of certain immunomodulators.

The expression of bovine IFN-γ, murine IFN-γ, or murine TNF-α by vaccinia virus recombinants has been shown to limit virus replication *in vitro* and increase survival of nude or X-irradiated mice, respectively.[103,104,117] It may be that lymphokines or cytokines (such as γ-interferon) that stimulate nonspecific host defense mechanisms can best be used as an additional safety feature in future live recombinant vaccines if placed behind a controllable promoter. Then if the vaccinated individual has any virus-induced complications, virus replication could be inhibited or limited following induction of expression of the lymphokine or cytokine. Thus, the incorporation of a lymphokine or cytokine that boosts nonspecific host defense mechanisms which have antiviral properties may show more promise in the attenuation of vaccinia virus than specific immune modulators do in amplifying the immune response.

D. A TRADE-OFF: IMMUNOGENICITY VS. ATTENUATION?

With the highly virulent WR strain of vaccinia virus, the immunogenicity levels which have been achieved in chimpanzees with certain recombinants have not been sufficient to protect completely on challenge, unlike results in certain rodent models. For example, in chimpanzees a TK minus recombinant vaccinia virus (WR strain) expressing the hepatitis B surface antigen did not elicit a detectable anti-HBs antibody response following a single i.d. injection.[118] The animals were primed, however, as subsequent challenges with live HBV resulted in an anamnestic anti-HBs antibody response and a mild inapparent infection with no severe liver pathology.[118] Similarly, a series of studies showed that TK minus recombinant vaccinia viruses (WR strain) expressing F or G proteins of respiratory syncytial virus (RSV) administered via a single i.d. inoculation into mice,[119] cotton rats,[120,121] and monkeys[122] induced high levels of RSV-specific serum antibodies and high resistance to RSV replication in the respiratory tract. In chimpanzees, however, a poor neutralizing antibody response was observed which did not reduce illness, and there was only a marginal reduction in the magnitude and duration of the RSV shedding from nose and throat.[123]

There are at least three possible explanations for the failure of a vaccinia virus recombinant to protect chimpanzees from challenge. First, the level of antigen production by the TK minus recombinant virus may not be high enough to optimally stimulate the immune system of the chimpanzees although, as indicated by lesion size and anti-vaccinia virus-neutralizing antibody production, vaccinia virus replication and, presumably, antigen expression appears similar among the tested nonhuman primates.[122,123] An increase in the absolute quantity of antigen produced during an immunization could be accomplished by increasing the levels of expressions of the gene, and the use of a TK positive recombinant, which would probably replicate *in vivo* better than TK minus virus. Second, due to the selective pressure of the immune system, human pathogens like HBV or RSV may have evolved proteins which are less immunogenic in their natural host (and, presumably, in chimpanzees) than in lower primates or rodents.[123] Third, RSV replicates to far higher levels of infectivity in the respiratory tract of chimpanzees than in lower primates or rodents. Thus, a level of immunity which is sufficient to protect a cotton rat may be insufficient to protect a chimpanzee from disease.[123]

V. CONSIDERATIONS FOR SAFE RECOMBINANT VACCINIA VIRUS VACCINES

Aside from increasing the level of antigen expression, changing the time of expression, or modifying the site of antigen accumulation within the infected cell, there are a number of additional parameters which should be evaluated on a case-by-case basis for their effect on recombinant vaccinia virus antigen immunogenicity and protection from challenge.

A. VIRUS REPLICATION AND SPREAD *IN VIVO*

Many studies with serial passaged vaccinia virus strains such as CVI have contributed to the view that immunogenicity correlated with the extent of virus replication. This was based on the observation that small lesions generally gave rise to lower anti-vaccinia virus antibody titers or a lower percentage of "takes".[17] More recently, TK minus recombinants have been shown to be attenuated by a large number of parameters (Table 7) and, in certain models, have been shown to induce significantly lower anti-vaccinia virus antibody and CTL responses than TK positive virus.[124] The conclusion from these studies appears straightforward — increased attenuation of the virus equates with decreased virus replication in the host, which results in lower immunogenicity of virus-expressed antigens. Recent studies, however, are suggesting that this observed linkage may not be absolute. Intravenous inoculation of a vaccinia virus recombinant expressing TNF-α into CBA/H mice resulted in infectivity levels in the ovary and lungs that were at least 100-fold lower than control virus. In comparison to control virus, the IgM and IgG anti-vaccinia virus responses were severely reduced, but the anti-vaccinia virus CTL responses were indistinguishable.[103]

Similarly, the infection of CBA/H mice by the i.p. route with a vaccinia virus recombinant expressing mu-IL-2 generated an anti-vaccinia virus CTL response similar to control virus at all times assayed,[109] although the vaccinia virus-mu-IL-2 recombinant replicated to significantly lower levels of infectivity in ovaries. As with the vaccinia virus expressing TNF-α, the anti-vaccinia virus antibody titers were usually lower than control virus. Although one cannot rule out that the CTL responses generated with the recombinants expressing IL-2 or TNF-α were somehow enhanced by the lymphokines, an alternative view would be that very little virus replication is needed for the optimal generation of CTL, and significantly more replication is required for optimal antibody responses.

It would seem reasonable that the optimal triggering of the host immune response is not solely linked to absolute quantities of antigen but, rather, to the production of antigen which is accessible to (and in the proper form for) the immune system. This line of reasoning is supported by the natural biology of a number of dermotropic viruses (such as *Molluscum contagiosum*) which replicate to high levels in the superficial layers of the skin but invoke a poor immune response. Taken together, these observations would suggest that vaccinia virus recombinants can be designed which possess the ability to infect certain cells, thereby insuring a good immune response, but not infect others which, in turn, will yield reduced virulence and fewer post-vaccinial complications.

The host range and VGF genes should be evaluated in this context, as it is likely they are important for virus replication only in certain cells *in vivo*. The vaccinia virus equivalent to the CPV 38-kDa protein, which apparently blocks the production of a chemoattractant during infection *in vivo*, provides an alternative way for restricting virus growth but, possibly, not reducing immunogenicity. Based on one study, CPV lacking this gene replicated normally in the epidermis for approximately 20 h until the host detected and responded to a chemoattractant(s) generated in its absence.[56] The ensuing inflammatory response may well increase the immunogenicity of the virus while at the same time restricting virus replication and spread through the epidermis. The TK gene, on the other hand, may not be the best choice as a target of attenuation, as it may too effectively restrict virus replication.

B. HOST RESPONSE TO IMMUNIZATION

Several vaccinia virus-encoded genes have been shown to block a host defense mechanism (e.g., VCBP). Should other such genes be identified, they would be attractive targets for genetic manipulation, alone or in combination with other virus genes, in attempts to increase the immunogenicity of the foreign protein without increasing the virulence of the recombinant virus. Also, the inclusion into the virus genome of a lymphokine or cytokine gene may be desirable in certain situations, as it would allow a period of unrestricted viral replication (antigen expression) until the appropriate host response(s) had been engaged and virus replication restricted.

C. ROUTE OF RECOMBINANT VACCINIA VIRUS IMMUNIZATION

Currently, recombinant vaccinia virus vaccines are evaluated for efficacy usually through the intradermal route of inoculation. This is due, in large part, to the prior experience with vaccinia virus in the smallpox eradication program. It may be, however, that depending on the pathogen of interest, the route of immunization may have to be changed. An i.d. immunization route may be ill-suited if the pathogen of interest infects the mucosal surfaces, as is the case with RSV, and an intranasal route of immunization may be indicated. If so, the selection of the vaccinia virus genes for maintaining good immunogenicity but lower virulence may differ depending on the route of immunization.

D. ANIMAL MODELS FOR TESTING THE EFFICACY OF RECOMBINANT VACCINIA VIRUS VACCINES

The experience with vaccinia virus recombinants expressing F or G of RSV has provided a very important lesson in vaccinia virus recombinant virus vaccine design: that the excellent protection from challenge obtained with vaccinia virus recombinants in mice, cotton rats, and lower primates does not guarantee protection in chimpanzees and, presumably, humans. These results again highlight the importance of the chimpanzee model in the evaluation of vaccinia virus recombinants for vaccine applications.

VI. SUMMARY

Currently, the LC16 or NYCBH strains of vaccinia virus appear to be reasonable choices as vaccinia virus strains for use as a universal vaccine vehicle in humans. They have well-documented pedigrees and have had extensive human and animal testing which suggest they would have lower frequenices of CNS complications over other strains commonly used in the smallpox eradication program. Using these parental viruses, recombinant vaccinia viruses should be constructed which are efficacious prior to further efforts at reducing virus virulence. Should further attenuation of virus virulence be required, the inactivation of virus virulence genes or the incorporation into the genome of lymphokine/cytokine genes could be considered.

ACKNOWLEDGMENTS

We thank Dr. B. Moss for continued support; Mr. C. Duarte for archival research; Dr. P. Earl for assistance with viral sequence and genomic map analyses; and Ms. B. R. Marshall for continued superb editorial assistance. We are also indebted to Drs. G. Karupiah and B. Murphy for their critical and helpful review of this manuscript.

REFERENCES

1. On the State of the Public Health. Annual Report of the Chief Medical Officer of the Ministry of Health for the Year 1922, England and Wales, Ministry of Health, H. M. Stationery Office, London, 1923, 83.
2. **Fenner, F., Henderson, D. A., Arita, I., Ježek, Z., and Ladnyi, I. D.,** Smallpox and its Eradication, World Health Organization, Geneva, 1988.
3. **Gins, H. A., Hackenthal, H., and Kamentzewa, N.,** Experimentelle Untersuchungen über die Generalisierung des Vaccine-virus beim Menschen und Versuchstier, *Z. Hyg. Infektionskr.,* 110, 429, 1929.
4. **Herzberg-Kremmer, H. and Kerzberg, K.,** Untersuchungen über postvakzinale Enzephalitis. Verhalten des Vakzinevirus im Menschen bei normalem und gestörtem Impfverlauf, *Zentralbl. Bakteriol. (Naturwiss.),* 115, 271, 1930a.
5. **Herzberg-Kremmer, H. and Herzberg, K.,** Untersuchungen über postvakzinale Enzephalitis. Weitere Untersuchungen über das Verhalten des Vakzinevirus im menschlichen Körper, *Zentralbl. Bakteriol. (Naturwiss.),* 119, 175, 1930b.
6. **Siegert, R. and Schulz, W.,** Über den Beginn und die Dauer der Virämie nach Pockenschutzimpfung, *Z. Hyg. Infektionskr.,* 137, 81, 1953.
7. **Vries, E.,** *Postvaccinial Perivenous Encephalitis,* Elsevier, Amsterdam, 1960.
8. **Landrigan, P. J. and Witte, J. J.,** Neurological disorders following live measles-virus vaccination, *JAMA,* 223, 1459, 1973.
9. **Weber, G. and Lange, J.,** Zur Variationsbreite der "Inkubationszeiten" postvakzinaler zerebraler Erkrankungen, *Dtsch. Med. Wochenschr.,* 86, 1461, 1961.
10. **Spillane, J. D. and Wells, C. E.,** The neurology of Jennerian vaccination. A clinical account of the neurological complications which occurred during the smallpox epidemic in South Wales in 1962, *Brain,* 87, 1, 1964.
11. **Lane, J. M., Ruben, F. L., Neff, J. M., and Millar, J. D.,** Complications of smallpox vaccination, 1968. National surveillance in the United States, *N. Engl. J. Med.,* 281, 1201, 1969.
12. **Lane, J. M., Ruben, F. L., Neff, J. M., and Millar, J. D.,** Complications of smallpox vaccination, 1968: results of ten statewide surveys, *J. Infect. Dis.,* 122, 303, 1970.
13. **Polak, M. F.,** Complications of smallpox vaccination in the Netherlands, 1959–1970, in *International Symposium on Smallpox Vaccine, Bilthoven, the Netherlands, 11-13 October 1972; Symposia Series in Immunobiological Standardization,* Vol. 19, S. Karger, Basel, 1973, 235.
14. **Berger, K. and Heinrich, W.,** Decrease in postvaccinial deaths in Austria after introducing a less pathogenic virus strain, in *International Symposium on Smallpox Vaccine, Bilthoven, the Netherlands, 11-13 October 1972; Symposia Series in Immunobiological Standardization,* Vol. 19, S. Karger, Basel, 1973, 199.
15. **Marennikova, S. S.,** Evaluation of vaccine strains by their behaviour in vaccinated animals and possible implications of the revealed features for smallpox vaccination practice, in *International Symposium on Smallpox Vaccine, Bilthoven, the Netherlands, 11-13 October 1972; Symposia Series in Immunobiological Standardization,* Vol. 19, S. Karger, Basel, 1973, 253.
16. **Polak, M. F., Beunders, B. J. W., Van der Werff, A. R., Sanders, E. W., Van Klaveren, J. N., and Brans, L. M.,** A comparative study of clinical reactions observed after application of several smallpox vaccines in primary vaccination of young adults, *Bull. W.H.O.,* 29, 311, 1963.
17. **Galasso, G. J., Karzon, D. T., Katz, S. L., Krugman, S., Neff, J., and Robbins, F. C., Eds.,** Clinical and serologic study of four smallpox vaccines comparing variations of dose and route of administration, *J. Infect. Dis.,* 135, 131, 1977.
18. **Perkus, M. E., Goebel, S. J., Davis, S. W., Johnson, G. P., Limbach, K., Norton, E. K., and Paoletti, E.,** Vaccinia virus host range genes, *Virology,* 179, 276, 1990.

19. **Drillien, R., Koehren, F., and Kirn, A.,** Host range deletion mutant of vaccinia virus defective in human cells, *Virology,* 111, 488, 1981.
20. **Gillard, S., Spehner, D., and Drillien, R.,** Mapping of a vaccinia host range sequence by insertion into the viral thymidine kinase gene, *J. Virol.,* 53, 316, 1985.
21. **Gillard, S., Spehner, D., Drillien, R., and Kirn, A.,** Localization and sequence of a vaccinia virus gene required for multiplication in human cells, *Proc. Natl. Acad. Sci. U.S.A.,* 83, 5573, 1986.
22. **Gillard, S., Spehner, D., Drillien, R., and Kirn, A.,** Antibodies directed against a synthetic peptide enable detection of a protein encoded by a vaccinia virus host range gene that is conserved within the *Orthopoxvirus* genus, *J. Virol.,* 63, 1814, 1989.

22a. **Drillien, R.,** personal communication.

23. **Tomley, G., Binns, M., Campbell, J., and Boursnell, M.,** Sequence analysis of an 11.2 kilobase, near-terminal, *Bam*HI fragment of fowlpox virus, *J. Gen. Virol.,* 69, 1025, 1988.
24. **Lux, S. E., John, K. M., and Bennett, V.,** Analysis of cDNA for human erythrocyte ankyrin indicates a repeated structure with homology to tissue-differentiation and cell-cycle control proteins, *Nature,* 344, 36, 1990.
25. **Mayr, A., Hochstein-Mintzel, V., and Stickl, H.,** Abstammung, Eigenschaften und Verwendung des attenuierten Vaccinia-Stammes MVA, *Infection,* 3, 6, 1975.
26. **Altenburger, W., Süter, C.-P., and Altenburger, J.,** Partial deletion of the human host range gene in the attenuated vaccinia virus MVA, *Arch. Virol.,* 105, 15, 1989.
27. **Drillien, R., Spehner, D., and Kirn, A.,** Host range restriction of vaccinia virus in Chinese hamster ovary cells: relationship to shutoff of protein synthesis, *J. Virol.,* 28, 843, 1978.
28. **Hruby, D. E., Lynn, D. L., Condit, R. C., and Kates, J. R.,** Cellular differences in the molecular mechanisms of vaccinia virus host range restriction, *J. Gen. Virol.,* 47, 485, 1980.
29. **Spehner, D., Gillard, S., Drillien, R., and Kirn, A.,** A cowpox virus gene required for multiplication in Chinese hamster ovary cells, *J. Virol.,* 62, 1297, 1988.
30. **Kotwal, G. J. and Moss, B.,** Analysis of a large cluster of nonessential genes deleted from a vaccinia virus terminal transposition mutant, *Virology,* 167, 524, 1988.
31. **Goebel, S. J., Johnson, G. P., Perkus, M. E., Davis, S. W., Winslow, J. P., and Paoletti, E.,** The complete DNA sequence of vaccinia virus, *Virology,* 179, 247, 1990.
32. **Blomquist, M. C., Hunt, L. T., and Barker, W. C.,** Vaccinia virus 19-kilodalton protein: relationship to several mammalian proteins, including two growth factors, *Proc. Natl. Acad. Sci. U.S.A.,* 81, 7363, 1984.
33. **Brown, J. P., Twardzik, D. R., Marquardt, H., and Todaro, G. J.,** Vaccinia virus encodes a polypeptide homologous to epidermal growth factor and transforming growth factor, *Nature,* 313, 491, 1985.
34. **Reisner, A. H.,** Similarity between the vaccinia virus 19K early protein and epidermal growth factor, *Nature,* 313, 801, 1985.
35. **Stroobant, P., Rice, A. P., Gullick, W. J., Cheng, D. J., Kerr, I. M., and Waterfield, M. D.,** Purification and characterization of vaccinia virus growth factor, *Cell,* 42, 383, 1985.
36. **Twardzik, D. R., Brown, J. P., Ranchalis, J. E., Todaro, G. J., and Moss, B.,** Vaccinia virus-infected cells release a novel polypeptide functionally related to transforming and epidermal growth factors, *Proc. Natl. Acad. Sci. U.S.A.,* 82, 5300, 1985.
37. **King, C. S., Cooper, J. A., Moss, B., and Twardzik, D. R.,** Vaccinia virus growth factor stimulates tyrosine protein kinase activity of A431 cell epidermal growth factor receptors, *Mol. Cell. Biol.,* 6, 332, 1986.
38. **Buller, R. M. L., Chakrabarti, S., Cooper, J. A., Twardzik, D. R., and Moss, B.,** Deletion of the vaccinia virus growth factor gene reduces virus virulence, *J. Virol.,* 62, 866, 1988.

39. **Buller, R. M. L., Chakrabarti, S., Moss, B., and Fredrickson, T.,** Cell proliferative response to vaccinia virus is mediated by VGF, *Virology,* 164, 182, 1988.
40. **Buller, R. M. L. and Moss, B.,** Genetic basis for vaccinia virus virulence, in *Vaccinia Viruses as Vectors for Vaccine Antigens,* Quinnan, G. V., Ed., Elsevier, New York, 1985, 37.
41. **Buller, R. M. L., Smith, G. L., Cremer, K., Notkins, A. L., and Moss, B.,** Decreased virulence of recombinant vaccinia virus expression vectors is associated with a thymidine kinase negative phenotype, *Nature,* 317, 813, 1985.
42. **Child, S. J., Palumbo, G. J., Buller, R. M. L., and Hruby, D. E.,** Insertional inactivation of the large subunit of ribonucleotide reductase encoded by vaccinia virus is associated with reduced virulence, *in vivo, Virology,* 174, 625, 1990.
43. **Mims, C. A.,** Aspects of the pathogenesis of virus diseases, *Bacteriol. Rev.,* 28, 30, 1964.
44. **Ada, G. L., Jackson, D. C., Blanden, R. V., Tha Hla, R., and Bowern, N. A.,** Changes in the surface of virus infected cells recognized by cytotoxic T cells. I. Minimal requirements for lysis of ectromelia infected P815, *Scand. J. Immunol.,* 5, 23, 1976.
45. **Jackson, D. C., Ada, G. L., and Tha Hla, R.,** Cytotoxic T cells recognize very early, minor changes in ectromelia virus-infected target cells, *Aust. J. Exp. Biol. Med. Sci.,* 54, 349, 1976.
46. **Hapel, A. J., Bablanian, R., and Cole, G. A.,** Inductive requirements for the generation of virus-specific T lymphocytes. II. Poxvirus and H-2 antigens associate without cellular or virus-directed protein synthesis, and remain immunogenic in cell membrane fragments, *J. Immunol.,* 124, 1990, 1980.
47. **Mallon, V. R. and Holowczak, J. A.,** Vaccinia virus antigens on the plasma membrane of infected cells. I. Viral antigens transferred from infecting virus particles and synthesized after infection, *Virology,* 141, 201, 1985.
48. **Natuk, R. J. and Holowczak, J. A.,** Vaccinia virus proteins on the plasma membrane of infected cells. III. Infection of peritoneal macrophages, *Virology,* 147, 354, 1985.
49. **Domber, E. and Holowczak, J. A.,** Vaccinia virus proteins on the plasma membranes of infected cells. IV. Studies employing L cells infected with ultraviolet-irradiated vaccinia virions, *Virology,* 152, 331, 1986.
50. **Travis, J. and Salvesen, G. S.,** Human plasma proteinase inhibitors, *Annu. Rev. Biochem.,* 52, 665, 1983.
51. **Pickup, D. J., Ink, B. S., Hu, W., Ray, C. A., and Joklik, W. K.,** Hemorrhage in lesions caused by cowpox virus is induced by a viral protein that is related to plasma protein inhibitors of serine proteases, *Proc. Natl. Acad. Sci. U.S.A.,* 83, 7698, 1986.
52. **Boursnell, M. E. G., Foulds, I. J., Campbell, J. I., and Binns, M. M.,** Non-essential genes in the vaccinia virus *Hin*dIII K fragment: a gene related to serine protease inhibitors and a gene related to the 37K vaccinia virus major envelope antigen, *J. Gen. Virol.,* 69, 2995, 1988.
53. **Kotwal, G. J. and Moss, B.,** Vaccinia virus encodes two proteins that are structurally related to members of the plasma serine protease inhibitor superfamily, *J. Virol.,* 63, 600, 1989.
54. **Pickup, D. J., Ink, B. S., Parsons, B. L., Hu, W., and Joklik, W. K.,** Spontaneous deletions and duplications of sequences in the genome of cowpox virus, *Proc. Natl. Acad. Sci. U.S.A.,* 81, 6817, 1984.
55. **Smith, G. L., Howard, S. T., and Chan, Y. S.,** Vaccinia virus encodes a family of genes with homology to serine proteinase inhibitors, *J. Gen. Virol.,* 70, 2333, 1989.
56. **Palumbo, G. J., Pickup, D. J., Fredrickson, T. N., McIntyre, L. J., and Buller, R. M. L.,** Inhibition of an inflammatory response is mediated by a 38-kDa protein of cowpox virus, *Virology,* 172, 262, 1989.

56a. **Palumbo, G. J.,** unpublished results.

57. **Kotwal, G. J. and Moss, B.,** Vaccinia virus encodes a secretory polypeptide structurally related to complement control proteins, *Nature,* 335, 176, 1988.

58. **Kotwal, G. J., Isaacs, S. N., McKenzie, R., Frank, M. M., and Moss, B.,** Inhibition of the complement cascade by the major secretory protein of vaccinia virus, *Science,* 250, 827, 1990.
59. **Friedman, H. M., Cohen, G. H., Eisenberg, R. J., Seidel, C. A., and Cines, D. B.,** Glycoprotein C of herpes simplex virus 1 acts as a receptor for the C3b complement component on infected cells, *Nature,* 309, 633, 1984.
60. **Seidel-Dugan, C., Ponce de Leon, M., Friedman, H. M., Fries, L. F., Frank, M. M., Cohen, G. H., and Eisenberg, R. J.,** C3b receptor activity on transfected cells expressing glycoprotein C of herpes simplex virus types 1 and 2, *J. Virol.,* 62, 4027, 1988.
61. **Whitaker-Dowling, P. and Younger, J. S.,** Vaccinia rescue of VSV from interferon-induced resistance: reversal of translation block and inhibition of protein kinase activity, *Virology,* 131, 128, 1983.
62. **Paez, E. and Esteban, M.,** Resistance of vaccinia virus to interferon is related to an interference phenomenon between the virus and the interferon system, *Virology,* 134, 12, 1984.
63. **Rice, A. P. and Kerr, I. M.,** Interferon-mediated, double-stranded RNA-dependent protein kinase is inhibited in extracts from vaccinia virus-infected cells, *J. Virol.,* 50, 229, 1984.
64. **Whitaker-Dowling, P. and Younger, J. S.,** Characterization of a specific kinase inhibitory factor produced by vaccinia virus which inhibits the interferon-induced protein kinase, *Virology,* 137, 171, 1984.
65. **Brown, G. E., Lebleu, B., Kawakita, M., Shaila, S., Sen, G. C., and Lengyel, P.,** Increased endonuclease activity in an extract from mouse Ehrlich ascites tumor cells which had been treated with a partially purified interferon preparation: dependence on double-stranded RNA, *Biochem. Biophys. Res. Commun.,* 69, 114, 1976.
66. **Ratner, L., Sen, G. C., Brown, G. E., Lebleu, B., Kawakita, M., Cabrer, B., Slattery, E., and Lengyel, P.,** Interferon, double-stranded RNA and RNA degradation. Characteristics of an endonuclease activity, *Eur. J. Biochem.,* 79, 565, 1977.
67. **Baglioni, C., Minks, M. A., and Maroney, P. A.,** Interferon action may be mediated by activation of a nuclease by pppA2′p5′A2′p5′A, *Nature,* 273, 684, 1978.
68. **Kerr, I. M. and Brown, R. E.,** pppA2′p5′A2′p5′A: an inhibitor of protein synthesis synthesized with an enzyme fraction from interferon-treated cells, *Proc. Natl. Acad. Sci. U.S.A.,* 75, 256, 1978.
69. **Paez, E. and Esteban, M.,** Nature and mode of action of vaccinia virus products that block activation of the interferon-mediated ppp(A2′p)$_n$A-synthetase, *Virology,* 134, 29, 1984.
70. **Rice, A. P., Kerr, S. M., Roberts, W. K., Brown, R. E., and Kerr, I. M.,** Novel 2′,5′-oligoadenylates synthesized in interferon-treated, vaccinia virus-infected cells, *J. Virol.,* 56, 1041, 1985.
71. **Esteban, M., Benavente, J., and Paez, E.,** Mode of sensitivity and resistance of vaccinia virus replication to interferon, *J. Gen. Virol.,* 67, 801, 1986.
72. **Youngner, J. S., Thacore, H. R., and Kelly, M. E.,** Sensitivity of ribonucleic acid and deoxyribonucleic acid viruses to different species of interferon in cell cultures, *J. Virol.,* 10, 171, 1972.
73. **Fenner, G. and Ratcliffe, F. N.,** *Myxomatosis,* Cambridge University Press, Cambridge, 1965.
74. **Strayer, D. S.,** Poxviruses, in *Viral-Induced Immunosuppression,* Specter, S., Bendinelli, M., and Friedmann, H., Eds., Plenum Press, New York, 1989, 173.
75. **Archard, L. C. and Mackett, M.,** Restriction endonuclease analysis of red cowpox virus and its white pock variant, *J. Gen. Virol.,* 45, 51, 1979.
76. **Dumbell, K. R. and Archard, L. C.,** Comparison of white pock (h) mutants of monkeypox virus with parental monkeypox and with variola-like viruses isolated from animals, *Nature,* 286, 29, 1980.

77. **Lake, J. R. and Cooper, P. D.,** Deletions of the terminal sequences in the genomes of the white pock (μ) and host-restricted (ρ) mutants of rabbitpox virus, *J. Gen. Virol.,* 48, 135, 1980.
78. **Moyer, R. W., Graves, R. L., and Rothe, C. T.,** The white pock (μ) mutants of rabbit poxvirus. III. Terminal DNA sequence duplication and transposition in rabbit poxvirus, *Cell,* 22, 545, 1980.
79. **Moyer, R. W. and Rothe, C. T.,** The white pock mutants of rabbit poxvirus. I. Spontaneous host range mutants contain deletions, *Virology,* 102, 119, 1980.
80. **Esposito, J. J., Cabradilla, C. D., Nakano, J. H., and Obijeski, J. F.,** Intragenomic sequence transposition in monkeypox virus, *Virology,* 109, 231, 1981.
81. **Cooper, J. A., Wittek, R., and Moss, B.,** Extension of the transcriptional and translational map of the left end of the vaccinia virus genome to 21 kilobase pairs, *J. Virol.,* 39, 733, 1981.
82. **Kotwal, G. J., Hügin, A. W., and Moss, B.,** Mapping and insertional mutagenesis of a vaccinia virus gene encoding a 13,800-Da secreted protein, *Virology,* 171, 579, 1989.
83. **Niles, E. G. and Seto, J.,** Vaccinia virus gene D8 encodes a virion transmembrane protein, *J. Virol.,* 62, 3772, 1988.
84. **Ichihashi, Y. and Dales, S.,** Biogenesis of poxviruses: interrelationship between hemagglutinin production and polykariocytosis, *Virology,* 46, 533, 1971.
85. **Shida, H.,** Nucleotide sequence of the vaccinia virus hemagglutinin gene, *Virology,* 150, 451, 1986.
86. **Shida, H., Tochikura, T., Sato, T., Konno, T., Hirayoshi, K., Seki, M., Ito, Y., Hatanaka, M., Hinuma, Y., Sugimoto, M., Takahashi-Nishimaki, F., Maruyama, T., Miki, K., Suzuki, K., Morita, M., Sashiyama, H., and Hayami, M.,** Effect of the recombinant vaccinia viruses that express HTLV-I envelope gene on HTLV-I infection, *EMBO J.,* 6, 3379, 1987.
87. **Flexner, C., Hügin, A., and Moss, B.,** Prevention of vaccinia virus infection in immunodeficient mice by vector-directed IL-2 expression, *Nature,* 330, 259, 1987.
88. **Hochstein-Mintzel, V., Hanichen, T., Huber, H. C., and Stickl, H.,** Vaccinia- und variolaprotektive Wirkung des modifizierten Vaccinia-Stammes MVA bei intramuskularer Immunisierung, *Zentralbl. Bakteriol. (Naturwiss.),* 230, 283, 1975.
89. **Tagaya, I., Kitamura, T., and Sano, Y.,** A new mutant of dermovaccinia virus, *Nature,* 192, 381, 1961.
90. **Hashizume, S., Morita, T., Yoshizawa, H., Suzuki, K., Arita, M., Komatsu, T., Amano, H., and Tagaya, I.,** Intracerebral inoculation of monkeys with several vaccinia virus strains: an approach to the comparison of different strains, in *International Symposium on Smallpox Vaccine, Bilthoven, the Netherlands, 11–13 October 1972; Symposia Series in Immunobiological Standardization,* Vol. 19, S. Karger, Basel, 1973, 325.
91. **Rivers, T. M.,** Cultivation of vaccine virus for Jennerian prophylaxis in man, *J. Exp. Med.,* 54, 453, 1931.
92. **Rivers, T. M. and Ward, S. M.,** Jennerian prophylaxis by means of intradermal injections of culture vaccine virus, *J. Exp. Med.,* 62, 549, 1935.
93. **Barker, L. F.,** Further attenuated vaccinia virus: a possible alternative for primary immunization, in *Sixth Annual Immunization Conference,* United States Communicable Disease Center, Atlanta, 1969, 55.
94. **Hashizume, S., Yoshizawa, H., Morita, M., and Suzuki, K.,** Properties of attenuated mutant of vaccinia virus, LC16m8 derived from Lister strain, in *Vaccinia Viruses as Vectors for Vaccine Antigens,* Quinnan, J., Ed., Elsevier, New York, 1985, 87.
95. **Sugimoto, M., Yasuda, A., Miki, K., Morita, M., Suzuki, K., Uchida, N., and Hashizume, S.,** Gene structures of low-neurovirulent vaccinia virus LC16m0, LC16m8, and their Lister original (LO) strains, *Microbiol. Immunol.,* 29, 421, 1985.

96. **Takahashi-Nishimaki, F., Suzuki, K., Morita, M., Maruyama, T., Miki, K., Hashizume, S., and Sugimoto, M.,** Genetic analysis of vaccinia virus Lister strain and its attenuated mutant LC16m8: production of intermediate variants by homologous recombination, *J. Gen. Virol.,* 68, 2705, 1987.
97. **Morita, M., Aoyama, Y., Arita, M., Amana, H., Yoshizawa, H., Hashizume, S., Komatsu, T., and Tagaya, I.,** Comparative studies of several vaccinia virus strains by intrathalamic inoculation into cynomolgus monkeys, *Arch. Virol.,* 53, 197, 1977.
98. **Morita, M., Suzuki, K., Yasuda, A., Kojima, A., Sugimoto, M., Watanabe, K., Kobayashi, H., Kajima, K., and Hashizume, S.,** Recombinant vaccinia virus LC16m0 or LC16m8 that expresses hepatitis B surface antigen while preserving the attenuation of the parental virus strain, *Vaccine,* 5, 65, 1987.
99. **Shida, H., Hinuma, Y., Hatanaka, M., Morita, M., Kidokoro, M., Suzuki, K., Maruyama, T., Takahashi-Nishimaki, F., Sugimoto, M., Kitamura, R., Miyazawa, T., and Hayami, M.,** Effects and virulences of recombinant vaccinia viruses derived from attenuated strains that express the human T cell leukemia virus type I envelope gene, *J. Virol.,* 62, 4474, 1988.
100. **Ramshaw, I. A., Andrew, M. E., Phillips, S. M., Boyle, D. B., and Coupar, B. E. H.,** Recovery of immunodeficient mice from a vaccinia virus/IL-2 recombinant infection, *Nature,* 329, 545, 1987.
101. **Ruby, J., Brinkman, C., Jones, S., and Ramshaw, I.,** Response of monkeys to vaccination with recombinant vaccinia virus which coexpress HIV gp160 and human interleukin-2, *Immunol. Cell Biol.,* 68, 113, 1990.
102. **Flexner, C., Moss, B., London, W. T., and Murphy, B. R.,** Attenuation and immunogenicity in primates of vaccinia virus recombinants expressing human interleukin-2, *Vaccine,* 8, 17, 1990.
103. **Ramshaw, I. A., Sambhi, S., Karupiah, G., Kohonen-Corish, M., Ruby, J., and Ramsay, A.,** Vaccinia-directed cytokine gene expression and infection, in *Immunology of Virus Diseases,* Blanden, R. V., Ed., Brolga Press, Curtin, Australia, 1989, 35.
104. **Kohonen-Corish, M. R. J., King, N. J. C., Woodhams, C. E., and Ramshaw, I. A.,** Immunodeficient mice recover from infection with vaccinia virus expressing interferon-γ, *Eur. J. Immunol.,* 20, 157, 1990.
105. **Ramshaw, I. A., Karupiah, G., Sambhi, S., Ruby, J., Ramsay, A., and Kohonen-Corish, M.,** The analysis of immune effector mechanisms using recombinant vaccinia viruses that express their own cytokines, in *Biological Agents in the Treatment of Cancer,* Hersey, P., Ed., NSW Government Publication Office, Australia, 1990, 1.
106. **Karupiah, G., Blanden, R. V., and Ramshaw, I. A.,** Interferon-γ is involved in the recovery of athymic nude mice from recombinant vaccinia virus/interleukin 2 infection, *J. Exp. Med.,* 172, 1495, 1990.
107. **Kohonen-Corish, M. R. J., Blanden, R. V., and King, N. J. C.,** Induction of cell surface expression of HLA antigens by human IFN-γ encoded by recombinant vaccinia virus, *J. Immunol.,* 143, 623, 1989.
108. **Andrew, M. E., Coupar, B. E. H., and Boyle, D. B.,** Use of recombinant vaccinia viruses to study MHC class I-restricted recognition and lymphokine function, in *Immunology of Virus Diseases,* Blanden, R. V., Ed., Brolga Press, Curtin, Australia, 1989, 47.
109. **Karupiah, G., Coupar, B. E. H., Andrew, M. E., Boyle, D. B., Phillips, S. M., Müllbacher, A., Blanden, R. V., and Ramshaw, I. A.,** Elevated nautral killer cell responses in mice infected with recombinant vaccinia virus encoding murine IL-2, *J. Immunol.,* 144, 290, 1990.
110. **Kempe, C. H.,** Studies on smallpox and complications of smallpox vaccination, *Pediatrics,* 20, 176, 1960.
111. **O'Connell, C. J., Karzon, D. T., Barron, A. L., Plaut, M. E., and Ali, V. M.,** Progressive vaccinia with normal antibodies. A case possibly due to deficient cellular immunity, *Ann. Intern. Med.,* 60, 282, 1964.

112. **Hansson, O., Johansson, S. G. O., and Vahlquist, B.,** Vaccinia gangrenosa with normal humoral antibodies, *Acta Paediatr. Scand.*, 55, 264, 1966.
113. **Fulginiti, V. A., Kempe, C. H., Hathaway, W. E., Pearlman, D. S., Serber, O. F., Jr., Eller, J. J., Joyner, J. J., Sr., and Robinson, A.,** Progressive vaccinia in immunologically deficient individuals, in *Immunologic Deficiency Diseases in Man. Birth Defects,* Original Article Ser., Vol. 4, Bergsma, D., Ed., National Foundation of the March of Dimes, New York, 1968.
114. **Redfield, R. R., Wright, D. C., James, W. D., Jones, T. S., Brown, C., and Burke, D. S.,** Disseminated vaccinia in a military recruit with human immunodeficiency virus (HIV) disease, *N. Engl. J. Med.*, 316, 673, 1987.
115. **van Eendenburg, J.-P., Yagello, M., Girard, M., Kieny, M.-P., LeCocq, J.-P., Muchmore, E., Fultz, P. N., Riviere, Y., Montagnier, L., and Gluckman, J.-C.,** Cell-mediated immune proliferative responses to HIV-1 of chimpanzees vaccinated with different vaccinia recombinant viruses, *AIDS Res. Human Retrovir.*, 5, 41, 1989.
116. **Müllbacher, A., Ramshaw, I. A., and Coupar, B. E. H.,** Vaccinia-interleukin 2 recombinant virus.or exogenous interleukin 2 does not alter the magnitude or immune response gene defects of the cytotoxic T-cell response to vaccinia virus *in vivo, Scand. J. Immunol.*, 29, 1, 1989.
117. **Yilma, T., Anderson, K., Brechling, K., and Moss, B.,** Expression of an adjuvant gene (interferon-γ) in infectious vaccinia virus recombinants, in *Vaccines 87,* Cold Spring Harbor Laboratory, Cold Spring Harbor, New York, 1987, 393.
118. **Moss, B., Smith, G. L., Gerin, J. L., and Purcell, R. H.,** Live recombinant vaccinia virus protects chimpanzees against hepatitis B, *Nature,* 311, 67, 1984.
119. **Wertz, G. W., Stott, E. J., Young, K. K.-Y., Anderson, K., and Ball, L. A.,** Expression of the fusion protein of human respiratory syncytial virus from recombinant vaccinia virus vectors and protection of vaccinated mice, *J. Virol.*, 61, 293, 1987.
120. **Elango, N., Prince, G. A., Murphy, B. R., Venkatesan, S., Chanock, R. M., and Moss, B.,** Resistance to human respiratory syncytial virus (RSV) infection induced by immunization of cotton rats with a recombinant vaccinia virus expressing the RSV G glycoprotein, *Proc. Natl. Acad. Sci. U.S.A.*, 83, 1906, 1986.
121. **Olmsted, R. A., Elango, N., Prince, G. A., Murphy, B. R., Johnson, P. R., Moss, B., Chanock, R. M., and Collins, P. L.,** Expression of the F glycoprotein of respiratory syncytial virus by a recombinant vaccinia virus: comparison of the individual contributions of the F and G glycoproteins to host immunity, *Proc. Natl. Acad. Sci. U.S.A.*, 83, 7462, 1986.
122. **Olmsted, R. A., Buller, R. M. L., Collins, P. L., London, W. T., Beeler, J. A., Prince, G. A., Chanock, R. M., and Murphy, B. R.,** Evaluation in non-human primates of the safety, immunogenicity and efficacy of recombinant vaccinia viruses expressing the F or G glycoprotein of respiratory syncytial virus, *Vaccine,* 6, 519, 1988.
123. **Collins, P. L., Purcell, R. H., London, W. T., Lawrence, L. A., Chanock, R. M., and Murphy, B. R.,** Evaluation in chimpanzees of vaccinia virus recombinants that express the surface glycoproteins of human respiratory syncytial virus, *Vaccine,* 8, 164, 1990.
124. **Andrew, M. E., Coupar, B. E. H., and Boyle, D. B.,** Humoral and cell-mediated immune responses to recombinant vaccinia viruses in mice, *Immunol. Cell Biol.*, 67, 331, 1989.
125. **Stott, E. J., Ball, L. A., Young, K. K., Furze, J., and Wertz, G. W.,** Human respiratory syncytial virus glycoprotein G expressed from a recombinant vaccinia virus vector protects mice against live-virus challenge, *J. Virol.*, 60, 607, 1986.
126. **Morrison, H. G., Bauer, S. P., Lange, J. V., Esposito, J. J., McCormick, J. B., and Auperin, D. D.,** Protection of guinea pigs from Lassa fever by vaccinia virus recombinants expressing the nucleoprotein or the envelope glycoproteins of Lassa virus, *Virology,* 171, 179, 1989.

127. **Earl, P. L. and Moss, B.,** Vaccinia virus, in *Genetic Maps: Locus Maps of Complex Genomes,* 5th ed., O'Brien, S. J., Ed., Cold Spring Harbor Laboratory Press, Cold Spring Harbor, New York, 1990, 1.138.
128. **Fredrickson, T. G., Sechler, J. M. G., Palumbo, G. J., Albert, J., Khairallah, L. H., and Buller, R. M. L.,** Acute inflammatory response to cowpox virus infection of the chorioallantoic membrane of the chick embryo, *Virology,* in press.
129. **Ramsay, A.,** personal communication.
130. **Hügin, A., Isaacs, S., Kotwal, G., and Moss, B.,** personal communication.

Chapter 8

AVIPOXVIRUS VECTORS

Michael E. G. Boursnell

TABLE OF CONTENTS

I. INTRODUCTION

As this book amply demonstrates, vaccinia virus has been widely used as a vector for expressing foreign genes, and many of these recombinant viruses have been used as prototype vaccines (see Reference 1 for review) and the possibility of using these recombinants in humans has been widely discussed. Although there have been complications associated with the use of vaccinia virus as a vaccine against smallpox,[2] these complications are well documented and the incidences are low. Hence when the benefit to the patient or community is great enough, it is reasonable to consider the use of recombinant vaccinia viruses in humans. Recombinant vaccinia viruses have also been tested and proved effective in protecting animals against diseases of veterinary importance.[3,4] However, because of the wide host range of vaccinia virus, there is a risk of spread of the recombinant outside the target species. In these cases it is not desirable or necessary to accept the risk of using vaccinia virus if an equivalent species-specific poxvirus is available. Such is the case in the poultry field, where the avipoxvirus, fowlpox virus, has a host range which is restricted to avian species.[5] Hence, it is a suitable candidate for the construction and use of recombinant poxvirus-based vaccines.

Avian poxviruses such as fowlpox, pigeonpox, canarypox, quailpox, etc., are all members of the genus *Avipox*.[5] They are similar in many ways to the orthopox viruses, but appear to have a considerably larger genome, with estimates of size ranging from 240 to 300 kb.[6,7] The most-studied member of the genus is fowlpoxvirus. Although the term "fowlpox" originally included all pox diseases of birds, it now refers to the disease of chickens. Although Mayr and Mahnel[8] have reported the isolation of the virus from a rhinocerous, there are no reports of its replication in nonavian species. It is a slow-spreading disease characterized by a pronounced transitory inflammatory process, hyperplasia of the epidermis and feather follicles with intracytoplasmic inclusions, and terminating with the formation of scabs and desquamation of degenerated epithelium. In infected dermal epithelium and chorioallantoic membranes of chicken embryos, the virus-containing intracytoplasmic inclusions can be seen surrounded by a membrane composed of a granular matrix. Infection occurs through mechanical transmission of the virus to injured or lacerated skin.[5] Transmission via mosquitoes can also occur, but there is no evidence of replication within the insect.[9]

In order for fowlpox virus to be developed as a vector it has been necessary to study several features of the virus and these will be discussed in turn. The next two sections deal with the main two components of a poxvirus vector system: (1) identification of nonessential regions of the genome for insertion of foreign genes, including methods of selecting for recombinants, and (2) investigation of promoters which would function well in fowlpox virus. In order for these features of the system to be well characterized, a certain amount of basic molecular biology of the virus has also been studied and this will also be discussed briefly. Finally the putting together of these components with genes from various pathogens into prototype recombinant vaccines against several diseases will be discussed.

II. NONESSENTIAL REGIONS AND SELECTION METHODS

By analogy with vaccinia virus, it was expected that fowlpox virus would have large areas of its genome which were not essential for growth in tissue culture. Two approaches have been used to identify such nonessential regions. The simplest of these approaches is to insert genes at randomly chosen sites within cloned fowlpox virus sequences and test empirically whether the site is nonessential. This approach has proved successful, suggesting that a fairly high proportion of the fowlpox virus genome is nonessential, although it has been found that expression levels of marker genes have depended on the particular site.[10] The second approach has been to identify and characterize particular regions of the genome which for one reason or another appear to be suitable for use as insertion sites, and to test those for nonessentiality. This second approach will be discussed in more detail here, not because it is more efficient, but because it has revealed more about the organization of the fowlpox virus genome.

The most widely used site of insertion in the vaccinia virus genome is the thymidine kinase (TK) gene. The TK gene in vaccinia, and in herpes-viruses, is not essential for replication in tissue culture. Hence it was one of the main targets of many groups to identify this gene in fowlpox virus and to investigate the possibility of using it as an insertion site. Several approaches were used to identify this gene.

A. HYBRIDIZATION

Hybridization of clones containing vaccinia TK sequences to fowlpox virus DNA was carried out by several groups, with mixed results. In general it was found that vaccinia DNA had little homology to fowlpox virus DNA. However the vaccinia virus *Hin*dIII J clone, containing the TK gene, was found to hybridize to a 13-kb fowlpox virus *Hin*dIII fragment.[11] A 3.1-kb section of this was sequenced and found to contain several genes which were clearly homologs of the vaccinia genes in the *Hin*dIII J fragment. Unfortunately one gene was missing from this conserved region, namely the TK gene itself.

B. RANDOM SEQUENCING

At the same time, another approach to identifying fowlpox virus genes was apparently closing in on the same region. Binns et al.[12] had developed a random sequencing strategy to identify large numbers of genes by homology with vaccinia sequences. This approach was to yield much information about the organization of the fowlpox virus genome, and was carried out as follows. Whole fowlpox virus genomic DNA was sonicated and cloned into an M13 vector. Clones were sequenced at random and compared in all six reading frames to libraries of amino acid sequences of published vaccinia genes. Several genes were quickly identified, including homologs of genes near to the TK gene in vaccinia. However when the region around these genes was sequenced in this strain of fowlpox virus the same translocation of the TK was found.[13]

C. COMPLEMENTATION

While these approaches at the nucleic acid level were being carried out, the fowlpox virus TK gene was identified by an elegant complementation technique by Boyle and Coupar.[14] Restriction fragments of fowlpox virus were shotgun-cloned into a TK^- vaccinia virus, and TK^+ recombinants selected. A 5.5-kb *Eco*RI fragment was found to confer the TK^+ phenotype on the vaccinia virus. When part of this fragment was sequenced the fowlpox virus TK gene was identified by its homology with the vaccinia virus TK.[15]

Although the TK seemed to offer a general site for insertion of foreign genes, in fact workers found mixed fortunes when attempting to insert genes at this site. Boyle and Coupar,[16] using the Arthur Webster mild vaccine strain of fowlpox virus growing in chickn embryo skin cells (CES), successfully constructed fowlpox virus recombinants with the human influenza hemagglutinin gene inserted into the *Nco*I site of the TK gene. Because of the fact that there are no good avian TK^- cell lines, it was not possible to select for TK^- recombinants as could be done for vaccinia virus. However, Boyle and Coupar[17] developed a positive selection technique whereby the *Escherichia coli* xanthine-guanine phosphoribosyl transferase gene (gpt) was inserted, under the control of a vaccinia promoter, alongside the foreign gene of interest. This allowed the recombinants to be selected by virtue of their resistance to mycophenolic acid. In some other labs, using chick embryo fibroblasts (CEFs), it was found that recombinants in the TK gene were unstable. Because recombinants elsewhere in the viral genome were successfully obtained by these groups, it seems clear that this is not merely a technical failure but is a genuine result, and it suggests that, in a particular cell-line or strain of virus, or both, the TK gene is not entirely dispensable for growth in tissue culture.

Several other sites in the fowlpox virus genome have been used for the insertion of foreign genes. Taylor et al.[18] used a pair of *Hinc*II sites within a 900-base pair *Pvu*II fragment of the FP-1 strain of fowlpox virus (a Duvette strain, O DCEP 25/CEP67/2309 October 1980: Institute Merieux) presumably identified empirically by random insertion into cloned fowlpox virus DNA. Boursnell et al.[19] used the terminal inverted repeat regions (TIR) which are present in avipoxviruses as well as in orthopoxviruses, and have been cloned and sequenced in fowlpox virus by Campbell et al.[20] Clones from the termini were found to comprise a short unique region followed by 3.8 kb of tandem repeats. Following the tandem repeats, but still within the TIR, three open reading frames were found. A marker gene, namely β-galactosidase under the control of a vaccinia promoter, was successfully inserted into each of these three open reading frames (into *Bgl*II, *Nco*I, and *Spe*I sites), showing that the sites were nonessential for replication in tissue culture. The terminal fragment has also been found to be nonessential in the FP-1 strain of fowlpox virus by Paoletti.[21] Binns et al.[22] have identified two nonessential genes in a region which contains genes equivalent to the vaccinia *Hin*dIII D fragment. They showed, by insertion of β-galactosidase marker genes, that the fowlpox virus equivalents of genes D8 and D9 were nonessential, the relevant insertion points being *Bcl*I and *Asu*II sites, respectively. Spehner et al.[23] have utilized the site of the missing TK gene which they had previously identified in the Salsbury strain of fowlpox virus.[11] The original size of the intergenic region in fowlpox virus, i.e., between the genes on either side of the site of the TK in vaccinia virus, was 32 bp. This was enlarged to 55 bp by site-directed mutagenesis in order to prevent problems arising from a possible overlap of the promoter elements of the downstream gene with the 3′ end of the upstream gene. The enlargement was achieved by replacing the 3′ end of the upstream gene by a different DNA sequence which encoded the same protein sequence. A unique *Bam*HI cloning site was introduced into the site at the same time for insertion of foreign genes. Spehner et al.,[23] in addition to testing the nonessentiality of this site with a β-galactosidase marker gene, used a screening system in which the β-galactosidase gene was used in combination with the foreign gene of interest. The use of β-galactosidase as a tandem marker gene has been reported by Chakrabarti et al.,[24] but Spehner et al.[23] introduced a new element in which the marker gene was placed upstream of, and in the same orientation as, the foreign gene, both genes being under the control of the same promoter. This meant that after using the marker gene to screen for recombinants, it could be allowed to "drop out" of the virus by virtue of intragenomic recombination across the duplicated promoter sequences, as described for vaccinia by Ball.[25]

As has been described above, a wide range of methods has been used to aid in identifying recombinant fowlpox viruses, including the *E. coli* gpt gene,[17] β-galactosidase as a marker,[19,23] and the use of temperature-sensitive fowlpox virus to boost the level of recombinants.[23] However, some groups have identified recombinants simply by screening viral plaques, lifted onto nitrocellulose filters, with radiolabeled DNA[19] and subsequent recovery of virus from the agar overlay. Because of the relatively high recombination rate (approximatley 0.1% in some cases) this has not proved too difficult.

III. PROMOTERS

As in the case of vaccinia virus recombinants, it was realized that it would be necessary to use poxvirus promoters to drive expression of foreign genes in fowlpox virus. However, it was not known whether vaccinia virus promoters would work in fowlpox. The first evidence that promoters from the two genera were similar was obtained when Boyle and Coupar[14] found that the fowlpox virus TK gene functioned in a vaccinia recombinant when it was cloned into vaccinia with its own promoter. Although it was possible that the fowlpox virus gene was being controlled by vaccinia virus promoters from flanking regions, this seemed unlikely, as the gene was expressed in both orientations. Subsequently, many workers found that vaccinia virus promoters such as the p11 promoter,[26] the p7.5 promoter,[27] and the H6 promoter[28] functioned in fowlpox virus recombinants[10,16–19,23,29] or in transient assays.[30] More recently, a comparative analysis of vaccinia virus promoter activity in fowlpox and vaccinia recombinants has been carried out.[31] These experiments showed that the vaccinia virus p11 promoter functioned better in vaccinia virus than in fowlpox virus, and provided supportive evidence for the fact that sequence elements involved in transcription of genes in vaccinia virus and in fowlpox virus were similar, as had been suggested by earlier sequence studies.[12,32]

Because these vaccinia virus promoters functioned in fowlpox virus, many of them have been used in fowlpox virus recombinants developed as prototype vaccines, particularly those that have early and late functions, namely the p7.5 and the H6 promoters. However there was interest in identifying fowlpox virus promoters, first, because they might function more efficiently in fowlpox virus than the heterologous vaccinia virus promoters, but also because it was felt that fowlpox virus recombinants designed to be used as veterinary vaccines would be preferable if they had no vaccinia virus sequences present in their genomes. Such fowlpox virus promoters have been identified in several ways.

By abundant RNA — Northern blot analysis has been used by Kumar and Boyle[33] to identify early fowlpox virus promoters. They found that the strongest signal within the sequences covered by four clones, representing some 17% of the fowlpox virus genome, was a 1-kb transcript within the *Pst*I F fragment.[34] The promoter element responsible was identified by sequencing and S1 mapping. The promoter turned out to be of the early/late type, and interestingly it appears that start sites of both early and late transcripts map to the same TAAAT sequence near to the translation start site. These authors have also described the late promoter of the gene adjacent to the early/late gene.[35] This was found to have approximately one tenth the activity of the vaccinia p11 promoter when comparisons in fowlpox virus recombinants were carried out.

Paoletti[21] had identified a fowlpox promoter element by making radiolabeled cDNA probes from RNA preparations, probing back onto blots of restriction digests of fowlpox virus DNA, and looking for DNA fragments which give the strongest signal. Using this method, a DNA fragment containing an open reading frame predicting a protein of 25.8-kDa has been found, which corresponded to an abundant protein of the same size. Preliminary results suggest that the promoter element can direct expression of the feline leukemia virus envelope gene in fowlpox and vaccinia recombinants.[21]

By analogy with vaccinia genes — Binns et al.[12] identified a large number of fowlpox virus genes which are homologous to genes previously identified in vaccinia virus. Some of these vaccinia genes were known to be expressed at a high level and presumably had strong promoters. Such a gene was the gene for the vaccinia 4b protein, which accounts for approximately 11% of vaccinia viral protein. Binns et al.[36] identified and sequenced the fowlpox virus 4b gene, and tested the promoter by fusing it to the β-galactosidase gene. In a transient assay system using fowlpox virus-infected cells, the fowlpox virus 4b promoter was found to function at a level somewhat higher than the vaccinia p11 promoter. The fowlpox virus 4a promoter appears to have a similar strength.[12a]

By random screening — Jenkins et al.[37] have reported the identification of fowlpox virus promoters by a random screening procedure. Restriction fragments of fowlpox virus genomic DNA were shotgun-cloned 5′ to a reporter gene, lacZ, in a translational fusion vector. The resulting plasmids were screened for β-galactosidase production in fowlpox virus-infected cells in a transient assay. A strong fowlpox virus promoter, designated C1, was identified and modified so that it could be used with genes containing their own initiation codons, by replacement of the ATG sequence within the promoter by ATA.

In addition to the use of vaccinia and fowlpox promoters, a promoter from the insect poxviruses (entomopoxviruses) has been used by Paoletti.[21] The entomopoxvirus concerned was AmEPV, isolated from *Amsacta moorei* and the promoter was identified by making cDNA probes from total RNA from cells infected with the virus, and probing back onto blots of restriction digests of AmEPV DNA. A fragment, presumably encoding a strongly expressed gene, was identified, sequenced, and found to contain an open reading frame of 42 K. The promoter was found to operate well in vaccinia virus, although in fowlpox and canarypox viruses the levels of expression were fairly low.

IV. GENOME ORGANIZATION

A considerable amount of sequence analysis of the fowlpox virus genome has been carried out, amounting to greater than 55 kb in all. The picture which seems to emerge is that there are regions at the termini of the virus which have little similarity to vaccinia virus.[20,32] However, other regions, which are not at the termini, have been found to have fairly high levels of homology to vaccinia virus. Fowlpox virus has been found to have regions homologous to the vaccinia *Hin*dIII J region,[11,13] the vaccinia *Hin*dIII region[22,38] and the vaccinia *Hin*dIII I region.[49] In addition, a large number of individual genes of interest have been identified, including the DNA polymerase,[12] the RNA polymerase, the DNA ligase, and the topoisomerase.[50] In the regions where there is homology to vaccinia, the overall gene organization often reflects that found in vaccinia virus. However the gene organization is only broadly similar and there are many exceptions. We have seen how the TK gene was missing from an otherwise similar block of genes. In the *Hin*dIII

D region the same is true, with occasional genes present in vaccinia missing in fowlpox virus, and vice versa. Genes missing from one region can appear in other regions. It has been noted that these ''wandering genes'' are largely nonessential, but it is not clear whether this is a general rule.

These basic studies of the fowlpox virus genome have given us information which is useful in the development of fowlpox virus as a vector. For example genes which might be expected to have strong promoters have been found.[36] ''Pseudogenes'', which are likely nonessential regions, have been found,[22] and subsequently shown to be sites for insertion which are neutral with respect to attenuation *in vivo*. In addition, genes have been found which could well be used to modify the immunological response of the host to the recombinant virus. Among such genes are the serine protease inhibitor (serpin) genes, a truncated form of which was identified in fowlpox virus by Tomley et al.[32] Another such gene is a highly immunogenic gene with repeating amino acid motifs,[39] alterations in which could alter the way the host responds to the virus.

Recently a complete *Bam*HI restriction map of the Arthur Webster strain of fowlpox virus has been published.[34] Restriction maps of other fowlpox virus strains are being constructed. This will allow the large number of genes which have been identified to be mapped onto the fowlpox virus genome. The fowlpox virus genome may be up to 100 kb larger than the vaccinia virus genome, and this sort of mapping will allow us to determine whereabouts in the fowlpox virus genome this extra DNA lies.

V. RECOMBINANT VACCINES

The information gained by the studies described above has been used in the construction of recombinant fowlpox viruses expressing genes of veterinary importance, and many of these have proved successful as prototype vaccines. The next sections will deal with some of these recombinant fowlpox virus vaccines.

A. INFLUENZA

Boyle and Coupar[16] constructed a recombinant fowlpox virus expressing the human influenza hemagglutinin protein. Expression of the protein was demonstrated by binding of specific antibodies to fowlpox virus plaques and by demonstration of hemagglutination inhibiting (HI) antibodies after inoculation into chickens. Taylor et al.[29] made fowlpox virus recombinants expressing the hemagglutinin protein from a highly virulent H5N8 avian strain of influenza. Expression was demonstrated by immunoprecipitation and immunofluorescence, the protein being found to be correctly cleaved into HA1 and HA2, and expressed on the surface of cells in tissue culture. Again HI antibodies were demonstrated after inoculation into chickens or turkeys. Chickens vaccinated by the wing web route using a double needle of the type used for commercial vaccination of poultry with fowlpox virus (6×10^5

plaque forming units (pfu)/bird) were completely protected against challenge by the homologous H5N8 strain of influenza, and also against a related but distinguishable H5N2 strain. In the case of the heterologous challenge, no influenza virus replication was detected, suggesting that the antibody had prevented replication and dissemination of the virus. It is interesting to note that antibodies induced by the fowlpox virus/H5N8 recombinants were not active in HI or in virus neutralization tests against the heterologous H5N2 strain, yet the chickens were protected against challenge by this virus for over nine months. In protection studies in turkeys no HI antibody was detected even against the homologous strain, but the majority of the turkeys were protected against challenge.

B. NEWCASTLE DISEASE VIRUS

In the poultry field, possibly the most important pathogen worldwide is Newcastle disease virus (NDV).[40] Several groups have reported the construction of fowlpox virus recombinants expressing NDV genes. Boursnell et al.[19] have inserted the fusion (F) gene from the Beaudette strain of NDV into the terminal regions of a vaccine strain of fowlpox virus. The promoter used was the vaccinia p7.5 promoter. Although no expression of the protein could be detected by immunofluorescence, immunoprecipitation, or Western blotting, chickens inoculated with the recombinant were protected against challenge by a virulent strain of NDV. Two vaccinations with 10^6 pfu/bird gave 100% protection, whereas a single vaccination with lower doses gave 70 to 92% protection. Taylor et al.[41] made a recombinant expressing the F gene from the Texas strain of NDV, driven by the vaccinia H6 promoter. In this case, expression of the F protein could be demonstrated in tissue culture by immunological techniques, and correct cleavage of the protein was found. Chickens inoculated by the wing web route with 10^4 $TCID_{50}$ doses of virus were completely protected after one or two vaccinations. Ogawa et al.[10] have also demonstrated protection against NDV using the F protein.

Recombinant fowlpox viruses expressing the HN gene of NDV have also been found to confer protective immunity,[10,42,43] but Edbauer et al.[43] found that vaccination with a mixture of fowlpox recombinants expressing the F and HN recombinants did not show added protection over that seen with the individual recombinants.

As in the case of influenza, the NDV work shows that protection is achieved in spite of relatively low levels of anti-NDV antibodies, but the relative contributions of humoral antibody and cell-mediated immunity have yet to be assessed.

C. INFECTIOUS BURSAL DISEASE VIRUS

In addition to the use of viral glycoproteins which are normally anchored in viral membranes, such as the influenza HA and the NDV F and HN proteins, protection against a viral disease using a recombinant fowlpox virus has also been achieved using components of the icosahedral shell of a virus. The VP2 protein of a birnavirus, infectious bursal disease virus (IBDV) has been expressed in fowlpox virus as a fusion protein with β-galactosidase.[44] Challenge studies showed that chickens were protected against mortality, either after challenge with the homologous strain, or with a new, highly virulent strain, but not against damage to the site of viral replication, the Bursa of Fabricius.

D. INFECTIOUS BRONCHITIS VIRUS

The coronavirus, infectious bronchitis virus (IBV), is an important pathogen of poultry, and one for which the use of an avipox vector would appear to be eminently suitable. However although the spike protein of IBV has been expressed in vaccinia virus[45] and there have been preliminary reports of its expression in fowlpox virus,[21] there have as yet been no reports of the successful use of a fowlpox/IBV recombinant to demonstrate protection of chickens against challenge by IBV.

E. *EIMERIA TENELLA*

Several groups are studying the expression of antigens from various species of the avian coccidial parasite *Eimeria*. Although no reports have as yet been published in the literature, Tomley[51] has expressed a major microneme antigen (Etp100) of *Eimeria tenella* in fowlpox virus and demonstrated the induction of specific antibodies when the recombinant is inoculated into chickens.

F. RABIES

Although there are no reports of fowlpox virus replicating in any other than avian species, there have been reports of its causing cytopathic effects in nonavian cell lines.[46] These reports have been followed up and extended by Taylor et al.,[18] using fowlpox virus recombinants, and have led to some extremely interesting results which suggest that fowlpox virus may be useful as a vector in nonavian species. They have constructed a fowlpox virus recombinant expressing the rabies virus glycoprotein. When mammalian cells are infected with the recombinant, although no infectious virus particles are produced, the rabies glycoprotein is expressed by the recombinant. In addition when this recombinant is used to inoculate a range of mammalian species, protective immunity was induced that was sufficient to protect against a challenge with live rabies virus. Controls involving β-propiolactone-inactivated recombinant virus showed that the immune response was due to *de novo* expression of the rabies glycoprotein and not to antigen bound to virus particles.

Taylor et al.[47] have extended these results by showing that the use of a canarypox virus vector to express the rabies virus glycoprotein significantly reduces the dose of recombinant virus needed to confer protection. In fact the canarypox virus was 100 times more effective in protection studies than the original fowlpox virus recombinant.

G. MEASLES

Following on these reports, Spehner et al.[23] and Wild et al.[48] have constructed a fowlpox virus recombinant expressing the fusion (F) protein of measles virus. They also demonstrate protective immunity in a model in which mice are protected against fatal measles encephalitis. The dose of recombinant was similar to that described in the rabies fowlpox virus experiments in mice. Wild et al[48] compared the protection offered by a vaccinia recombinant and a fowlpox virus recombinant both expressing the measles F protein. Although both recombinants could confer 100% protection at suitable doses, the authors found that with the fowlpox virus recombinant protection fell off dramatically as the vaccination dose was decreased, whereas with vaccinia, presumably because of the fact that it replicates productively in mice, even at lower doses there was still significant protection. In both the fowlpox virus and vaccinia cases, protection was achieved even though antibodies against measles virus were not detected.

These demonstrations of the use of fowlpox virus in mammalian systems opens up the possibility of the use of fowlpox virus-based vaccines, not only in veterinary species but also in humans. Assuming that the safety of fowlpox virus could be tested, such a vaccine could be considered, particularly in cases when there are contraindications as to the use of vaccinia virus, for example, with immunocompromised individuals, such as AIDS patients. Although no reports of fowlpox virus-based HIV vaccines have been published, recently, the expression of the gag-pol and env genes of simian immunodeficiency virus (SIV) in fowlpox virus has been reported.[37] Expression of SIV-encoded polypeptides was demonstrated in avian and primate cells. Defective lentivirus-like particles could be seen in cells infected with a fowlpox virus-gag-pol recombinant alone, or in combination with a fowlpox virus-env recombinant. Such recombinants could be used as a system to produce purified lentivirus particles which will be free from any infectious poxvirus contamination. In addition the recombinant itself could be used as a prototype vaccine in animal studies to determine whether protection against a lentivirus infection can be achieved.

VI. CONCLUSION

Fowlpox virus has been developed as a vector to an extent that it is now fairly routine to make recombinants containing avian or other genes. Although the longer life cycle and the fact that chicken primary or secondary cells are used for growth of the virus mean that it is still easier to make recombinants in vaccinia virus, in other respects the use of fowlpox virus has advantages over vaccinia virus. For instance, when the pathogen is of avian origin, the recombinants can be tested for efficacy in the target species, rather than in model systems, as has most often been the case with vaccinia/human recombinants. In addition, when the pathogen is of nonavian origin, the recombinant

TABLE 1

Virus	Gene	Protection	Ref.
Influenza (human)	Hemagglutinin (HA)	n.a.[a]	16
Influenza (avian)	Hemagglutinin (HA)	+	29
Influenza (avian)	Nucleoprotein (NP)	n.a.	21
Infectious bronchitis virus	Spike (S)	n.a.	21
Infectious bronchitis virus	Membrane (M)	n.a.	21
Feline leukemia virus	Envelope	n.a.	21
Bovine leukemia virus	Envelope	n.a.	21
Rous-associated virus (RAV)	Envelope	n.a.	21
Herpes simplex virus	Glycoprotein D	n.a.	21
Newcastle disease virus (NDV)	Fusion protein (F)	+	10,19,41
Newcastle disease virus	Hemagglutinin-neuraminidase (HN)	+	10,42,43
Infectious bursal disease virus	VP2	+/−	44
Rabies virus	Glycoprotein (G)	+	18,47
Measles virus	Fusion protein (F)	+	23,48
Simian immunodeficiency virus (SIV)	Gag-pol, Env	n.a.	37

[a] n.a., Not applicable, not tested, or data not available.

can be safely tested with no risk of spread to humans. In the same way that a range of sites have been identified in vaccinia, there are now a wide range of insertion sites available in fowlpox virus, some of which can be used without further attenuating suitable vaccine strains, and doubtless some which do attenuate. In addition there are several fairly well characterized fowlpox promoters which can be used if regulatory authorities are uneasy about recombinant viruses containing vaccinia virus sequences. Several prototype vaccines against diseases of veterinary importance have now been developed using recombinant avipoxviruses. Table 1 shows a list, mainly based on those that have been reported in the literature, but including some recombinants that have been described in patents. Because of the wide commercial interest in fowlpox-based vaccines, data is not available on some of this work, and in fact much other exciting work is being carried out that has not yet been published in the literature. Many of these recombinants are highly effective, and because they are based on existing licensed fowlpox vaccine strains, it should be possible for these to be used safely in the field before long. Indeed, because of the safety factor conferred by the fact that avipoxviruses have a host range restricted to avian species, it may be that the first widely used recombinant poxvirus vaccines are based on avipoxviruses.

REFERENCES

1. **Mackett, M. and Smith, G. L.,** Vaccinia virus expression vectors, *J. Gen. Virol.*, 67, 2067, 1986.
2. **Lane, J. M., Ruben, F. L., Neff, J. M., and Millar, J. D.,** Complications of smallpox vaccination, 1968. National surveillance in the United States, *N. Engl. J. Med.*, 281, 1201, 1969.
3. **Boyle, D. B., Coupar, B. E. H., Parsonson, I. M., Bagust, T. J., and Both, G. W.,** Responses of cattle, sheep and poultry to a recombinant vaccinia virus expressing a swine influenza haemagglutinin, *Res. Vet. Sci.*, 41, 40, 1986.
4. **Yilma, T., Hsu, D., Jones, L., Owens, S., Grubman, M., Mebus, C., Yamanaka, M., and Dale, B.,** Protection of cattle against rinderpest with vaccinia virus recombinants expressing the HA or F gene, *Science,* 242, 1058, 1988.
5. **Tripathy, D. N. and Cunningham, C.,** Avian pox, in *Diseases of Poultry,* Hofstad, M. S., Barnes, J. H., Calnek, B. W., Reid, W. M., and Yoder, H. W., Eds., Iowa State University Press, 1984, 524.
6. **Gaffard, L. G. and Randall, C. C.,** The high molecular weight of the fowlpox virus genome, *J. Mol. Biol.*, 26, 303, 1967.
7. **Muller, H. K., Wittek, R., Schaffner, W., Schumperli, A., Menna, A., and Wyler, R.,** Comparison of five poxvirus genomes by analysis with restriction endonucleases *Hind*III, *Bam*HI and *Eco*RI, *J. Gen. Virol.*, 38, 135, 1977.
8. **Mayr, A. and Mahnel, H.,** Charakterisierung eines vom Rhinozeros isolierten Huhnerpockenvirus, *Arch. Gesammte Virusforsch.*, 31, 51, 1970.
9. **Da Massa, A. J.,** The role of *Culex tarsalis* in the transmission of fowlpox virus, *Avian Dis.*, 10, 57, 1966.
10. **Ogawa, R., Yanagida, Y., Saeki, S., Saito, S., Ohkawa, S., Gotoh, H., Kodama, K., Kamogawa, K., Sawaguchi, K., and Iritani, Y.,** Recombinant fowlpox viruses inducing protective immunity against Newcastle disease and fowlpox viruses, *Vaccine,* 8, 486, 1990.
11. **Drillien, R., Spehner, D., Villeval, D., and Lecocq, J.-P.,** Similar genetic organisation between a region of fowlpox virus DNA and the vaccinia virus *Hind*III J fragment despite divergent location of the thymidine kinase gene, *Virology,* 160, 203, 1987.
12. **Binns, M. M., Stenzler, L., Tomley, F. M., Campbell, J., and Boursnell, M. E. G.,** Identification by a random sequencing strategy of the fowlpoxvirus DNA polymerase gene, its nucleotide sequence and comparison with other viral DNA polymerase, *Nucleic Acids Res.*, 15, 6563, 1987.

12a. **Binns, M. M.,** personal communication.

13. **Binns, M. M., Tomley, F. M., Campbell, J., and Boursnell, M. E. G.,** Comparison of a conserved region in fowlpox virus and vaccinia virus genomes and the translocation of the fowlpox virus thymidine kinase gene, *J. Gen. Virol.*, 69, 1275, 1988.
14. **Boyle, D. B. and Coupar, B. E. H.,** Identification and cloning of the fowlpox virus thymidine kinase gene using vaccinia virus, *J. Gen. Virol.*, 67, 1591, 1986.
15. **Boyle, D. B., Coupar, B. E. H., Gibbs, A. J., Seigman, L. J., and Both, G. W.,** Fowlpox virus thymidine kinase: nucleotide sequence and relationships to other thymidine kinases, *Virology,* 156, 355, 1987.
16. **Boyle, D. B. and Coupar, B. E. H.,** Construction of recombinant fowlpox viruses as vectors for poultry vaccines, *Virus Res.*, 10, 343, 1988.
17. **Boyle, D. B. and Coupar, B. E. H.,** A dominant selectable marker for the construction of recombinant poxviruses, *Gene,* 65, 123, 1988.
18. **Taylor, J., Weinberg, R., Languet, B., Desmettre, P., and Paoletti, E.,** Recombinant fowlpox virus inducing protective immunity in non-avian species, *Vaccine,* 6, 497, 1988.

19. **Boursnell, M. E. G., Green, P. F., Campbell, J. I. A., Deuter, A., Peters, R. W., Tomley, F. M., Samson, A. C. R., Chambers, P., Emmerson, P. T., and Binns, M. M.,** Insertion of the fusion gene from Newcastle disease virus into a non-essential region in the terminal repeats of fowlpox virus and demonstration of protective immunity induced by the recombinant, *J. Gen Virol.,* 71, 621, 1990.
20. **Campbell, J. I. A., Binns, M. M., Tomley, F. M., and Boursnell, M. E. G.,** Tandem repeated sequences within the terminal region of the fowlpox virus genome, *J. Gen. Virol.,* 70, 145, 1989.
21. **Paoletti, E.,** Recombinant avipox virus. Patent, International Publ. No. WO 89/03429, Health Research Inc., 1989.
22. **Binns, M. M., Britton, B. S., Mason, C., and Boursnell, M. E. G.,** Analysis of the fowlpox virus genome region corresponding to the vaccinia virus D6 to A1 region: location of, and variation in, non-essential genes in poxviruses, *J. Gen. Virol.,* 71, 2873, 1990.
23. **Spehner, D., Drillien, R., and Lecocq, J.-P.,** Construction of fowlpox virus vectors with intergenic insertions: expression of the β-galactosidase gene and the measles virus fusion protein, *J. Virol.,* 64, 527, 1990.
24. **Chakrabarti, S., Brechling, K., and Moss, B.,** Vaccinia virus expression vector: coexpression of β-galactosidase provides visual screening of recombinant virus plaques, *Mol. Cell. Biol.,* 5, 3403, 1985.
25. **Ball, L. A.,** High-frequency homologous recombination in vaccinia virus DNA, *J. Virol.,* 61, 1788, 1987.
26. **Bertholet, C., Drillien, R., and Wittek, R.,** One hundred base pairs of 5′ flanking sequence of a vaccinia virus late gene are sufficient to temporally regulate late transcription, *Proc. Natl. Acad. Sci. U.S.A.,* 82, 2096, 1985.
27. **Venkatesan, S., Baroudy, B. M., and Moss, B.,** Distinctive nucleotide sequences adjacent to multiple initiation and termination sites of an early vaccinia virus gene, *Cell,* 25, 805, 1981.
28. **Rosel, J. L., Earl, P. L., Weir, J. P., and Moss, B.,** Conserved TAAATG sequence at the transcriptional and translation initiation sites of vaccinia virus late genes deduced by structural and functional analyses of the *Hin*dIII H genome fragment, *J. Virol.,* 60, 436, 1986.
29. **Taylor, J., Weinberg, R., Kawaoka, Y., Webster, R. G., and Paoletti, E.,** Protective immunity against avian influenza induced by a fowlpox virus recombinant, *Vaccine,* 6, 504, 1988.
30. **Tripathy, D. N. and Wittek, R.,** Regulation of foreign gene in fowlpox virus by a vaccinia virus promoter, *Avian Dis.,* 34, 218, 1990.
31. **Prideaux, C. T., Kumar, S., and Boyle, D. B.,** Comparative analysis of vaccinia virus promoter activity in fowlpox and vaccinia virus recombinants, *Virus Res.,* 16, 43, 1990.
32. **Tomley, F. M., Binns, M. M., Campbell, J., and Boursnell, M. E. G.,** Sequence analysis of an 11.2 kilobase, near-terminal, BamHI fragment of fowlpox virus, *J. Gen. Virol.,* 69, 1025, 1988.
33. **Kumar, S. and Boyle, D. B.,** Mapping of a major early/late gene of fowlpox virus, *Virus Res.,* 15, 175, 1990.
34. **Coupar, G. E. H., Teo, T., and Boyle, D. B.,** Restriction endonuclease mapping of the fowlpox virus genome, *Virology,* 179, 159, 1990.
35. **Kumar, S. and Boyle, D. B.,** Activity of a fowlpox virus late gene promoter in vaccinia and fowlpox virus recombinants, *Arch. Virol.,* 112, 139, 1990.
36. **Binns, M. M., Boursnell, M. E. G., Tomley, F. M., and Campbell, J.,** Analysis of the fowlpoxvirus gene encoding the 4b core polypeptide and demonstration that it possesses efficient promoter sequences, *Virology,* 170, 288, 1989.
37. **Jenkins, S., Gritz, L., Fedor, C. H., Cormier, N. A., McGuigan, L. C., O'Neill, E. M., Osborne, M., Cohen, L. K., Payne, L. G., Mazzara, G. P., and Panicali, D. L.,** Formation of lentivirus particles in mammalian cells infected with recombinant virus, 1990, submitted.

38. **Tartaglia, J., Winslow, J., Goebel, S., Johnson, G. P., Taylor, J., and Paoletti, E.,** Nucleotide sequence analysis of a 10.5kbp *Hind*III fragment of fowlpox virus: relatedness to the central portion of the vaccinia virus *Hind*III D region, *J. Gen. Virol.*, 71, 1517, 1990.
39. **Binns, M. M., Mason, C., and Boursnell, M. E. G.,** A 39 000 Mr immunodominant protein of fowlpox virus contains multiple copies of a 12 amino acid repeat sequence, *J. Gen. Virol.*, 71, 2883, 1990.
40. **Lancaster, J. E. and Alexander, D. J.,** Newcastle Disease Virus and Spread, Canada Department of Agriculture Monograph No. 11, 1975.
41. **Taylor, J., Edbauer, C., Rey-Senelonge, A., Bouquet, J.-F., Norton, E., Goebel, S., Desmettre, P., and Paoletti, E.,** Newcastle disease virus fusion protein expressed in a fowlpox virus recombinant confers protection in chickens, *J. Virol.*, 64, 1441, 1990a.
42. **Boursnell, M. E. G., Green, P. F., Samson, A. C. R., Campbell, J. I. A., Deuter, A., Peters, R. W., Millar, N. S., Emmerson, P. T., and Binns, M. M.,** A recombinant fowlpox virus expressing the hemagglutinin gene of Newcastle disease virus (NDV) protects chickens against challenge by NDV, *Virology*, 178, 297, 1990.
43. **Edbauer, C., Weinberg, R., Taylor, J., Rey-Senelonge, A., Bouquet, J.-F., Desmettre, P., and Paoletti, E.,** Protection of chickens with a recombinant fowlpox virus expressing the Newcastle disease virus hemagglutinin-neuraminidase protein, *Virology*, 179, 901, 1990.
44. **Bayliss, C. D., Peters, R. W., Cook, J. A., Reece, R. L., Howes, K., Binns, M. M., and Boursnell, M. E. G.,** A recombinant fowlpox virus that expresses the VP2 antigen of Infectious Bursal Disease Virus induces protection against mortality caused by the virus, *Arch. Virol.*, 120, 193, 1991.
45. **Tomley, F. M., Mockett, A. P. A., Boursnell, M. E. G., Binns, M. M., Cook, J. K. A., Brown, T. D. K., and Smith, G. L.,** Expression of the infectious bronchitis virus spike protein by recombinant vaccinia virus and induction of neutralizing antibodies in vaccinated mice, *J. Gen. Virol.*, 68, 2291, 1987.
46. **Burnett, J. W. and Frothingham, T. E.,** The cytotoxic effect of fowlpox virus on primary human amniotic cell cultures, *Arch. Gesammte Virusforsch.*, 24, 137, 1968.
47. **Taylor, J., Trimarch, C., Weinberg, R., Languet, B., Guillemin, F., Desmettre, P., and Paoletti, E.,** Efficacy studies on a Canarypox-Rabies Recombinant Virus, *Vaccine*, 9, 190, 1991.
48. **Wild, F., Giraudon, P., Spehner, D., Drillien, R., and Lecocq, J.-P.,** Fowlpox virus recombinant encoding the measles virus fusion protein: protection of mice against fatal measles encephalitis, *Vaccine*, 8, 441, 1990.
49. **Binns, M. M. and Boursnell, M.,** manuscript in preparation.
50. **Binns, M. M.,** personal communication.
51. **Tomley, F. M.,** personal communication.

Chapter 9

PARAPOXVIRUSES: THEIR BIOLOGY AND POTENTIAL AS RECOMBINANT VACCINES

Anthony J. Robinson and David J. Lyttle

TABLE OF CONTENTS

I. INTRODUCTION

During the past ten years, the staff in this laboratory have been investigating the molecular biology of the parapoxvirus, orf virus. In sheep the virus causes a disease called orf, contagious pustular dermatitis, or, colloquially, scabby mouth. The disease can be transmitted to humans, where it is usually referred to as orf. Our goal is to identify the viral genes sufficient to induce a protective immune response to this disease and for this purpose we have cloned segments of the orf virus genome into vaccinia virus to dissect out the relevant genes. The widespread use of vaccinia virus as a gene vector and the development of techniques associated with its use prompted us to consider the possibility of using orf virus and other parapoxviruses as gene vectors themselves. The biology of the parapoxviruses and their potential as recombinant vaccines is reviewed.

II. OVERVIEW OF THE GENUS *PARAPOXVIRUS*

A. CLASSIFICATION AND NOMENCLATURE

The classification of the parapoxviruses has evolved slowly and has been accompanied by some degree of confusion, particularly with regard to nomenclature. The first attempt at a formal classification of the poxviruses was by Goodpasture in 1933,[1] when he referred to the diseases of fowlpox, molluscum contagiosum, and variola-vaccinia as Borreliotoses, acknowledging the contribution of Borrel, who first described the minute elementary bodies in smears of fowlpox lesions. In an earlier review of fowlpox, Goodpasture had noted the similarities between these three diseases.[2] The group we now call the parapoxviruses were not included, although the diseases of orf, pseudocowpox, and bovine papular stomatitis were recognized as distinct clinical entities at that time and their similarity to other poxvirus diseases had been noted (see Section II.B.)

Parapoxviruses were first classified in 1956,[3] when it was stated that the virus of pustular dermatitis of sheep and goats and that of milkers' nodes should be included with the poxviruses. In 1957, a classification scheme based on a latin binomial system was proposed[4] and under the genus *Poxvirus,* contagious pustular dermatitis was mentioned as being within the group and related immunologically to goatpox virus. Seven years later, Andrewes' *Viruses of Vertebrates*[5] included for the first time the three viruses which we now recognize as the main members of the parapoxvirus group, although the name parapoxvirus was not used. The diseases of "contagious pustular dermatitis" and "bovine pustular dermatitis" were listed as Group II poxviruses, "viruses related to CPD". In the text, the diseases of orf (synonyms: contagious pustular dermatitis; contagious ecthyma of sheep; sore mouth; scabby mouth; contagious pustular stomatitis; infectious labial dermatitis), milkers' nodes (synonyms: pseudocowpox; paravaccinia), and stomatitis papulosa (of cattle) were included and the "ball of yarn" appearance of their virions was described. The similarity between the appearance of these virions and those of molluscum contagiosum was also noted. In 1966,[6] a similar classification appeared with "orf-like viruses" placed in subgroup II of the poxvirus group and including "bovine papular stomatitis, contagious pustular dermatitis (orf virus) and milker's nodules (human)".

The International Committee on Nomenclature of Viruses (now the International Committee on Taxonomy of Viruses [ICTV]), in 1971,[7] consolidated the classification of these viruses as a distinct group within the poxviruses but the nomenclature remained confused. Listed under "subgroup B" of the poxviruses were "bovine papular dermatitis virus, chamois contagious ecthyma virus, milkers' node (paravaccinia) virus, and orf virus". Why chamois contagious ecthyma virus suddenly appeared is not clear, although an orf-like virus isolated from chamois had been reported in 1963.[8] This nomenclature was perpetuated in the ICTV report of 1976,[9] but by then the current scheme of genera within families of viruses had emerged and for the first time the orf subgroup became the parapoxvirus genus within the family *Poxviridae,* the name presumably being derived from "paravaccinia".[10] Orf virus was chosen as the type species, with "bovine pustular stomatitis virus, chamois contagious ecthyma virus, and milker's node virus" being the other members. The description of the genus that emerged at that time was, "viruses of ungulates that may infect man. Infectivity is ether sensitive. Virion ovoid 220 × 300 nm, external coat and filaments are thicker than in vaccinia virions and are arranged in a regular spiral coil consisting of a single thread. Species show serological cross-reactivity, infected cells do not produce hemagglutinin."

Prior to the publication of the 1976 scheme, a brief report appeared which announced the new scheme for classifying poxviruses and included a description of the genus *Parapoxvirus.*[11] It stated, "No multiplication in embryonated eggs or common experimental animals; probable immunological relationships between members of the genus but distinct from all other poxviruses. Mass of virion (bovine papular stomatitis) 2—3.7 × 10^{-15} g. Virus specific antigen common to all members of the genus and probably to all other poxviruses; no haemagglutinating properties; non-genetic reactivation between all members and probably with other poxviruses; ether-sensitive. Includes viruses of orf, ulcerative dermatosis of sheep, bovine papular stomatitis, milker's nodule''. This was the first and last mention of the virus of ulcerative dermatosis of sheep as a member and, as will be discussed later, it could well be that isolates of this virus were orf virus.

The ICTV reports of 1979[12] and 1982[13] gave identical descriptions of the parapoxviruses and little had changed from the report of 1971, except that the cryptogram scheme was omitted and it was reported that the DNA of parapoxviruses was double stranded with a molecular weight of 85×10^6.

Currently, the Poxvirus Study Group of the ICTV is reviewing this classification[14] and the new list is likely to include the viruses of **orf** (*syn.* contagious pustular dermatitis, contagious ecthyma), stomatitis papulosa (*syn.* **bovine papular** (pustular) **stomatitis**), and **pseudocowpox** (*syn.* milker's node, milkers' nodule, paravaccinia), together with three new members, listed as probable members; a parapoxvirus of red deer, a parapoxvirus of seals, and a parapoxvirus of camels (the agent of contagious ecthyma of camels or Ausdyk disease). The virus of chamois contagious ecthyma is still included in the list of probable members but as orf virus is recorded as affecting chamois, it could be that this virus is none other than orf virus.

In this review we will use the names emphasized above in bold type. We prefer the name bovine papular stomatitis to stomatitis papulosa, as the former is the name used in most veterinary texts and was the name given to the disease by Plowright and Ferris[15] and Griesemer and Cole,[16] who described the disease definitively. We also consider that the synonym ''bovine pustular stomatitis'' should be dropped altogether as it is not a name used to describe bovine papular stomatitis.

B. PARAPOXVIRUS DISEASES — HISTORICAL PERSPECTIVE

1. Orf

The disease of contagious pustular dermatitis or orf in sheep has been reviewed.[17] Shepherds have long recognized scabby diseases of sheep and an early record has been attributed by Oláh and Elek[18] to Steeb (1787),[19] who wrote what is probably the first description of contagious pustular dermatitis or orf. The contagious nature of orf was reported by Walley in 1890[20] and this confirmed what had been suspected by local shepherds for many years. Humans are susceptible to infection[21] and there is an early report of the disease by Hansen (1879).[22] Since 1937, the disease in humans has usually been referred to as orf.[23]

In 1923, Aynaud[24] reported that the disease could be transmitted between sheep by inoculation with preparations of scab material that had been passed, albeit with difficulty, through Berkfield V (coarse, 8 to 12 μm) and Chamberland L1 (passes all bacteria) and L1 *bis* (retains all but the smallest bacteria) candles. This proved the virus was smaller than most bacteria and justified its inclusion in the group of ''filterable'' viruses. These experiments were only a few of many on orf virus recorded by Aynaud in this monumental work. He distinguished the infection from that caused by vaccinia virus by showing that orf infection did not prevent vaccinia virus infection but did prevent reinfection with orf virus. He also described the natural disease, its histological appearance in sections of tongue and soft palate, the susceptibility of the virus to various solvents and chemicals, the development and duration of immunity in inoculated animals, the lack of protection afforded by serum from immune animals, and the development of a vaccine which, essentially, is still in use today. He also described the results of inoculations of goats, calves, a foal, pigs, cats, dogs, rabbits, white rats, guinea pigs, mice, chickens, pigeons, frogs, and a 15-year-old donkey. Only sheep and goats were fully susceptible to infection, although the calves and the foal developed mild lesions on the muzzle.

Orf virus was first isolated in cell cultures by Greig in 1957.[25] The strains of virus isolated were from cases of contagious ecthyma (orf) in sheep in Canada and the cultures used were primary embryonic sheep skin cells. There are many reports in the literature of growth of the virus in primary bovine and human cells, but field strains do not grow readily in continuous cell cultures or in eggs.[17] However, three North American strains have been adapted to growth in vero cells[26] and a European strain will propagate in MRC-5 cells.[27]

A question remains about the status of the disease called ulcerative dermatosis of sheep,[28,29] as to whether or not it is a manifestation of infection with orf virus. The disease has been given different names over the years, including lip and leg ulceration, anovulvitis, ulcerative vulvitis, and, in rams, balanoposthitis, infectious balanoposthitis, foul sheath, sheath infection, and pizzle rot.[30,31] Undoubtedly, the etiology of the condition, at least in rams, is multifactorial and a number of bacteria have been implicated, in particular *Fusobacterium necrophorum*. However, a parapoxvirus resembling orf virus has been isolated on a number of occasions and the disease reproduced with bacteria-free filtrates.[32] In 1963, the earlier literature was reviewed and the viruses of ulcerative dermatosis and orf compared.[33,34] The two viruses survived treatment with 20% ether and fluorocarbons, and mice, rabbits, guinea pigs, and hamsters were refractory to infection with either virus. Significantly, however, the viruses would not cross-protect sheep. In a later publication, it was shown that antisera prepared in rabbits against virus-infected chicken embryo fibroblast cultures failed to distinguish between the two viruses in a cross-neutralization test but, in a precipitin test, a line was seen with orf virus antiserum against orf virus antigen but not with an antigen prepared from the agent of ulcerative dermatosis. It was concluded that the virus of ulcerative dermatosis was likely to be a variant of orf virus.[35] As far as we are aware, there has been no further work done on viruses associated with ulcerative dermatosis.

2. Pseudocowpox

Prior to the 1930s, the disease we now recognize as pseudocowpox in cattle was generally not distinguished from cowpox, although the disease has been present in cattle for a long time. Edward Jenner came to recognize "pustulous sores" with no "bluish or livid tint"[36] and other "spurious"[37] forms of cowpox on cattle teats during his historic studies in the latter half of the eighteenth century. It seems likely that one of these diseases was caused by pseudocowpox virus. Bonnevie (1937)[38] reviewed the early literature describing attempts to distinguish "vaccine rouge" from true cowpox. It was recognized that some smallpox vaccinations, called "vaccine rouge", produced a different form of lesions from that caused by true cowpox and that no protection was elicited to subsequent vaccination with true cowpox. Such infections were called "false cowpox", "paravaccine", "milkers' warts", and "milkers' nodules". In a review of milkers' nodules published in 1940,[39] the first report that led to the differentiation between the two viruses was attributed to Winternitz (1899).[40] Clearly, by the beginning of the 20th century it was established that there were two distinct forms of cowpox. Lipschütz (1932)[41] referred to the disease in humans as paravaccinia ("Strongyloplasma paravaccinia"[10,42]) and confirmed, contrary to reports from some earlier workers, that material from milkers' nodules did not confer immunity to vaccinia after transmission to humans and calves. Lipschütz and others could also distinguish the agent of milkers' nodules from that causing cowpox by the application of Paul's test. The latter test involved the inoculation of the infectious material onto the cornea of rabbits. Unlike vaccinia, the agent of milkers' nodules failed to produce a lesion.

In 1963, the virus was isolated in cell cultures from teat lesions in cattle[43] and from milkers' nodules.[44] The virions were shown by electron microscopy to resemble those of orf virus and bovine papular stomatitis virus (see Section II.E). As with orf virus, growth was restricted to primary cells of bovine or human origin and the virus would not propagate in embryonated hen eggs.[43,44]

3. Bovine Papular Stomatitis

The early literature regarding bovine papular stomatitis was reviewed by Griesemer and Cole in 1960.[16] The disease was first described by Degive in Belgium in 1884,[45] who gave it the name of "la stomatite papillaire ou papillomateuse" (papular or papillomatous stomatitis). Subsequently, a number of diseases of cattle with similarities to bovine papular stomatitis were described and included infectious stomatitis,[46] Armagh's disease,[47] erosive stomatitis,[48] proliferative stomatitis,[49] muzzle disease,[50] parotido-stomatitis,[51] and ulcerative stomatitis.[52] These diseases could be transmitted to calves with varying degrees of success. Some reports of bovine hyperkeratosis (X disease), which can be reproduced by feeding cattle feed contaminated with highly chlorinated naphthalenes may also have been complicated by intercurrent infection with the virus of bovine papular stomatitis. The transmissible proliferative stomatitis reported by Olsen and Paliones in 1953, which arose in cattle being fed hyperkeratosis-producing feedstuffs, was most likely bovine papular stomatitis.[49] The agent of proliferative stomatitis was reported to have been transmitted to two people handling infected calves.[49]

In 1959, Plowright and Ferris,[53] and one year later Griesemer and Cole,[54] reported the isolation of the virus in cell culture and the reproduction of the disease in calves and it was not until then that the disease of bovine papular stomatitis was characterized definitively. In 1967, Carson and Kerr[55] reported a case of apparent transmission of bovine papular stomatitis virus to humans and, later, it was shown that the virus derived from lesions in humans could be transmitted back to calves.[56]

4. Other Parapoxviruses

Parapoxvirus infections in red deer, seals, and camels have been described relatively recently, although there is every reason to believe they have been in existence for as long as the more familiar parapoxviruses of cattle and sheep. The infection in red deer has been reported only from New Zealand[57,58] and was thought initially to be orf virus infection resulting from deer being grazed on pastures previously grazed by sheep. However, restriction endonuclease analysis of viral DNA,[57,59] DNA/DNA hybridization studies,[59] and animal inoculation tests[60] have shown that the virus is unique. The recent practice of intensive farming of red deer may have created the conditions for the emergence of the disease from previously unnoticed infections in wild herds.

From 1969 onward, a number of parapoxvirus infections have been recorded in seal species, including Californian sea lions (*Zalophus californianus*),[61] South Amerian sea lions (*Otaria byronia*),[62] harbor seals (*Phoca vitulina*),[63] northern fur seals (*Callorhinus ursinus*),[64] and grey seals (*Halichoerus grypus*)[65] and human infection with the virus from grey seals has been reported.[65] The extent and importance of these infections in wild seal populations is not known.

A disease of camels called Ausdyk or camel contagious ecthyma is caused by a parapoxvirus and was first reported in the Soviet Union by Rosylakov in 1972.[66] Subsequently, the disease has been recorded in Mongolia,[67,68] Somalia,[69] and Kenya.[70] Over the years, a number of cases of Ausdyk have been referred from Middle Eastern countries to the Agricultural and Food Research Council (AFRC) Institute for Animal Health at Pirbright and parapoxvirus isolates have been obtained.[71] A report of contagious ecthyma of sheep occurring in camels,[72] and the fact that the Kenyan outbreak occurred at the same time as outbreaks occurring in sheep, begs the question of whether or not the camel parapoxvirus is orf virus. Animal inoculation tests and restriction endonuclease analysis may provide the answer (see Section II.H).

Records of two other parapoxvirus infections are to be found in the literature. Parapoxvirus virions have been found in scabby lesions on the eyelids of a red squirrel[73] and a parapoxvirus is recorded as the cause of contagious papular dermatitis of Japanese serow.[74,75] The relationship between the virus causing the latter condition and orf virus is not known.

5. Criteria for Subgeneric Classification

It is appropriate at this point to address the question of how viruses are classified within the parapoxvirus genus. What criteria should be used to decide whether a new isolate of a parapoxvirus is a new member of the genus or is an existing member discovered in a new host? There are no rules at present but, historically, the three main members of the genus were separated on the basis of the main species affected and the pathological character of the disease. The separation of these viruses into three distinct groups has generally been borne out by animal inoculation tests,[76] serology,[77] DNA/DNA hybridization,[78] and restriction endonuclease analysis,[78,79] despite some early disagreement particularly concerning the relationship between the viruses causing pseudocowpox and bovine papular stomatitis.[80,81] Initial confusion was reasonable at the time, however, as there seems little doubt that pseudocowpox virus will occasionally cause mouth lesions in calves and bovine papular stomatitis virus, teat lesions in cows. With this proviso in mind, any new isolate can be tentatively assigned to one of the described member groups by taking into account the pathological appearance of the disease and the taxonomic relationship between the species affected by the newly described virus and the species affected by known virus types. To obtain a definitive answer restriction endonuclease analysis is useful. Using certain endonucleases such as *Bam*HI or *Kpn*I the established members of the genus can be readily distinguished[78,79,82] (see Section II.H).

Nevertheless, it is possible that as more isolates are examined the distinction will become less clear. In Figure 1, the restriction endonuclease patterns of bovine papular stomatitis virus isolates show that isolates 4 and 5 have a distinctly orf virus-like pattern. Discounting obvious errors such as mislabeling, it is possible that recombination events between different restriction types could confuse the picture. It could be considered a circular argument to say that a virus is orf virus by restriction pattern even though it was isolated from a case of bovine papular stomatitis. Subtle changes in gene arrangement due to recombination or mutation could alter the host range of a virus without altering significantly its restriction pattern.

Another way of looking at subgeneric classification would be, therefore, to lump all isolates as merely different strains of parapoxvirus rather than to attempt to split them. In some ways, though, this would tend to blur the fact that there are distinct strains of virus that appear to be maintained in some hosts and not others and that, generally they have very different restriction patterns and nonhomologous terminal regions. Those that do not fit the pattern might be useful tools for delineating what it is that constitutes host range.

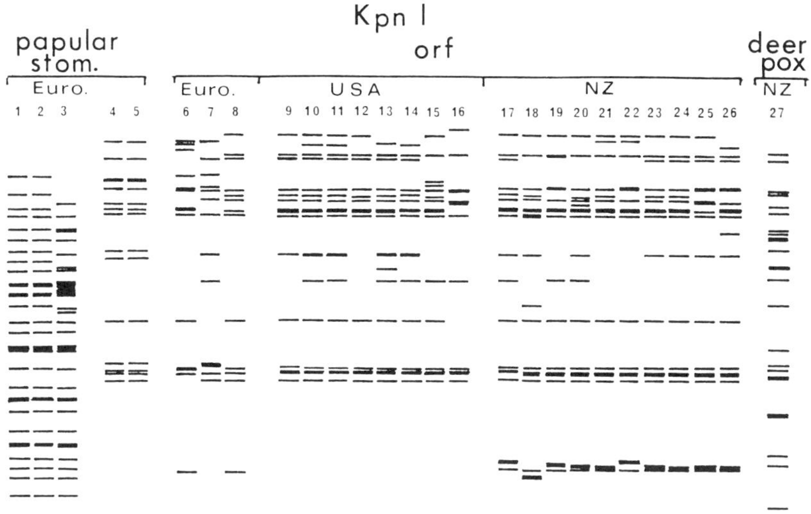

FIGURE 1. Comparison of parapoxvirus strains by KpnI restriction endonuclease analysis. Bovine papular stomatitis strains 1 to 5 and orf virus strains 6 to 8 are redrawn from Reference 79; orf virus strains 9 to 16 from Reference 178; orf virus strains 17 to 26 from Reference 170, and deer parapoxvirus strain 27 from Reference 59. Strain designations are 1, V660; 2, V619; 3, V380; 4, V590; 5, V256; 6, D1701/8; 7, Düsseldorf; 8, 6/36; 9, CE GP11; 10, CE MW30; 11, CE Vacc; 12, CE LW32; 13, CE Shoe; 14, CE Musk ox; 15, CE CO18; 16, CE Dall; 17, NZ2; 18, NZ3; 19, NZ7; 20, NZ10; 21, NZ48; 22, NZ52; 23, NZ113; 24, NZ116; 25, NZ125; 26, NZ125; 27, Halden.

C. HOST RANGE OF PARAPOXVIRUSES

The range of hosts recorded as being affected by any one member of the parapoxvirus genus is much narrower than the range recorded for vaccinia virus and in this respect resembles most other poxviruses. This is summarized in Table 1. A feature of the parapoxviruses is their lack of pathogenicity for the commonly used laboratory animals, which has to some extent hindered studies on the biology of the group.

Just how widespread parapoxviruses are in nature is not known. Thus far they have been found as naturally occurring infections in a limited number of genera within the orders Artiodactyla and Carnivora, and one in Rodentia. Generally, they have been found in animals that are managed in some manner either in the wild, in captivity, or as a farmed or domesticated species. This might indicate that their distribution in wild populations is wider than generally recognized and that their presence in additional species would be revealed by closer observation or by the introduction of more intensive management practices. The recent discovery of a parapoxvirus in a red squirrel and in red deer serve as examples.

TABLE 1
Natural and Experimental Infection with Parapoxviruses

Natural Infection

Species	Ref.
Orf Virus	
Domestic sheep	17[a], 20
Bighorn sheep (*Ovis canadensis*)	237, 238
Thinhorn sheep (Dall sheep, *Ovis dalli*)	239, 240
Domestic goat	17[a], 89, 241
Rocky Mountain goat (*Oreamnos americanus*)	237
Chamois (*Rupicapra rupicapra*)	8, 242
Himalayan tahr (*Hemitragus jemlahicus*)	8, 243
Musk-ox (*Ovibos moschatus*)	97, 239, 244, 245
Reindeer (*Rangifer tarandus*)	97, 245
Steenbok (*Raphicerus campestris*)	242
Alpaca (*Lama guanicoe*)	72, 247
Dog	72, 248
Human	17[a], 22, 23
Pseudocowpox Virus	
European cattle	39[a], 43
Human	38[a], 39[a], 44
Bovine Papular Stomatitis Virus	
European cattle	16[a], 54
Human	55
Parapoxvirus of Seals	
Californian sea lion (*Zalophus californianus*)	61
South American sea lion (*Otaria byronia*)	62
Harbor seal (*Phoca vitulina*)	63
Northern fur seal (*Callorhinus ursinus*)	64
Grey seal (*Halichoerus grypus*)	65
Human	65
Parapoxvirus of Camels	
Camel (*Camelus dromedarius*)	66, 67, 68, 69, 70
Parapoxvirus of Red Deer	
Red deer (*Cervus elaphus*)	57

TABLE 1 (continued)
Natural and Experimental Infection with Parapoxviruses

Species	Ref.
Parapoxvirus (Unspecified)	
Red squirrel (*Sciurus vulgaris*)	73
Japanese serow (*Capricornis crispus*)	74, 75

Experimental Infection

Orf Virus (Excluding Sheep and Goats)

Species	Susceptibility	Ref.
Mouse	Refractory	24, 33, 88, 92, 249, 250
White rat	Refractory	24
Rabbit	Refractory	24, 33, 88, 90, 124, 251, 252
Guinea pig	Mild lesions, refractory	92, 250, 253, 254, 24, 88, 90, 249, 250, 251
Hamster	Moderate lesions, refractory	255, 33
Dog	Refractory	24, 90, 251
Cat	Refractory	24
Pigeon	Moderate lesions, refractory	255, 24, 88
Domesticated chicken	Refractory	24, 88
Embryonated hen egg (CAM)[b]	Refractory	92, 112, 249
Frog	Limited growth, refractory	250, 256, 257, 24
Cow	Refractory	90
Calf	Moderate lesions	24, 76, 250, 258
Horse	Mild lesions	24
Red deer (*Cervus elaphus*)	Mild lesions	60
Wapiti (*Cervus elaphus nelsoni*) Mule deer (*Odocoileus hemionus*) White-tailed deer (*Odocoileus virginianus*)	Mild oral lesions	259
Pronghorn (*Antilocapra americana*)	Moderate lesions	259
Moose (*Alces alces*)	Mild lesions	260
Caribou (*Rangifer tarandus*)	Mild lesions	260
Monkey	Susceptible	258
Human	Susceptible	96, 261

Pseudocowpox Virus (Excluding Cattle)

Species	Susceptibility	Ref.
Mouse	Refractory	43, 44
Rabbit (cornea)	Refractory	39, 41, 43, 44
Rabbit (skin scarification)	Refractory	262
Guinea pig	Refractory	39
Embryonated hen egg (CAM)[b]	Refractory	43, 44, 262
Sheep	Refractory	76, 115

TABLE 1 (continued)
Natural and Experimental Infection with Parapoxviruses

Species	Susceptibility	Ref.
Lamb	Susceptible	263
Human	Susceptible	41, 96
Bovine Papular Stomatitis Virus (Excluding Cattle)		
Mouse	Refractory	16
Rabbit	Refractory	16
Guinea pig	Refractory	16
Embryonated hen egg (CAM)[b]	Refractory	16
Dog	Limited growth ?	54
	(1/6 developed papilloma)	49
Pig	Refractory	16
Sheep	Refractory	16
Goat	Refractory	16
Horse	Refractory	16
Human	Susceptible	96
Parapoxvirus of Red Deer		
Sheep	Mild lesions	228

[a] Review of the early literature.
[b] CAM, Chorioallantoic membrane.

D. PATHOLOGY OF PARAPOXVIRUS DISEASES IN ANIMALS AND HUMANS

Descriptions of the pathological changes occurring in animals (for convenience we will use this term to refer to nonhuman animals) and humans after infection with the parapoxviruses are covered in standard texts[30,83–85] or in specific articles[57,61,70,86,87] and only the briefest outline is given here. Generally, the pathology in animals is confined to the epithelium and oral mucosa and the incubation period for experimental inoculation is three to four days.[24,54,88–90] Skin lesions begin as reddening and swelling around the sites of inoculation and these develop into small vesicles over 24 hours. These vesicles develop a pustular appearance over the next few days. Histologically, the vesicles and pustules can be seen to be of a multiloculated nature due to swelling and degeneration of epithelial cells. Early in infection, eosinophilic, intracytoplasmic inclusion bodies are seen within degenerating epithelial cells but these can be difficult to find in more advanced lesions. Epithelial cells at the periphery of the lesion undergo proliferation and many mitotic figures can be found. The pustular appearance is predominantly due to a massive infiltration of polymorphonucleocytes. As the disease progresses, adjacent lesions may coalesce and scabs form over the surface. The underlying dermis becomes edematous and proliferative and if the scabs are removed a reddened, swollen, and moist surface is revealed, which gives a granulomatous appearance to the lesion. In orf virus infection, lesions around the mouth tend to develop extensive thickened scab formation presumably due to lifting and cracking of scabs with resulting ''weeping'' of blood and serum. Resolution occurs over a period of 4 to 6 weeks,[24,53,90] usually without complication. Mucosal lesions appear, grossly, as punched out ulcers but, at least in the case of bovine papular stomatitis, on histological examination the lesions can be seen to be areas of epithelial proliferation.[86]

There is little evidence that the parapoxviruses can spread systemically in animals. In a report of an outbreak of bovine papular stomatitis in neonatal calves that had been thymectomized and treated with anti-bovine lymphocyte globulin, an increase in incidence of the disease was recorded but systemic spread of the virus did not appear to occur.[91] In another experiment, intravenous administration of 10^9 plaque forming units (pfu) of orf virus failed to produce lesions in sham-scarified skin in a susceptible sheep.[26] On the other hand, rare severe outbreaks of orf virus infection in sheep have been reported where pathological changes to internal organs were found.[92,93] Whether or not this was systemic spread was not resolved.

The effects on the animal are determined by the site and extent of the lesions but fever is not a feature.[24,53,90] Lesions around or in the mouth interfere with feeding and thus retard growth in young animals. Teat lesions interfere with milking or suckling. In the case of red deer, lesions on the epithelium of the growing antler (velvet) can affect antler growth and marketability.[57]

The parapoxviruses are capable of infecting humans, as has been noted. Milkers' nodules have been recognized for a hundred years or more[38] as have infections caused by orf virus.[18] The viruses of bovine papular stomatitis[55] and seal pox[65] can also be readily transmitted to humans. The lesions these viruses cause are seldom serious although they can be painful and, in the healing stages, very itchy. The progression of the disease caused by orf virus from infection to resolution has been divided into six stages:[94] a maculopapular stage (1 to 7 days), characterized by vacuolization of the cells of the upper one third of the epidermis; a target stage (7 to 14 days), macroscopically having a red center surrounded by a white ring of cells in the maculopapular stage, in turn surrounded by a red halo of inflammation; an acute stage (14 to 21 days), where the epidermis has completely disappeared in some areas and involved hair follicles are dilated and full of pycnotic cells; a regenerative stage (21 to 28 days), where the epithelium is growing back and black spots appear where hair follicle cells have been extruded; a papilloma stage (28 to 35 days), characterized by an elevated epidermal lesion with finger-like projections of epidermis extending down into the dermis; and a regressive phase (35 days to resolution), in which the skin returns to its normal thickness and appearance, usually without scarring. Crust formation occurs at the acute and regenerative stages and we have seen a series of ''stacks'' forming a crust

5 to 10 mm thick on facial lesions. Second infections have been recorded within 14 months of the first, indicating that immunity is short-lived.[95] Descriptions of milkers' nodules are similar, with lesions beginning as reddish-purple, raised nodules turning bullous or pustular and being surrounded by a red halo of inflammation.[39] The incubation period is 5 to 7 days and lesions resolve in 5 to 6 weeks.[96]

An unusual manifestation of parapoxvirus infection in humans was recorded in handlers of musk-oxen and reindeer in Norway.[97] The lesions seen in these people were markedly granulomatous in nature and took many months to heal. Orf virus is known to affect reindeer and musk-oxen, but in this case the authors suggested that the virus might not have been orf virus but a new type of parapoxvirus. We and others[98,99] have noted that orf infections on the face in humans can elicit a marked proliferative response which could be considered unusual. It would be of interest to subject isolates from reindeer- and musk-ox-related human infections to restriction endonuclease analysis to determine whether or not the virus was orf virus.

E. THE UNIQUE MORPHOLOGY OF PARAPOXVIRUS VIRIONS

The viruses in the parapoxvirus group display a unique morphology when viewed in the electron microscope and, early on, formed the basis for their inclusion in a group separate from other members of the poxvirus family. Electron microscopy of orf virus reveals a virion that is oval in outline, with a long axis of 260 nm on average and a short axis of 160 nm.[100–105] This contrasts with virions of the other poxvirus genera which are brick-shaped and 350 nm × 250 nm on average. On closer examination of negatively stained preparations, parapoxviruses, in common with vaccinia virus, appear in two forms. Where the stain has penetrated the virion a finely crenulate membrane appears to surround an inner amorphous core (C or "capsular" form) whereas virions that remain impervious to the stain reveal a regular array of tubule-like structures arranged in a criss-cross manner along the length of the particle (M or "mulberry" form).[103] In addition, a layer of regular saucer-shaped subunits spanning the interval between the tubules and 1.5 to 2 nm thick has been seen closely associated with the outer surface of some intracellular forms of the virus.[77] A proportion of virions isolated from infected cell culture media (extracellular virions) are surrounded by a membranous structure 9 to 18 nm thick.[77,106] By analogy with vaccinia virus, this membrane is likely to be derived from the Golgi. Shadowing methods have revealed that the criss-cross appearance is due to the superimposition of images of the tubule-like structure as it winds its way in spiral fashion around the particle. The other parapoxviruses have been shown to exhibit a similar morphology[43,44,57,66,107,108] and electron microscopy still provides the most practical criterion for the confirmatory diagnosis of parapoxvirus infections.

Studies characterizing parapoxvirus virion proteins have shown that up to 40 polypeptide bands can be resolved by sodium dodecyl sulfate (SDS)/polyacrylamide gel electrophoresis of ^{35}S methionine-labeled milkers' node (pseudocowpox) virus and that a 42- to 45-kDa polypeptide, thought to be the characteristic virion surface tubule protein, can be released after detergent treatment, sonication, and differential centrifugation.[109] Similar results have been obtained with purified virions of orf virus, although only 30 polypeptide bands could be resolved in gels.[110,111] The size of the putative surface tubule protein was smaller and in the range of 37 to 44 kDa.[110,111] A study of the temporal appearance of pulse-labeled orf virus-specific polypeptides showed that they could be first detected at 8 to 10 hours post infection (p.i.) with host cell protein shut off occurring at 16 to 18 hours p.i.[111]

F. IMMUNOLOGICAL RELATIONSHIPS WITH OTHER POXVIRUS GENERA

Serological studies have shown that the parapoxviruses share common antigens with other poxvirus genera. Shared antigens can be demonstrated in complement fixation tests and in agar gel diffusion tests done in tubes, between orf virus, vaccinia virus, and ectromelia virus, the latter two viruses being members of the genus *Orthopoxvirus.*[112] Complement fixation tests have also revealed shared antigens between orf virus and sheep pox and goatpox viruses, members of the genus *Capripoxvirus.*[113] A common antigen shared between bovine papular stomatitis virus and vaccinia virus and myxoma virus (genus *Leporipoxvirus*) was shown in ring precipitin tests to be a nucleoprotein antigen.[114] On the other hand, fowlpox virus, a member of the genus *Avipoxvirus,* showed no cross-reactivity in tube gel diffusion tests with either orf virus or the orthopoxviruses, suggesting that the avipoxviruses are only distantly related at the antigenic level to the other poxvirus genera.[112]

The relatedness of the parapoxviruses to the orthopoxviruses has not always been demonstrated in agar gel diffusion studies. Tests done in a plate format using orf virus antiserum revealed precipitin lines of identity between orf virus and pseudocowpox virus, but not between orf virus and cowpox or vaccinia viruses.[76,115] In serum neutralization tests, antisera raised in rabbits against the orthopoxvirus, vaccinia virus, did not neutralize orf virus infectivity, nor did orf virus antisera raised in sheep neutralize vaccinia virus.[112] Two groups have reported neutralization of orf virus with goatpox virus antiserum,[113,116] but in another study cross-neutralization could not be demonstrated.[117]

Despite the apparent serological cross-reactivity and genetic relatedness between orf virus and vaccinia virus, cross-protection in sheep has not been found.[24,118] Cross-protection has not been demonstrated in calves[119] or humans[40,120] between vaccinia virus and pseudocowpox virus nor has cross-protection been demonstrated between goatpox virus[113,117] or sheep pox virus[113] and orf virus. Studies demonstrating cross-reactivity have invariably involved the interaction of sera, generated either by natural infection or artificially by inoculation of virion preparations into rabbits or other animals, with virion preparations. Therefore the serological cross-reactivity observed is most likely between epitopes on virion proteins (mostly late) and these serological responses seem not to be involved in the protective immune response. Perhaps not surprisingly, cross-protection has been demonstrated between orf virus and bovine papular stomatitis virus[121] and between orf virus and the parapoxvirus of red deer,[60] suggesting that all members of the parapoxvirus genus will cross-protect.

G. EPIDEMIOLOGY

The parapoxviruses of cattle and sheep have a worldwide distribution and it is believed that the spread and maintenance of the infections in each species is related to the resistant nature of the virions in the environment and the short-lived immunity to reinfection.

The conventional view of the epidemiology of orf virus infection is that the virus, being resistant to desiccation and being protected within the scab material that falls from infected animals, persists in the environment and acts as the source of infection for other animals.[122–125] It is believed that orf virus in scab material can remain infective for many years in heavily contaminated areas such as yards, barns, and sheep camps. Under experimental conditions, orf virus will remain infective for years, provided it is kept cool, dry, and out of direct sunlight.[126,127] However, from a study done in Scotland, if the virus is dispersed from the scab material and allowed to remain under field conditions over winter, the infectivity of the virus falls off rapidly.[128] Orf infections have also been found to persist for many months on the heads of rams and these persistent infections are thought to contribute to the maintenance and spread of the disease.[129,130] Another factor in the epidemiology is the short-lived immunity to reinfection, particularly on the udder of ewes.[131] Second and even third infections can be induced on the udder in animals which are otherwise immune to reinfection on the mouth, coronet, or leg.[131]

This view of the epidemiology of orf virus has been challenged and it has been suggested that the virus is maintained in some animals as a latent infection and that it is these animals which act as the reservoir of infection.[132] This idea appears to have arisen from experiments where scabs were harvested from lambs on successive occasions throughout the course of experimental infection and found to contain few virus particles at the stage of resolution of the lesion.[133] These experiments bear repeating with scab that has been undisturbed from the time of the initial lesion and that has been well dispersed by trituration. In our experience, large amounts of virus can be found in the dry scabs that have fallen off naturally infected animals, providing the scabs are well triturated before examination.

The epidemiology of pseudocowpox in milking herds has been well documented.[134] Of 20 herds with the disease, less than 5% of cases occurred in the summer, increasing to 50 to 100% in the autumn or spring. In one herd of 60 animals that was undergoing a primary infection, 95% of animals were infected in 6 weeks. After that, chronic infection was established in the herd. In herds chronically infected, lesions appeared on teats two weeks after calving, individual cows had recurrence of lesions at six-monthly intervals and some animals had lesions that persisted for 6 months. Lesions caused by reinfection were generally of a milder nature than those of primary infection. Poor hygiene was a feature of these herds and spread from cow to cow was via milking machine teat clusters, teat cloths, and milkers hands. The corollary to this study is that pseudocowpox can be controlled in milking herds by good hygienic practice and the reduction of factors causing teat damage.

There are no reports of epidemiological studies with other parapoxvirus infections but it appears from the studies of orf and pseudocowpox that the parapoxviruses are maintained in a population of animals by direct spread from acute and chronically infected animals and from virus persisting in the environment. There seems to be no evidence of a latent state.

H. MECHANISM OF PROTECTIVE IMMUNITY

The mechanisms involved in the generation of protective immunity, or apparent lack of it in some cases, after infection with parapoxviruses have not been satisfactorily resolved.

Our hypothesis is that the protective immune response to parapoxvirus infections is entirely cell mediated, although the evidence for this is incomplete and indirect. Only the response to orf virus infection has been studied in any detail.

Evidence that antibody is not involved in protection from orf virus infection is compelling, but not proven. Early work showed that serum from immune animals was not protective[24] and although neutralizing antibody present in colostrum is passed onto lambs from their immune dams, it was considered to be ineffective in protecting lambs from infection.[135] This is consistent with field evidence that lambs born to immune dams can succumb to natural infection in the first few days of life.[124,136,137] However, this view has been contradicted in other experiments where the titer of neutralizing antibody in colostrum correlated directly with the degree of protection transferred to lambs.[138,139]

Further evidence that antibody, or at least antibody directed against virion antigens, is not involved in protection comes from the observation that killed orf virus preparations do not induce a protective immune response. Unfortunately, we have not located any published data to verify this and have had to rely on personal communication with manufacturers of orf virus vaccine[140] (see Section IV.E). However, there seems little doubt that if it was possible to produce effective killed vaccines using standard protocols they would be in use, but this is not the case. Virus preparations inactivated at 60°C for 30 minutes are capable of inducing a delayed-type hypersensitivity (DTH) response in immune animals but it would seem that the antigen involved is not one that confers a protective response.[59,140] A report of a heat-killed orf virus vaccine being effective is probably not valid, as the heated preparations gave rise to mild lesions and it is likely that the heat treatment was not sufficient to inactivate all the virus.[137]

The observation that inactivated orf virus does not induce a protective effect is similar to the situation found when attempts were made to produce an effective, killed vaccinia virus vaccine during the lead-up to the smallpox eradication campaign. Although the inactivated vaccinia virus preparations induced neutralizing antibody, such vaccines were not protective.[141–147]

Leaving aside parapoxviruses for the moment, there is evidence that circulating antibody has a moderating effect on systemic poxvirus disease. Hyperimmune vaccinial γ-globulin is effective in treating generalized vaccinia[148,149] and has a moderating effect in smallpox.[150] Likewise, hyperimmune antifibroma serum is effective in preventing secondary fibromas in rabbits infected with Shope rabbit fibroma virus[151] and anti-rabbitpox serum will prevent fatal disease in rabbits infected with rabbitpox virus and mice infected with the neurovirulent vaccinia virus strain, WR.[147] In the latter case, it was shown that extracellular virus but not intracellular virus could induce the appropriate immunity and that the antigens involved are probably virally coded protein components of the extra envelope. This envelope is acquired from the inner membrane of the Golgi as the virions leave the cell.[152] Perhaps the most relevant point pertaining to immunity to parapoxvirus infection arising from these studies was that the primary lesion at the site of inoculation was little affected by the antiserum. As parapoxvirus infections are confined to the site of inoculation, the fact that neutralizing antibody has little effect on parapoxvirus growth is consistent with results observed in other poxvirus infections.

Another approach to understanding the mechanism of the protective immune response to parapoxvirus infection has been to study the humoral and cell-mediated responses induced after infection. Early work showed that viral neutralizing,[135,138,139,153,154] agglutinating,[154] precipitating,[154] and complement-fixing[136,154,155] antibodies can be detected after orf virus infection. Also, cytotoxic antibody, as measured in a complement-mediated cytotoxicity assay, can be detected in orf-infected sheep,[156,157] peaking 28 d after infection.[157] More recently, it has been shown that circulating antibody can be detected as early as 6 d after inoculation using anti-sheep Ig and purified virus antigen in an enzyme-linked immunosorbent assay (ELISA).[158] Western blotting showed that the antibody from infected animals would bind to virion antigens, but marked differences were noted between animals undergoing natural infection and those experimentally infected. Naturally infected animals showed responses to at least 13 virion components, although there was extensive individual variation in responses shown by different animals. Interestingly, there was a positive correlation between the number of antigens recognized by serum antibody and the severity and duration of clinical disease. Experimentally infected animals responded consistently to only three or four antigens, one of which appeared to be the characteristic surface tubule antigen. IgG could be detected at 14 d post-inoculation by an ELISA test using a hypotonic glycine extract of cells harvested 24 h after infection as a source of antigen.[158]

Cellular responses to infection in sheep have been studied using cells from cannulated efferent ducts of draining lymph nodes. For both orf virus and pseudocowpox virus infections, the numbers of cells emerging from the draining popliteal node rapidly decreased in the first day then increased 10-fold over the next 3 d. Large blastic lymphocytes increased 15- to 25-fold over the next 3 d. Antibody, detected in an antibody-mediated cytotoxicity assay, followed the appearance of blastic cells by 24 h. Perhaps surprisingly, no qualitative differences in cell movement and morphology were noted between infections with the two viruses. In these experiments no virus was recovered from the draining lymph.[159] In another study of cells emerging from draining supramammary nodes, similar timings and morphological changes were seen, although cell numbers increased only 2- to 3-fold. Most blast cells contained immunoglobulin of the IgG class, with a few containing IgM and antibody, could be detected in lymph 5 d after infection. In two sheep, the proportion of cells staining with T-cell-specific monoclonal antibody decreased as the response developed.[160]

The proliferative responses of peripheral lymphocytes to orf virus antigens was the subject of a third study. Little proliferative response was seen in a primary infection, although five of six animals responded on at least one occasion to a purified virion antigen but not to extracts of infected cells. A greater response was seen after a second inoculation.[161]

The conclusions to be drawn from these experiments are that antibody responses and draining node responses to orf virus infection are relatively normal and are within the normal range expected following an acute, cutaneous viral infection. Unfortunately, they tell us little about the mechanism of the protective response. There have been no studies reported on the cytotoxic T-cell response to parapoxvirus infection and it is likely such studies would be informative. It is considered that cytotoxic T-cells are an important component in recovery and protection from poxvirus infections[162] and it could be that in parapoxvirus infections where the virus is confined to the epithelium, cytotoxic T-cells play the major role in recovery and protection from disease. For orf virus and pseudocowpox virus infections the role of cytotoxic T-cells remains to be established.

It has been reported that purified inactivated vaccinia virus, avipoxvirus, and orf virus act as immunological potentiating agents and promote a generalized nonspecific stimulation of the immune response in the mouse.[163–165] Whether or not this phenomenon is a significant factor in the development of a protective immunity to poxviruses during infection in the natural host has yet to be determined.

I. PARAPOXVIRUS GENOMES

Molecular studies on the DNA of the parapoxviruses were first reported in 1979 with the publication of a restriction endonuclease cleavage site map of bovine papular stomatitis virus DNA.[166] The DNA was shown to be double stranded and linear, with a molecular weight of approximately 85 $\times$ 10^6 (135 kb), and with cross-linked termini. The overall structure of the parapoxvirus genome resembled that of vaccinia virus, although the size was smaller by some 40 kbp. At about the same time, it was reported that the DNAs of orf virus, milkers' node virus, and the virus of bovine papular stomatitis had G+C contents of 63% compared with 36 to 37% for vaccinia virus and rabbitpox virus.[167]

Subsequently, orf virus, bovine papular stomatitis virus, and pseudocowpox virus were compared by restriction endonuclease analysis and DNA/DNA hybridization and, generally, the results were in agreement with their classification based on host range and pathology.[78,79] However, analysis with *Eco*RI and *Hin*dIII showed considerable heterogeneity in the distribution of cleavage sites along the DNA, both between and within isolates of the three main virus groups.[78,79] Cleavage-site maps confirmed major differences between nine isolates of bovine papular stomatitis virus and between the bovine papular stomatitis virus isolates and two orf virus and two pseudocowpox virus isolates. The orf virus isolates had similar maps as did the pseudocowpox virus isolates, but the two groups differed markedly from each other. DNA/DNA hybridization studies showed that the three groups of viruses shared strong homology in the central regions of their genomes but not at their termini. The extent of this terminal nonhomologous region was not determined precisely but was at least 15.9 kb between a bovine papular stomatitis virus isolate and the other two virus groups. Hybridization of terminal fragments revealed further that these parapoxvirus genomes were terminally redundant.

Work in this laboratory on orf virus DNA generated restriction endonuclease maps of strains of orf virus isolated in New Zealand.[168] The DNA of orf virus obtained from scab material from clinical cases of the disease in sheep was shown by restriction endonuclease analysis to have a molecular weight of 88 $\times$ 10^6 (140 kb), similar to that found for the DNA of bovine papular stomatitis virus. All isolates examined contained DNAs of a similar size but there was variation in restriction fragment patterns between isolates.

A number of orf virus isolates have been plaque purified in cell culture and one, NZ2, selected for further study. Plaque-purified NZ2 has been amplified on sheep skin, viral DNA prepared, and maps produced for the restriction endonucleases *Eco*RI, *Hin*dIII, *Bam*HI, and *Hpa*I.[169] Restriction endonuclease-derived fragments representing the complete genome, except for 100 bp at the termini, have been cloned into various vectors[169] and with the aid of these clones, a *Kpn*I map of NZ2 and restriction maps for 16 other isolates have been deduced.[170] This analysis revealed that restriction endonuclease cleavage sites were conserved between isolates, except for a region 3 to 35 kbp in from the left end.[170] Sites in this region were variably present or absent between isolates and some sites were unique. A section of non-homology in this region, extending for at least 2.75 kbp, was detected in a number of isolates.

In addition to the 30-kbp variable region it was shown that one isolate had undergone a 1-kbp deletion near the junction of the inverted terminal repetition and unique sequences at the left end.[170] This observation is reminiscent of the orthopoxvirus genus where some members have been shown to undergo terminal rearrangements of their genomes.[171–176] Terminal sequences are duplicated in an inverted fashion at the opposite end of the genome and this is sometimes accompanied by a loss of sequences at the recipient end. Since the discovery of the deletion in naturally occurring orf virus strains we have observed a spontaneous terminal rearrangement of the NZ2 genome occurring during cell culture passage.[177] The rearrangement appeared as an increase in the size of the ITR by some 10 kbp resulting from the duplication of what were left-end unique sequences and the deletion of 2.5 kbp from a region 3.5 kbp from the right end.

We have compared restriction endonuclease cleavage site maps and gel patterns of New Zealand orf virus strains with European[78,79] and North American[178] strains. There are obvious differences, but by comparing *Kpn*I patterns they conform to a common type (Figure 1). It is interesting to note here that we would have classified two European parapoxvirus strains, V590 and V256, as orf virus on the basis of restriction patterns although they had been isolated from cases of bovine papular stomatitis in calves. Similarly, the European parapoxvirus strains Ch 32/H and 60898 had been designated as orf virus strains based on their clinical origin but were grouped with the bovine papular stomatitis virus isolates on the basis of restriction maps and hybridization data. It suggests the possibility that orf virus can give rise to natural clinical disease in calves and bovine papular stomatitis virus can affect sheep. Certainly some strains of orf virus can cause lesions in calves after experimental inoculation.[24,76]

*Kpn*I and *Bam*HI enzymes appear to be useful in grouping parapoxviruses. With these enzymes, DNA from the parapoxvirus of red deer can be distinguished easily from orf virus and the cattle parapoxviruses. DNA/DNA hybridization has shown that the terminal regions of the deer virus do not hybridize to those of orf virus.[59]

J. PARAPOXVIRUS GENES AND TRANSCRIPTIONAL CONTROLS

The genome of orf virus is the only parapoxvirus genome for which transcriptional and sequence data is available. Portions of the orf virus genome have been sequenced and a number of open reading frames identified.[179,180] Both early and late vaccinia virus-like promoter sequences have been found

5′ to the first ATG in many of these putative genes and the early transcriptional stop motif, T_5NT, has been found 3′ to those preceded by early promoter sequences. Transcriptional analysis of one early gene has shown that the 5′ end of the message begins 12 to 13 nucleotides downstream from the promoter-like sequence and ends 24 to 32 nucleotides downstream from each of two T_5NT motifs.[181] This close similarity between orf virus and vaccinia virus transcriptional control sequences[182] has been analyzed further using vaccinia virus recombinants containing this and other early genes. The orf virus genes are faithfully transcribed from their respective promoters.[183] Other vaccinia/orf virus recombinants also express orf virus proteins, as determined by immunofluorescence, Western blotting, and lymphocyte transformation assays.[184]

In addition to similarities in transcriptional signals, it is becoming clear that some orf virus genes bear a strong homology to those of vaccinia virus at the amino acid level and occupy similar positions on the genome. Homologies have been found between the pseudoprotease[180,185] (dUTPase-like[186]) gene, the 14K fusion gene[187,188] and the vaccinia virus H4, H5, H6, and H7 (topoisomerase) genes.[189,190] Despite the disparate G+C content of the two viruses, their common evolutionary origin is evident. This is probably not unexpected from their serological cross-reactivity. Just how many orf virus genes are related to those of vaccinia has yet to be determined.

Based on these findings, the orf virus genome map and consequently the maps for the other parapoxviruses can be oriented relative to the vaccinia virus map (Figure 2). The orf virus genome appears to be the equivalent of that region of the vaccinia virus genome lying betwen *Hin*dIII M and the serine protease inhibitor genes (B13R, B14R) in *Hin*dIII B.[191] It is, therefore, reasonable to conclude that the two viruses have drifted to the extremes of the G+C spectrum from a common ancestor and that vaccinia virus has gained, or orf virus has lost, nonessential terminal genes. The discovery that the physical arrangements of the vaccinia virus and orf virus genomes are similar should accelerate the gene mapping of orf virus and other parapoxviruses, although it must be kept in mind that some regions of the genome might be rearranged, as has been shown for fowlpox virus[192] and the capripoxviruses,[193] when compared with vaccinia virus.

III. RECOMBINANT PARAPOXVIRUSES

What can be gleaned from this information relevant to gene vectors and recombinant vaccines? The methodology needed to produce poxvirus recombinants has been pioneered using vaccinia virus[194,195] and it should be possible to adapt these methods to the parapoxviruses. As orf virus is the parapoxvirus for which most knowledge is available, we will describe progress toward a parapoxvirus gene vector using that virus system. It is reasonable to assume that any system developed for orf virus would be readily adaptable to the other parapoxviruses.

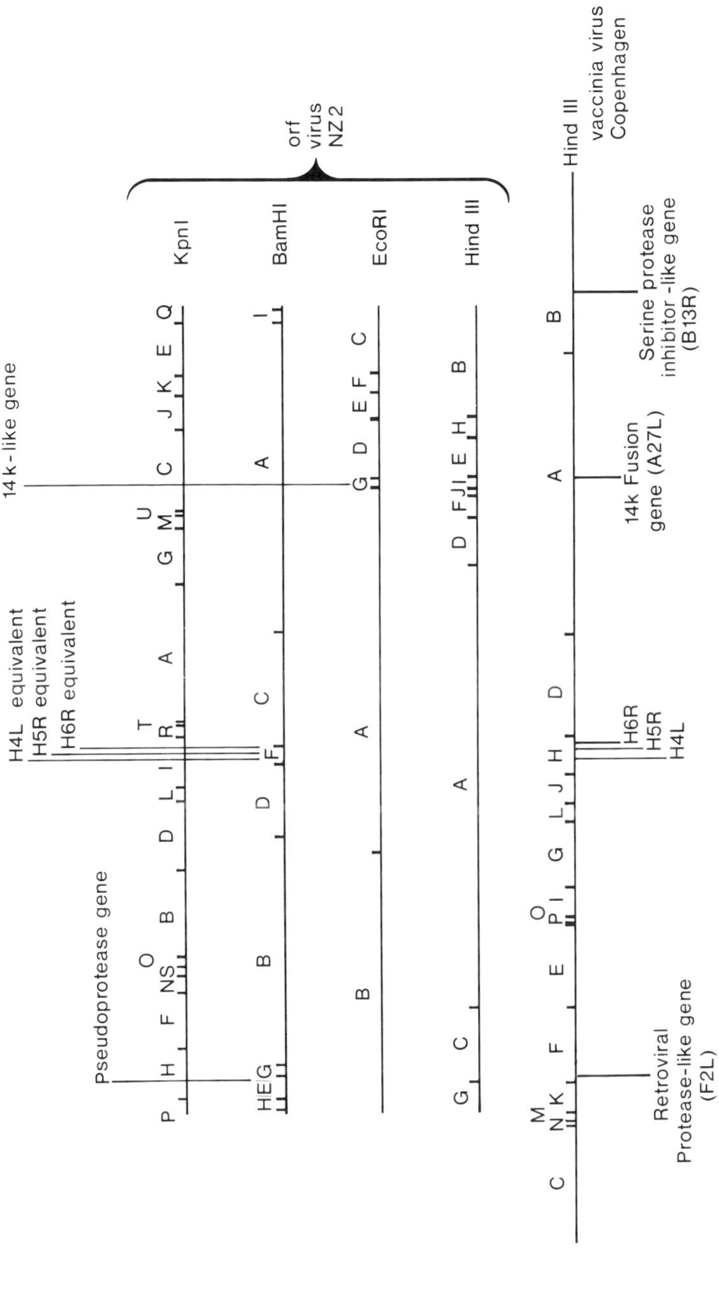

FIGURE 2. Orientation of the orf virus NZ2 genome map with that of the Copenhagen strain of vaccinia virus. The orf virus genome map has been drawn in the reverse orientation from that reported in Reference 169. The data for the vaccinia virus map is from Reference 191.

A. CELL CULTURE SYSTEMS

The basic requirements for producing poxvirus recombinants are cell culture systems capable of supporting viral growth, a transfection system for introducing DNA into cells, and a means of selecting recombinants. Until recently, the only cell culture systems available routinely for orf virus work were primary cells of ovine or bovine origin. The virus will grow in primary cells of human origin[196] and there is a report of propagation of the virus in HeLa cells[197] but in our laboratory, using New Zealand strains, we have not been successful in adpating the virus to continuous cell lines of rabbit, monkey, ovine, bovine, or human origin. For this reason we have done all our work in primary bovine or ovine testis cells. More recently, there has been a report of North American orf virus strains adapted to vero cells[26] and we have learned that New Zealand and European strains grow in MRC5 cells.[27] It might be that these viruses and cells will be useful in developing systems for producing orf virus recombinants.

Orf virus replicates more slowly than vaccinia virus in cell cultures, and to lower titers. One-step growth experiments have shown that in primary bovine testis cells, orf virus DNA replication can be detected at 10 h p.i., whereas with vaccinia virus it is 1 to 2 h.[111] New infectious orf virus particles appear at 20 to 24 h p.i., whereas with vaccinia virus it is 5 to 6 h.[198] In another study, in bovine fetal spleen cells, orf virus particles were detected intracellularly at 8 h p.i. and extracellularly at 12 h p.i.[199] Whether cell type or virus strain could account for these different results has not been determined. In our experience the yield of New Zealand strains of orf virus from recently confluent bovine testis cells in a culture flask with a 75-cm^2 surface area is between 10^7 and 10^8 pfu. Similar results have been obtained for pseudocowpox virus (milkers' node virus).[109] The yields of vaccinia (WR) virus from similar cultures is 10- to 100-fold higher. However high-titer orf virus stocks can be prepared by concentrating cell culture lysates by centrifugation.

B. TRANSFECTION PROTOCOL

As parapoxviruses replicate more slowly than vaccinia virus in cell cultures, it is possible that the optimal parameters for generating recombinants will differ from those which have been worked out for vaccinia virus. In attempts to generate orf virus recombinants, we have opted to use liposome-mediated transfection[200] rather than calcium phosphate precipitates,[201] as in our hands, primary bovine testis cells will not tolerate calcium phosphate for periods longer than a few hours. Cells will survive for at least 24 h in the medium recommended for liposome transfection and this allows the cells to be exposed to DNA for a greater time, thus extending the period available for transfection to occur.

C. SELECTION OF RECOMBINANTS

In order to recover orf virus recombinants from transfection experiments, the recombinants need to be enriched and cloned. The method that has become a standard for vaccinia virus recombinant selection is based on insertional inactivation of the viral thymidine kinase (*tk*) gene.[202] Results obtained from experiments designed to measure *tk* activity in orf virus-infected BT cells and abortively infected tk^- cells have been equivocal and we have been unable to find a *tk* gene in the NZ2 strain of orf virus by hybridization with redundant sets of radiolabeled oligonucleotides. The oligonucleotide sets were based on the relatively conserved region in the mammalian *tk* polypeptide and included sequences with an added G + C bias.[203] As the virus will not grow in continuous cell lines that have a tk^- phenotype, we have not been able to select for tk^+ or tk^- orf virus. The strains of orf virus that have been adapted to vero cells may offer a solution to this dilemma if a tk^- vero cell line is available or can be selected. From sequence data we have located a gene in orf virus that shows strong homology to a vaccinia virus gene that lies adjacent to the vaccinia virus *tk* gene.[204] If the *tk* gene is in the same relative position on the orf virus genome, further sequencing through this region should locate it. However, a negative result would not prove the absence of a *tk* gene in orf virus as in fowlpox virus the *tk* gene is in a different position in the genome relative to the equivalent genes flanking the vaccinia virus *tk* gene.[192] To date, we have not been able to locate a *tk* gene in orf virus to use as a nonessential site for gene insertion or a selectable marker.

The approach we have adopted has been to utilize the positive selection system based on the *Escherichia coli* guanine phosphoribosyltransferase (*gpt*) gene.[205] This selection has been used successfully for generating both vaccinia virus[206] and fowlpox virus recombinants[207] and is based on the observation that the mammalian *gpt* gene can transfer a phosphoribosyl pyrophosphate group to guanine but not xanthine, whereas the bacterial enzyme can utilize both substrates. Viruses that are expressing a bacterial *gpt* gene protect the cell in which they are replicating from the effects of the DNA synthesis inhibitor mycophenolic acid (MPA). MPA inhibits inosine monophosphate (IMP) dehydrogenase, preventing the formation of xanthine monophosphate (XMP), the precursor to guanine monophosphate (GMP), in the normal *de novo* synthesis pathway. In the mammalian cell, if guanine is supplied, then the mammalian *gpt* can overcome the block by directly synthesizing GMP in the salvage pathway. If the cells are starved of guanine then GMP becomes depleted and nucleic acid synthesis stops even in the presence of xanthine. If, however, the bacterial *gpt* is present and xanthine is supplied in the absence of guanine, GMP is synthesized via XMP and the cell can support nucleic acid synthesis. The block induced by MPA can be made more efficient by supplementing the medium with aminopterin which prevents the *de novo* synthesis of IMP. In this case hypoxanthine or adenine must be added to allow the formation of hypoxanthine or adenosine monophosphate by hypoxanthine or adenine phosphoribosyltransferase in the salvage pathway.

D. INSERTION OF FOREIGN GENES

For the selection system to work, the *gpt* gene needs to be expressed early in the virus cell cycle from a nonessential region of the viral genome. We have partially characterized an orf virus early promoter which appears to be suitable for this purpose and have constructed hybrid molecules that allow expression of the *gpt* gene under control of this promoter.[208] Subsequent work has suggested that it might not be necessary to use orf virus promoters because vaccinia virus polymerase recognizes orf virus early and late promoters[183] and it is very likely that the reverse situation will apply where orf virus polymerase recongizes vaccinia virus promoters, such as 7.5K or H6.

Despite our inability to locate an orf virus *tk* gene, we have mapped other regions on the genome which appear to be nonessential and provide suitable alternatives as insertion sites. Two sites are in regions where deletions have been observed[170,177] and the other is a region of the orf genome which appears to have undergone substitution in some strains.[170] As the mapping of orf virus genes progresses it is likely that better-defined, nonessential genes or intergenic sites will be identified, which will offer suitable alternative insertion sites.

A number of plasmids have been constructed which contain the *gpt* gene under control of an orf virus promoter and flanked by sequences from these nonessential regions.[208] We have constructed other plasmids where both a β-galactosidase gene under the control of an orf virus late promoter and a *gpt* transcriptional unit are flanked by nonessential sequences. The constructs follow the general principles of recombinant poxvirus construction as pioneered for vaccinia virus. It is interesting to note that the late promoter/β-galactosidase construct on its own is active in transient assays using vaccinia virus- or orf virus-infected bovine testis cells. Perhaps significantly, the β-galactosidase activity is expressed maximally at 48 h p.i. in the orf virus-infected cells, compared with 24 h in those infected with vaccinia virus. At the time of writing, we have selected a stable orf virus recombinant using the *gpt*-MPA selection system and this recombinant expresses the β-galactosidase gene.

IV. ADVANTAGES AND DISADVANTAGES OF RECOMBINANT PARAPOXVIRUS VACCINES

Having demonstrated experimentally that parapoxivurs gene vectors can be constructed, what advantage would such vectors have over vectors based on vaccinia virus? As experimental tools they may offer little when compared with vaccinia virus, but a case can be made for the use of parapoxviruses as recombinant vaccines based on their biological properties, which differ significantly from those of vaccinia virus. Some arguments elaborating these differences are presented in the sections which follow.

A. RECOMBINANT VACCINIA VIRUS VACCINES

The use of recombinant vaccinia viruses as vaccines has been promoted for some years[209–211] and briefly the arguments for their use are: recombinants are easy to produce and propagate; they can deliver protective antigens from a variety of pathogenic organisms into an animal in such a way as to stimulate both the humoral and cell-mediated effector arms of the immune system; and in numerous cases can be shown to induce protection. There is a long history of the safe use of vaccinia virus in the human population and the logistics of its delivery throughout the world were established and could be reestablished if necessary.

On the other hand, there have been a number of arguments against the use of vaccinia virus recombinants as vaccines.[212] First, there could be an unacceptable number of vaccine-related illnesses or deaths relative to that caused by the target disease. Vaccinia virus, although not a highly virulent virus, is still capable of causing a systemic febrile illness. The fact that the viruses are recombinants adds an extra element of uncertainty into this equation, although recombinants appear to be less virulent than the parent vaccinia. This has been the case at least in experimental animals using recombinants derived from the WR strain which was originally selected for neurovirulence in mice.[213] Second, vaccinia virus has a wide host range and it has been argued that there might be unforeseen effects in species other than the target species. Third, the widespread use of vaccinia virus recombinants against one disease might preclude the use of another vaccinia virus-based recombinant against other diseases at a later date because of the increase of immunity to vaccinia virus in the population. There is evidence, however, from animal experiments that individuals immune to vaccinia virus will still be capable of being infected with the virus and mounting an immune response against a foreign antigen introduced in a recombinant vaccina virus.[214–216]

B. ADVANTAGES OF RECOMBINANT PARAPOXVIRUS VACCINES

The advantage of using parapoxviruses rather than vaccinia virus as recombinant vaccines in humans are threefold. First, they are relatively avirulent compared with vaccinia virus. In general, both orf virus and pseudocowpox virus cause mild, self-limiting lesions without systemic febrile reactions. This is supported by numerous well-documented human case histories. Second, they have narrow host ranges, thereby reducing their possible impact outside the target species. As orf virus is endemic in the sheep and goat populations of the world and other parapoxviruses are endemic in cattle, the impact of recombinant parapoxviruses on these species is likely to be small. Third, there is little or no cross-immunity between the parapoxviruses and vaccinia virus, indicating that a target species immune to vaccinia virus might be more effectively immunized with the first or a second antigen with a parapoxvirus recombinant than with the equivalent vaccinia virus recombinant.

C. ADVERSE REACTIONS ASSOCIATED WITH PARAPOXVIRUSES IN HUMANS

An advantage that vaccinia virus has over the parapoxviruses in their potential use as recombinant vaccines, at least in the human population, is the 200 years of experience gained in its use to eradicate smallpox. The rate of adverse reactions is well documented and the impact of these reactions can be balanced against the impact of an uncontrolled target disease. Records of adverse reactions in humans to parapoxvirus infections are scanty but some have been recorded for orf virus. Erythma multiformae reactions characterized by reddened, papular rashes on the backs of the hands, wrists, and on the feet and ankles have been noted.[217–219] These reactions appear some 7 days after the appearance of the primary viral lesion and persist for 1 to 2 weeks. An unusual response to orf infection was reported in England, where two women, mother and daughter, contracted orf infection and subsequently developed papulo-vesicular eruptions on the skin and mucosae accompanied by pyrexia, malaise, and lymphadenopathy.[220] Both recovered but the symptoms persisted for 4 to 5 weeks. No virus could be isolated from these secondary lesions. One severe reaction attributed to orf virus infection was recorded in the U.S., where the outcome was embolism of a retinal artery and subsequent blindness.[221] In New Zealand, a farmer developed Stevens-Johnson syndrome (a severe form of erythma multiformae affecting skin and mucous membranes) after an orf virus infection.[222] The patient recovered but scarring permanently occluded the lachrymal ducts. The number of these types of reactions in proportion to the number of uneventful recoveries is unknown. A survey carried out in New Zealand meatworks showed that up to 500 meat workers out of a population of 20,000 at risk were infected with orf virus in the year of the survey.[95] Infections are also common in the farming community but generally go unrecorded because of the mild nature of the disease. It is likely that there are in excess of 1000 cases of orf in the New Zealand population each year and to date no deaths and only one severe reaction have been recorded.

It has been observed that orf virus can cause markedly enlarged lesions in humans suffering from burns. Two cases in New Zealand have been brought to our attention, one an electrician who had fallen into a sheep yard following electrocution, and another in a man who had suffered petrol burns to the face and had been handling sheep.[222] Both cases recovered uneventfully, but before healing, the lesions had reached 2 to 3 cm in diameter. It is interesting to note that severe lesions have been recorded in a sheep following burns.[223]

Vaccinia virus can cause severe disease in the immunocompromized host, characterized by systemic spread of the virus. This has not been recorded for the parapoxviruses but there are reports in the literature of orf infection in immunocompromised individuals. A patient undergoing treatment with cyclophosphamide, methotrexate, and vincristine for malignant lymphoma contracted orf virus on the finger.[224] The lesion persisted for some weeks and proliferation of the underlying dermis resulted in what was described as a ''giant orf'' resembling pyogenic granuloma. It is interesting to note that although the lesion became excessively large and did in fact lead, probably unnecessarily, to surgical amputation of the finger, there was no evidence of systemic spread of the virus or development of secondary lesions. Another case of a ''giant'' orf on the hand was reported in a patient undergoing treatment with chlorambucil and prednisone for chronic lymphocytic leukemia.[225] The lesion healed after excision of the main mass of the growth. Other ''giant'' proliferative lesions with resemblance to pyogenic granuloma have been recorded in otherwise normal individuals following parapoxvirus infection. In one case, the lesion developed at the site of a rope burn injury[98] to the face and in another, in sunburnt facial skin,[99] emphasizing the effect burns have on the ability of the skin at the site to contain virus growth. In none of these cases was there any evidence of systemic spread of the virus.

A case of orf has been reported in a 16-month-old child with atopic dermatitis.[226] A large granulomatous lesion appeared over a period of 10 d under the child's chin and within 2 d multiple satellite lesions had developed around the periphery of the main lesion. Small lesions also developed on the face, wrists, and the backs of the hands. Cervical and maxillary lymph nodes were enlarged and the child was feverish but the latter was attributed to a concurrent upper respiratory tract infection. Initially the lesion was diagnosed as either a pyogenic granuloma or bromoderma and was treated with topical corticosteroids. Subsequently, when no improvement was noted and a history of contact with sheep with contagious ecthyma was forthcoming, the lesion was treated with antibiotics to reduce secondary infection and the lesions healed uneventfully 2 weeks later. The satellite lesions that developed in this case could be interpreted as evidence of systemic spread, but their distribution made it more likely that they were secondary infections from the primary lesion.

In summary it appears that parapoxviruses are incapable of producing severe viremic disease in humans, such as is seen, for instance, in generalized vaccinia, but can cause rare severe erythma multiformae reactions and large granulomatous lesions where the individual is immunocompromised. It might have been expected that, for such a common infection in those handling cattle or sheep, some cases of systemic spread of the virus would have been recorded.

D. RECOMBINANT PARAPOXVIRUS VACCINES IN ANIMALS

Perhaps a more obvious use of parapoxvirus recombinants would be as vaccines in production animals. The parapoxviruses cause mild skin infections in their respective hosts and the insertion of foreign genes into the genome might be expected to reduce the virulence of the virus even further. However, this is an uncertain prediction and each recombinant would need to be carefully checked. Potentially, a more serious problem could turn out to be an unexpected increase in the prevalence of the parapoxvirus disease itself. For instance, if an orf virus recombinant was used in sheep, the recombinant itself might become a problem in the flock. Orf virus is a hardy virus and, if kept cool and out of direct sunlight, can maintain infectivity for many years. Virions are present in large numbers in the scabs that form on the animal during the course of the disease[168] or after so-called vaccination (for a description of vaccination against orf virus, see Section IV.E). There is no reason to expect that a recombinant virus would not be maintained in the environment in a similar manner and that it could be responsible for naturally occurring clinical disease.

There are a number of possible solutions to this problem. One could be that as knowledge of the molecular biology of the parapoxviruses accumulates it will be possible to engineer a parapoxvirus such that its virulence is reduced to an acceptable level and/or is rendered incapable of maintaining itself in the environment, while still maintaining its ability to deliver an immunizing dose of a particular antigen. Another possibility is to use a recombinant parapoxvirus in a species in which it is not normally endemic, for instance, an orf virus recombinant in cattle or a sealpox virus recombinant in sheep. Experience with fowlpox virus recombinants has shown that, despite the apparent absence of replication at the site of inoculation in hosts considered refractory to infection with fowlpox virus, immune responses can be elicited to the products of the inserted genes.[227] There is some evidence to suggest that although a particular parapoxvirus does not cause an apparent lesion when inoculated into an animal in which it is not normally endemic, there is still sufficient viral replication to induce a DTH response and/or protective immunity against the parapoxvirus which is endemic to that species. This was the case when the parapoxvirus of red deer was inoculated onto the skin of sheep[228] and when orf virus was inoculated onto the skin of red deer.[60] However, if this strategy was introduced, one would have to be aware of the possibility, albeit unlikely, that the virus might adapt to the target species and evolve toward an endemic disease problem in that species.

E. RECOMBINANT ORF VIRUS VACCINES

In the sheep population, where it is common to vaccinate against orf, the impact of the introduction of a recombinant orf virus vaccine might not be as great as might occur if a recombinant vaccine of one of the other parapoxviruses was used in its natural host species. Orf is the only parapoxvirus disease where vaccination is practiced routinely as a control measure. Clinical orf virus infection on the mouth during a lamb's growing period interferes with the lamb's ability to feed and is a cause of ill-thrift. At the present time, the means of controlling orf in lambs is to inoculate the animal on the leg with unattenuated virus when the animal is 3 to 4 weeks of age.[24,88–90,125] A typical orf lesion occurs at the site of inoculation and after 12 to 13 d, the animal is immune to reinfection.[131] The immunity to reinfection on the mouth persists for about 5 months, after which time lesions can be elicited but their resolution is accelerated.[131] The effect of this procedure is to protect the animals from developing clinical disease during their period of rapid growth and so is an effective and relatively cheap means of preventing production losses due to orf in the lamb-meat industry, although vaccines are not invariably protective.[229]

Orf virus preparations for use in lambs are available commercially[230] and although they are called vaccines the procedure is more akin to variolation than to vaccination. This distinction noted, one consequence of a vaccination program is to perpetuate the virus on the sheep farm or ranch. Virus that is formed following vaccination contaminates the environment just as surely as virus from naturally occurring clinical cases and, for obvious reasons, it is recommended that the vaccine be used only on properties where the disease is known to exist. If that caution is adhered to, then vaccination might not be expected to affect the level of environmental contamination any more than would be expected from clinical cases of the disease on a farm were the vaccine was not used. In the absence of an effective vaccine that avoids environmental contamination, the use of live unattenuated virus that prevents clinical disease is still considered acceptable animal husbandry, but vaccination must be done annually to prevent the appearance of clinical cases of orf in lambs in subsequent years.

Taking the current orf vaccination strategy into account, it could be argued that the introduction of an orf virus recombinant vaccine into a sheep flock would not alter the orf disease situation to any significant extent, provided the disease already existed as a problem on the farm. In fact, contrary to the argument presented previously, it might be considered an advantage if an orf virus recombinant that afforded protection against another pathogen in addition to itself became endemic in a sheep flock. The problem of uncontrolled cross-infection of a recombinant virus into humans would still exist, however, and that might be a reason for not using such a vaccine in sheep. To overcome this it might be possible to engineer a host-range mutant of the virus such that humans are not susceptible. Host-range mutants of vaccinia virus that grow poorly in human cell-lines have been isolated.[231,232]

F. OTHER CONSIDERATIONS

If it was possible to make parapoxvirus recombinant vaccines and if approval for their use in humans or other animals was likely, then one would need to decide what recombinants to make and when to use them. Any recombinant would need to be more effective, as safe or safer, and cheaper to produce than competitive products. Using the example of orf virus, it is likely that recombinants could be produced cheaply on sheep skin, in a similar manner to production methods for the current orf virus vaccines, which are similar to the methods used to produce smallpox vaccine (vaccinia virus). Alternatively, it might be possible to develop a cell-culture-derived vaccine.[233] Presumably, recombinants of other parapoxviruses could be produced in the same way in the skin of young animals of the appropriate species. The safety aspect has been mentioned briefly in relation to human infections and any specified recombinant would have to go through a comprehensive series of trials to gain registration. As regards effectiveness, each recombinant would have to be assessed separately, but as a general rule it would be unlikely that parapoxvirus recombinants would be any more effective than conventional vaccines where the protective effect was mediated by antibody. Exceptions to this might be where access to the target species is difficult, such as when wanting to vaccinate wild animals. In cases where antibody is the protective mechanism against a particular disease but where conventionally prepared vaccines are difficult or impossible to produce, then recombinant parapoxvirus vaccines might be useful, provided the antigen target is known and can be cloned. Even then, the production of a defined purified protein in fermenters by genetically engineered organisms might be the preferred method of production and this route has been followed in the production of an effective hepatitis B vaccine.[234] An advantage that a recombinant parapoxvirus could have in this situation might be the ability to process and present antigens in a more effective way and to be able to present the animal with a number of different antigens, the latter as a consequence of the capacity of poxviruses to accept large inserts of DNA.[235]

Poxvirus recombinant vaccines would seem to have their main advantage over conventionally prepared vaccines against diseases where the protective effect is cell-mediated cytotoxicity. Generally, viral replication in the host is necessary to stimulate such an immunity and, with conventional vaccines, this is achieved by administering a living, attenuated form of the virus.[236] A problem with the conventional approach can be that the degree of attenuation required to reduce the virulence of the virus down to acceptable levels reduces the ability of the vaccine to stimulate a protective immunity. Another problem can be the difficulty of growing the virus in quantity. Poxvirus recombinants offer the alternative of being able to deliver an effective dose of the appropriate antigen separate from the parent virus and its attendant pathogenicity. The argument then becomes whether or not the recombinant is more or less virulent than the attenuated form of the parent virus for an equivalent immunizing effect.

V. CONCLUSION

We have demonstrated that it is possible to construct parapoxvirus gene vectors. The mild nature of parapoxvirus infections and their narrow host ranges suggest that they could have an advantage over vaccinia virus vectors for use as recombinant vaccines in both animal and human populations. The factors that will determine whether or not they are used will be the same as for any other vaccine; efficacy, economy, and safety.

REFERENCES

1. **Goodpasture, E. W.,** Borreliotoses. Fowlpox, molluscum contagiosum, variola-vaccinia, *Science,* 77, 119, 1933.
2. **Goodpasture, E. W.,** Virus diseases of fowls as exemplified by contagious epithelioma (fowl-pox) of chickens and pigeons, in *Filterable Viruses,* Rivers, T. M., Ed., Williams & Wilkins, Baltimore, 1928, 233.
3. **Downie, A. W. and Dumbell, K. R.,** Pox Viruses, *Annu. Rev. Microbiol.,* 10, 237, 1956.
4. **Fenner, F. and Burnett, F. M.,** A short description of the poxvirus group (vaccinia and related viruses), *Virology,* 4, 305, 1957.
5. **Andrewes, C.,** *Viruses of Vertebrates,* Bailliere, Tindall and Cox, London, 1964, chap. 12.
6. **Melnick, J. L. and McCombs, R. M.,** Classification and nomenclature of animal viruses, *Prog. Med. Virol.,* 8, 400, 1966.
7. **Wildy, P.,** Classification and nomenclature of viruses. First report of the International Committee on Nomenclature of Viruses, *Monogr. Virol.,* 5, 28, 1971.
8. **Daniel, M. J. and Christie, A. H. C.,** Untersuchungen über krankheiten der gemse (Rupicapra rupicapra) und des Thars (Hemitragus jemlaicus Smith) in den südalpen von Neuseeland, *Schweiz. Arch. Tierheilkd.,* 105, 399, 1963.
9. **Fenner, F.,** Classification and nomenclature of viruses. Second report of the International Committee on Taxonomy of Viruses, *Intervirology,* 7, 28, 1976.
10. **Lipschütz, B.,** Untersuchungen über paravakzine, *Arch. Dermatol. Syph.,* 127, 193, 1919.
11. **Fenner, F., Pereira, H. G., Porterfield, J. S., Joklik, W. K., and Downie, A. W.,** Family and generic names for viruses approved by the International Committee on Taxonomy of Viruses, *Intervirology,* 3, 193, 1974.
12. **Matthews, R. E. F.,** Classification and nomenclature of viruses. Third report of the International Committee on Taxonomy of Viruses, *Intervirology,* 12, 160, 1979.
13. **Matthews, R. E. F.,** Classification and nomenclature of viruses. Fourth report of the International Committee on Taxonomy of Viruses, *Intervirology,* 17, 1, 1982.
14. **Esposito, J. J.,** personal communication, 1990.
15. **Plowright, W. and Ferris, R. D.,** Papular stomatitis of cattle in Kenya and Nigeria, *Vet. Rec.,* 71, 718, 1959.
16. **Griesemer, R. A. and Cole, C. R.,** Bovine papular stomatitis. I. Recognition in the United States, *J. Am. Vet. Med. Assoc.,* 137, 404, 1960.
17. **Robinson, A. J. and Balassu, T. C.,** Contagious pustular dermatitis (orf), *Vet. Bull.,* 51, 771, 1981.
18. **Oláh, P. and Elek, P.,** Immunisation against the contagious pustular dermatitis (sore mouth) of sheep, *Acta Vet.,* 3, 35, 1953.
19. **Steeb,** Von der Schaf Raude (Grind), 1787; cited in **Nurnberg,** *Diss. Berlin,* 1942.
20. **Walley, T.,** Contagious dermatitis: "orf" in sheep, *J. Comp. Pathol. Ther.,* 3, 357, 1890.
21. **Newsom, I. E. and Cross, F.,** Sore mouth in sheep transmissible to man, *J. Am. Vet. Med. Assoc.,* 84, 799, 1934.
22. **Hansen,** *Tidsskr. Vet.,* 9, 298, 1879; cited in **Hodgson-Jones, I.S.,** Orf in London, *Br. Med. J.,* 1, 795, 1951.
23. **Peterkin, G. A. G.,** The occurrence in humans of contagious pustular dermatitis of sheep ("orf"), *Br. J. Dermatol. Syph.,* 49, 492, 1937.
24. **Aynaud, M.,** La stomatite pustuleuse contagieuse des ovins (chancre du mouton), *Ann. Inst. Pasteur (Paris).,* 36, 498, 1923.
25. **Greig, A. S.,** Contagious ecthyma of sheep, *Can. J. Comp. Med.,* 21, 304, 1957.
26. **Hussain, K. A. and Burger, D.,** In vivo and in vitro characteristics of contagious ecthyma virus isolates: host response mechanisms, *Vet. Microbiol.,* 19, 23, 1989.

27. **Drillien, R.,** personal communication, 1990.
28. **Flook, W. H.,** An outbreak of venereal disease among sheep (ulcerative dermatosis), *J. Comp. Pathol.*, 16, 374, 1903.
29. **Tunnicliffe, E. A.,** Ulcerative dermatosis of sheep, *Am. J. Vet. Res.*, 10, 240, 1949.
30. **Jubb, K. V. F., Kennedy, P. C., and Palmer, N.,** *Pathology of Domestic Animals,* Vol. 1, 3rd ed., Academic Press, Orlando, 1985, 464.
31. **Roberts, R. S. and Bolton, J. F.,** A venereal disease of sheep, *Vet. Rec.*, 57, 686, 1945.
32. **Tunnicliffe, E. A. and Matisheck, P. H.,** A filterable virus demonstrated to be the infective agent in ovine balanoposthitis, *Science,* 94, 283, 1941.
33. **Trueblood, M. S. and Chow, T. L.,** Characterization of the agents of ulcerative dermatosis and contagious ecthyma, *Am. J. Vet. Res.*, 24, 47, 1963.
34. **Trueblood, M. S., Chow, T. L., and Griner, L. A.,** An immunological study of ulcerative dermatosis and contagious ecthyma, *Am. J. Vet. Res.*, 24, 42, 1963.
35. **Trueblood, M. S.,** Relationship of ovine contagious ecthyma and ulcerative dermatosis, *Cornell Vet.*, 56, 521, 1966.
36. **Jenner, E.,** *An Enquiry into the Causes and Effects of the Variolae Vacciniae, a Disease Discovered in Some of the Western Counties of England, Particularly Gloucestershire, and Known by the Name of Cowpox,* London, 1798; reprinted in **Camac, L. N. B., Ed.,** *Classics of Medicine and Surgery,* Dover, New York, 1959, 213.
37. **Jenner, E.,** *Further Observations on the Variolae Vacciniae,* Samson and Low, London, 1799.
38. **Bonnevie, P.,** Milker's warts: infection from "false cowpox" with a paravaccinal virus, *Br. J. Dermatol.*, 49, 165, 1937.
39. **Becker, F. T.,** Milker's nodules, *JAMA,* 115, 2140, 1940.
40. **Winternitz, R.,** Knotenbildungen bei melkerinnen, *Arch. Dermatol. Syph.*, 49, 195, 1899.
41. **Lipschütz, B.,** Paravaccine, in *Handbuch der Haut- und Geschlechtskrankheiten,* Vol. 2, Jadassohn, J., Ed., Julius Springer, Berlin, 1932, 80.
42. **Cowdry, E. V.,** Intracellular pathology in virus diseases, in *Filterable Viruses,* Rivers, T. M., Ed., Williams & Wilkins, Baltimore, 1928, 113.
43. **Moscovici, C., Cohen, E. P., Sanders, J., and DeLong, S. S.,** Isolation of a viral agent from pseudocowpox disease, *Science,* 141, 915, 1963.
44. **Friedman-Kein, A. E., Rowe, W. P., and Banfield, W. G.,** Milker's nodules: isolation of a poxvirus from a human case, *Science,* 140, 1335, 1963.
45. **Degive, A.,** Une affection-type ou maladie inedite — la stomatite papillaire ou papillomateuse — observee sur quatre genesis, *Ann. Med. Vet.*, 33, 369, 1884.
46. **Haag,** Infektiöse stomatitis, *Wochenschr. Tierheilk. Viehz.*, 51, 906, 1907.
47. **Mettam, A. E. and Norris, J. H.,** Report on Foot-and-Mouth Disease in Ireland in the Year 1912, Appendix B, HMS Office (Cd. 7103), 1913.
48. **Mason, J. H. and Neitz, W. O.,** Erosive stomatitis of cattle, *Onderstepoort J. Vet. Sci. Anim. Ind.*, 15, 159, 1940.
49. **Olson, C. and Paliones, T.,** The transmission of proliferative stomatitis of cattle, *J. Am. Vet. Med. Assoc.*, 123, 419, 1953.
50. **Hollister, C. J., Fagan, R., and Arnold, M. A.,** Muzzle disease, *J. Am. Vet. Med. Assoc.*, 128, 70, 1956.
51. **Pande, P. G. and Krishnamurty, D.,** Parotido-stomatitis in calves, *J. Infect. Dis.*, 98, 142, 1956.
52. **Pritchard, W. R., Claffin, R. M., Gustafson, D. P., and Ristic, M.,** An infectious ulcerative stomatitis of cattle, *J. Am. Vet. Med. Assoc.*, 132, 273, 1958.
53. **Plowright, W. and Ferris, R. D.,** Papular stomatitis of cattle. II. Reproduction of the disease with culture-passaged virus. *Vet. Rec.*, 71, 828, 1959.
54. **Griesemer, R. A. and Cole, C. R.,** Bovine papular stomatitis. II. The experimentally produced disease, *Am. J. Vet. Res.*, 22, 473, 1961.

55. **Carson, C. A. and Kerr, K. M.,** Bovine papular stomatitis with apparent transmission to man, *J. Am. Vet. Med. Assoc.*, 151, 183, 1967.
56. **Carson, C. A., Kerr, K. M., and Grumbles, L. C.,** Bovine papular stomatitis: experimental transmission from man, *Am. J. Vet. Res.*, 29, 1783, 1968.
57. **Horner, G. W., Robinson, A. J., Hunter, R., Cox, B. T., and Smith, R.,** Parapoxvirus infections in New Zealand farmed red deer, *N.Z. Vet. J.*, 35, 41, 1987.
58. **Robinson, A. J.,** Deer parapoxvirus: a new virus, Proc. Deer Course for Veterinarians, Course No. 4, Deer Branch of the New Zealand Veterinary Association, Dunedin, 1987.
59. **Robinson, A. J.,** unpublished data, 1988.
60. **Mackintosh, C. G. and Smith, D. R.,** Preliminary results of experimental deer parapox and orf virus infection trials in red deer, Proc. Deer Course for Veterinarians, Course No. 4, Deer Branch of the New Zealand Veterinary Association, Dunedin, 1987.
61. **Wilson, T. M., Cheville, N. F., and Karstad, L.,** Seal pox, *Bull. Wildl. Dis. Assoc.*, 5, 412, 1969.
62. **Wilson, T. M. and Poglayen-Neuwall, I.,** Pox in South American sea lions, *Otaria byronia, Can. J. Comp. Med.*, 35, 174, 1971.
63. **Wilson, T. M., Dykes, R. W., and Tsai, K. S.,** Pox in young, captive harbor seals, *J. Am. Vet. Med. Assoc.*, 161, 611, 1972.
64. **Hadlow, W. J., Cheville, N. F., and Jellison, W. L.,** Occurrence of pox in a northern fur seal on the Pribilof Islands in 1951, *J. Wildl. Dis.*, 16, 305, 1980.
65. **Hicks, B. D. and Worthy, G. A. J.,** Sealpox in captive grey seals (Halichoerus grypus) and their handlers, *J. Wildl. Dis.*, 23, 1, 1987.
66. **Roslyakov, A. A.,** Comparative ultrastructure of viruses of camel pox, pox-like disease of camels (AUZDUK) and contagious ecthyma of sheep, *Voprosi Virusol.*, 17, 26, 1972.
67. **Dashtseren, T. S., Solovyev, B. V., Varejka, F., and Khokhoo, A.,** Camel contagious ecthyma (Pustular Dermatitis), *Acta Virol.*, 28, 122, 1984.
68. **Jezek, Z., Kriz, B., and Rothbauer, V.,** Camelpox and its risk to the human population, *J. Hyg. Epidemiol. Microbiol. Immunol.*, 27, 29, 1983.
69. **Maollin, A. S. and Zessin, K. H.,** Outbreak of camel contagious ecthyma in central Somalia, *Trop. Anim. Health Prod.*, 20, 185, 1988.
70. **Munz, E., Schillinger, D., Reimann, M., and Mahnel, H.,** Electron microscopical diagnosis of *ecthyma contagiosum*, in camels (camelus dromedarius). First report of the disease in Kenya, *J. Vet. Med. Ser. B*, 33, 73, 1986.
71. **Kitching, P.,** personal communication, 1988.
72. **Hartung, J.,** Contagious ecthyma of sheep (cases in man, dog, alpaca and camel), *Tieraerztl. Prax.*, 8, 435, 1980.
73. **Scott, A. C., Keymer, I. F., and Labram, J.,** Parapoxvirus infection of the red squirrel (Sciurus vulgaris), *Vet. Rec.*, 109, 202, 1981.
74. **Okada, H. M., Okada, K., Numakunai, S., and Ohshima, K.,** Histopathological studies on mucosal and cutaneous lesions in contagious papular dermatitis of Japanese serow, *Jpn. J. Vet. Sci.*, 46, 257, 1984.
75. **Okada, H. M., Okada, K., Numakunai, S., and Ohshima, K. -I.,** Electron microscopy on mucosal and cutaneous lesions in contagious papular dermatitis of Japanese serow (Capricornis crispus), *Jpn. J. Vet. Sci.*, 46, 297, 1984.
76. **Huck, R. A.,** A paravaccinia virus isolated from cows' teats, *Vet. Rec.*, 78, 503, 1966.
77. **Rosenbusch, R. F. and Reed, D. E.,** Reaction of convalescent bovine antisera with strain-specific antigens of parapoxviruses, *Am. J. Vet. Res.*, 44, 875, 1983.
78. **Gassmann, U., Wyler, R., and Wittek, R.,** Analysis of parapoxvirus genomes, *Arch. Virol.*, 83, 17, 1985.
79. **Wittek, R., Herlyn, M., Schumperli, D., Bachmann, P. A., Mayr, A., and Wyler, R.,** Genetic and antigenic heterogeneity of different parapoxvirus strains, *Intervirology*, 13, 33, 1980.
80. **Nagington, J., Lauder, I. M., and Smith, J. S.,** Bovine papular stomatitis, pseudocowpox and milker's nodules, *Vet. Rev.*, 18, 306, 1967.

81. **Liebermann, H.,** Relationships between milkers' nodules, udder pox, papular stomatitis, and contagious ecthyma, *Z. Aertzl. Fortbild.*, 61, 447, 1967.
82. **Robinson, A. J. and Roulston, J.,** unpublished data, 1989.
83. **Lever, W. F. and Schaumburg-Lever, G.,** *Histopathology of the Skin,* 6th ed., Lippincott, Philadelphia, 1983, 369.
84. **Ackerman, A. B.,** *Histological Diagnosis of Inflammatory Skin Diseases,* Lea & Febiger, Philadelphia, 1978, 508.
85. **Blood, D. C. and Radostits, O. M.,** *Veterinary Medicine,* 7th ed. Bailliere and Tindall, London, 1989, 945.
86. **Griesemer, R. A. and Cole, C. R.,** Bovine papular stomatitis. III. Histopathology, *Am. J. Vet. Res.*, 22, 482, 1961.
87. **Cheville, N. F.,** Cytopathology of viral diseases, in *Monographs in Virology,* Vol. 10, Melnick, J. L., Ed., S. Karger, Basel, 1975, 9.
88. **Glover, R. E.,** Contagious pustular dermatitis of the sheep, *J. Comp. Pathol. Ther.*, 41, 318, 1928.
89. **Schmidt, H. and Hardy, W. T.,** Soremouth (contagious ecthyma) in sheep and goats, *Tex. Agric. Exp. Stn. Bull.*, No. 457, 1932.
90. **Howarth, J. A.,** Infectious pustular dermatitis of sheep and goats, *J. Am. Vet. Med. Assoc.*, 75, 741, 1929.
91. **Snider, T. G., McConnell, S., and Pierce, K. R.,** Increased incidence of bovine papular stomatitis in neonatal calves, *Arch. Virol.*, 71, 251, 1982.
92. **Darbyshire, J. H.,** A fatal ulcerative mucosal condition of sheep associated with the virus of contagious pustular dermatitis, *Br. Vet. J.*, 117, 97, 1961.
93. **Valder, W.-A., Straub, O. C., Theil, W., Wachendorfer, G., and Zettl, K.,** Ecthyma contagiousum des schafes — wandel des klinischen bildes, *Tieraerztl. Umsch.*, 34, 828, 1979.
94. **Leavell, U. W., McNamara, J., Muelling, R. J., and Landrum, F.,** Ecthyma contagiosum (orf), *South. Med. J.*, 58, 239, 1965.
95. **Robinson, A. J. and Petersen, G. V.,** Orf virus infection of workers in the meat industry, *N.Z. Med. J.*, 96, 81, 1983.
96. **Liebermann, H. and Jung, H.,** Experimentelle infection des menschen mit stomatitis-papulosa-, enterpocken-, dermatitis-pustulosa- und melkerknotenvirus-stammen, *Arch. Exp. Veterinaermed.*, 31, 603, 1977.
97. **Falk, E. S.,** Parapoxvirus infections of reindeer and musk ox associated with unusual human infections, *Br. J. Dermatol.*, 99, 647, 1978.
98. **Pether, J. V., Guerrier, C. J., Jones, S. M., Adam, A. E., and Kingsbury, W. N.,** Giant orf in a normal individual, *Br. J. Dermatol.*, 115, 497, 1986.
99. **Rogers, M., Bale, P., De Silva, L. M., Glasson, M. J., and Collins, E.,** Giant parapox infection in a two year old child, *Australas. J. Dermatol.*, 30, 87, 1989.
100. **Abdussalam, M. and Cosslett, V. E.,** Contagious pustular dermatitis. I. Studies on morphology, *J. Comp. Pathol.*, 67, 145, 1957.
101. **Reczko, E.,** Electronenmikroskopische untersuchungen am virus der stomatitis papulosa, *Zentralbl. Bakteriol. Abt. 1 Orig. B.*, 169, 425, 1957.
102. **Knocke, K.-W.,** Electronenmikroskopische untersuchungen am virus des ecthyma contagiosum der schafe, *Zentralbl. Bakteriol.*, 185, 304, 1962.
103. **Nagington, J. and Horne, R. W.,** Morphological studies of orf and vaccinia viruses, *Virology,* 16, 248, 1962.
104. **Peters, D., Müller, G., and Büttner, D.,** The fine structure of paravaccinia viruses, *Virology,* 23, 609, 1964.
105. **Nagington, J., Newton, A. A., and Horne, R. W.,** The structure of orf virus, *Virology,* 23, 461, 1964.
106. **Mitchner, M. B.,** The envelope of vaccinia and orf viruses: an electron-cytochemical investigation, *J. Gen. Virol.*, 5, 211, 1969.

107. **Nagington, J., Plowright, W., and Horne, R. W.,** The morphology of bovine papular stomatitis virus, *Virology,* 17, 361, 1962.
108. **Wilson, T. M. and Sweeny, P. R.,** Morphological studies of seal poxvirus, *J. Wildl. Dis.,* 6, 94, 1970.
109. **Thomas, V., Flores, L., and Holowczak, J. A.,** Biochemical and electron microscopic studies of the replication and composition of milker's node virus, *J. Virol.,* 34, 244, 1980.
110. **Buddle, B. M., Dellers, R. W., and Schurig, G. G.,** Heterogeneity of contagious ecthyma virus isolates, *Am. J. Vet. Res.,* 45, 75, 1984.
111. **Balassu, T. C. and Robinson, A. J.,** Orf virus replication in bovine testis cells: kinetics of viral DNA, polypeptide, and infectious virus production and analysis of virion polypeptides, *Arch. Virol.,* 97, 267, 1987.
112. **Webster, R. G.,** The immunological relations of the contagious pustular dermatitis virus to the mammalian pox group, *Aust. J. Exp. Biol. Med. Sci.,* 36, 267, 1958.
113. **Sharma, S. N. and Dhanda, M. R.,** Studies on sheep- and goat-pox viruses: relationship with contagious pustular dermatitis and vaccinia viruses, *Indian J. Anim. Sci.,* 41, 864, 1971.
114. **Woodroofe, G. M. and Fenner, F.,** Serological relationships within the poxvirus group: an antigen common to all members of the group, *Virology,* 16, 334, 1962.
115. **Papadopoulos, O. A., Dawson, P. S., Huck, R. A., and Stuart, P.,** Agar gel diffusion studies of paravaccinia viruses, *J. Comp. Pathol.,* 78, 219, 1968.
116. **Dubey, S. C. and Sawhney, A. N.,** Serological relationship between viruses of goatpox and contagious pustular dermatitis, *Indian J. Anim. Sci.,* 49, 135, 1979.
117. **Renshaw, H. W. and Dodd, A. G.,** Serological and cross-immunity studies with contagious ecthyma and goatpox virus isolates from the western United States, *Arch. Virol.,* 56, 201, 1978.
118. **Robinson, A. J. and Mercer, A. A.,** Orf virus and vaccinia virus do not cross-protect sheep, *Arch. Virol.,* 101, 255, 1988.
119. **Berger, K.,** Infectionsversuche mit dem virus des melkerknotens, *Zentralbl. Bakteriol. Parasitenkd. Infektionskrankh. Hyg. Abt. 1 Orig. Reihe B.,* 162, 363, 1955.
120. **Salkan, P. M.,** *Acta Derm. Venereol. Stockh.,* 14, 342, 1933; cited in **Davies, J. H. T., Janes, L. R., and Downie, A. W.,** Cowpox infection in farmworkers, *Lancet,* ii, 1534, 1938.
121. **Mayr, A.,** Control of acute virus disease of calves in the Federal Republic of Germany, *Vet. Sci. Commun.,* 3, 3, 1979.
122. **Theiler, A.,** Ecthyma contagiosum of sheep and goats, *Onderstepoort Laboratory 13th and 14th Annual Report,* Part 1, Onderstepoort, Union of South Africa, 1928, 7.
123. **Glover, R. E.,** Contagious pustular dermatitis. (A) Cross-immunity experiments with various strains. (B) Serological tests, *Rep. Dir. Inst. Anim. Pathol.,* 1932–1933, 4.
124. **Boughton, I. B. and Hardy, W. T.,** Contagious ecthyma (sore mouth) of sheep and goats, *J. Am. Vet. Med. Assoc.,* 85, 150, 1934.
125. **Boughton, I. B. and Hardy, W. T.,** Immunization of sheep and goats against soremouth (contagious ecthyma), *Tex. Agric. Exp. Stn. Bull.,* No. 504, 1935.
126. **Livingston, C. W. and Hardy, W. T.,** Longevity of contagious ecthyma virus, *J. Am. Vet. Med. Assoc.,* 137, 651, 1960.
127. **Buxton, A. and Fraser, G.,** *Animal Microbiology,* Vol. 2, *Rickettsias and Viruses,* Blackwell, Oxford, 1977, 690.
128. **McKeever, D. J. and Reid, H. W.,** Survival of orf virus under British winter conditions, *Vet. Rec.,* 118, 613, 1986.
129. **Greig, A., Linklater, K. A., and Clark, W. A.,** Persistent orf in a ram, *Vet. Rec.,* 115, 149, 1984.
130. **McKeever, D.,** Persistent orf, *Vet. Rec.,* 115, 334, 1984.

131. **Schmidt, D.,** Experimentelle beiträge zur kenntnis der dermatitis pustulosa des schafes V. Untersuchungen über die ausbildung der immunität gegen das virus der dermatitis pustulosa an verschiedenen stellen der körperoberfläche, *Arch. Exp. Veterinaermed.*, 21, 937, 1967.
132. **Robertson, A.,** *Handbook on Animal Diseases in the Tropics*, 3rd ed., British Veterinary Association, London, 1976, 9.
133. **Romero-Mercado, C. H., McPherson, E. A., Laing, A. H., Lawson, J. B., and Scott, G. R.,** Virus particles and antigens in experimental orf scabs, *Arch. Gesammte Virusforsch.*, 40, 152, 1973.
134. **Gibbs, E. P. J. and Osborne, D.,** Observations on the epidemiology of psuedocowpox, *Br. Vet. J.*, 130, 150, 1974.
135. **Buddle, B. M. and Pulford, H. D.,** Effect of passively-acquired antibodies and vaccination of the immune response to contagious ecthyma virus, *Vet. Microbiol.*, 9, 515, 1984.
136. **Schmidt, D.,** Experimentelle beiträge zur kenntnis der dermatitis pustulosa des schafes. V. Die specifität der komplementbindungs-reaction mit dermatitis-pustulosa-antigen, *Arch. Exp. Veterinaermed.*, 21, 947, 1967.
137. **Richter, J. and Jansen, J.,** Over de actieve immunisatie van schapen tegen ecthyma. Een praktijkproef met een onverwarmd en een verwarmd autovaccin, *Tijdschr. Diergeneeskd.*, 93, 757, 1968.
138. **Poulain, J., Gourreau, J. -M., and Dautigny, A.,** Ecthyma contagieux du mouton: anticorps sériques neutralisants, *Ann. Rech. Vet.*, 3, 571, 1972.
139. **Jan, C. Le., L'Harridon, R., Madelaine, M. F., Cornu, C., and Asso, J.,** Transfer of antibodies against the CPD virus through colostrum and milk, *Ann. Rech. Vet.*, 9, 343, 1978.
140. **Fastier, L.,** personal communication, 1980.
141. **Weil, A. J.,** Studies on the immunization of rabbits with formalinized vaccine virus, *J. Immunol.*, 38, 1, 1940.
142. **Collier, L. H., McClean, D., and Vallet, L.,** The antigenicity of ultra-violet irradiated vaccinia virus, *J. Hyg.*, 53, 513, 1955.
143. **Kaplan, C.,** The antigenicity of gamma-irradiated vaccinia virus, *J. Hyg.*, 58, 391, 1960.
144. **Beunders, B. J. W., Driessen, J. H., and van den Hoek, C.,** Clinical picture and serological response to vaccination with formalinized vaccinia virus followed by scarification with active vaccine, in military personnel, *Arch. Gesammte Virusforsch.*, 10, 382, 1960.
145. **Dostal, V.,** Advances in the production of smallpox vaccine, *Prog. Med. Virol.*, 4, 259, 1962.
146. **RamanaRao, A. V.,** The immunogenicity of inactivated vaccinia virus, *J. Pathol. Bacteriol.*, 84, 367, 1962.
147. **Boulter, E. A., Zwartouw, H. T., Titmus, D. H. J., and Maber, H. B.,** The nature of the immune state produced by inactivated vaccinia virus in rabbits, *Am. J. Epidemiol.*, 94, 612, 1971.
148. **Kempe, C. H.,** Hyperimmune vaccinial gamma globulin, *Pediatrics*, 18, 177, 1956.
149. **Kempe, C. H.,** Studies on smallpox and complications of smallpox vaccination, *Pediatrics*, 26, 176,. 1960.
150. **Pierce, E. R., Melville, F. S., Downie, A. W., and Duckworth, M. J.,** Antivaccinial gamma-globulin in smallpox prophylaxis, *Lancet*, ii, 635, 1958.
151. **Allison, A. C. and Friedman, R. M.,** Effects of immunosuppressants on Shope rabbit fibroma, *J. Natl. Cancer Inst.*, 36, 859, 1966.
152. **Ichihashi, Y., Matsumoto, S., and Dales, S.,** Biogenesis of pox-viruses: role of A-type inclusions and host cell membranes in virus dissemination, *J. Virol.*, 46, 507, 1971.
153. **Nagington, J. and Whittle, C. H.,** Human orf: isolation of the virus by tissue culture, *Br. Med. J.*, ii, 1324, 1961.

154. **Abdussalam, M.,** Contagious pustular dermatitis. IV. Immunological reactions, *J. Comp. Pathol.,* 68, 23, 1958.
155. **Rottgardt, A. A., Aramburu, H. G., and Garcia-Pirazzi, A. J.,** Complement fixation test in contagious ecthyma, *Nature (London),* 163, 219, 1949.
156. **DeMartini, J. C., Pearson, L. D., and Fiscus, S. A.,** Chromium-51-release assay of antibody and complement-mediated cytotoxicity for contagious ecthyma virus-infected cells, *Am. J. Vet. Res.,* 39, 1922, 1978.
157. **Koptopoulos, G., Reid, H. W., and Pow, I.,** Cytotoxic antibodies in orf virus infection of sheep, *Zentralbl. Veterinaermed. Reihe B,* 29, 284, 1982.
158. **McKeever, D. J., Reid, H. W., Inglis, N. F., and Herring, A. J.,** A qualitative and quantitative assessment of the humoral antibody response of the sheep to orf virus infection, *Vet. Microbiol.,* 15, 229, 1987.
159. **Pearson, L. D., DeMartini, J. C., and Fiscus, S. A.,** Cellular responses in efferent lymph during parapox virus infection in sheep, *Fed. Proc., Fed. Am. Soc. Exp. Biol.,* 38, 1158, 1979.
160. **McKeever, D. J. and Reid, H. W.,** The response of the supramammary lymph node of the sheep to secondary infection with orf virus, *Vet. Microbiol.,* 14, 3, 1987.
161. **Yirrel, D. L., Reid, H. W., Norval, M., and Howie, S. E.,** Immune response of lambs to experimental infection with orf virus, *Vet. Immunol. Immunopathol.,* 22, 321, 1989.
162. **Cole, G. A. and Blanden, R. V.,** Immunology of poxviruses, in *Comprehensive Immunology,* No. 9., Nahmias, A. J. and O'Reilly, R. J., Eds., Plenum Press, New York, 1982, 1.
163. **Buttner, M. and Mayr, A.,** Tests on protection against viral diseases, *Comp. Immunol. Microbiol. Infect. Dis.,* 9, 205, 1986.
164. **Muller-Brunecker, G., Erfle, V., and Mayr, A.,** Comparison of the effect of viral paraimmunity inducers PIND-AVI and PIND-ORF with that of BCG, *Corynebacterium parvum* and Levamisole on the growth of radiation-induced murine osteosarcoma, *J. Vet. Med. Ser. B,* 33, 188, 1986.
165. **Mayr, A., Büttner, M., Wolf, G., Meyer, H., and Czerny, C.,** Experimenteller nachweis paraspezifischer wirkungen von gereinigten und inaktivierten pockenviren, *J. Vet. Med. Ser. B,* 36, 81, 1989.
166. **Menna, A., Wittek, R., Bachmann, P. A., Mayr, A., and Wyler, R.,** Physical characterisation of a stomatitis papulosa virus genome: a cleavage map for the restriction endonucleases *Hind*III and *Eco*RI, *Arch. Virol.,* 59, 145, 1979.
167. **Wittek, R., Kuenzle, C. C., and Wyler, R.,** High C + G content in parapoxvirus DNA, *J. Gen. Virol.,* 43, 231, 1979.
168. **Robinson, A. J., Ellis, G., and Balassu, T.,** The genome of orf virus: restriction endonuclease analysis of viral DNA isolated from lesions of orf in sheep, *Arch. Virol.,* 71, 43, 1982.
169. **Mercer, A. A., Fraser, K., Barns, G., and Robinson, A. J.,** The structure and cloning of orf virus DNA, *Virology,* 157, 1, 1987.
170. **Robinson, A. J., Barns, G., Fraser, K., Carpenter, E., and Mercer, A. A.,** Conservation and variation in orf virus genomes, *Virology,* 157, 13, 1987.
171. **Archard, L. C. and Mackett, M.,** Restriction endonuclease analysis of red cowpox virus and its white pock variant, *J. Gen. Virol.,* 45, 51, 1979.
172. **Moyer, R. W. and Rothe, C. T.,** The white pock mutants of rabbit poxvirus. I. Spontaneous host range mutants contain deletions, *Virology,* 102, 119, 1980.
173. **Dumbell, K. R. and Archard, L. C.,** Comparison of white pock (h) mutants of monkeypox virus with parental monkeypox and with variola-like viruses isolated from animals, *Nature (London),* 286, 29, 1980.
174. **Drillien, R., Koehren, F., and Kirn, A.,** Host range deletion mutant of vaccinia virus defective in human cells, *Virology,* 111, 488, 1981.

175. **Panicali, D., Davis, S. W., Mercer, S. R., and Paoletti, E.,** Two major DNA variants present in serially propagated stocks of the WR strain of vaccinia virus, *J. Virol.,* 37, 1000, 1981.
176. **Pickup, D. J., Bastia, D., Stone, H. O., and Joklik, W. K.,** Sequence of terminal regions of cowpox virus DNA: arrangement of repeated and unique sequence elements, *Proc. Natl. Acad. Sci. U.S.A.,* 79, 4927, 1982.
177. **Robinson, A. J. and Lyttle, D. J.,** unpublished data, 1990.
178. **Rafii, F. and Burger, D.,** Comparison of contagious ecthyma virus genomes by restriction endonucleases, *Arch. Virol.,* 84, 283, 1985.
179. **Fraser, K. M., Hill, D. F., Mercer, A. A., and Robinson, A. J.,** Sequence analysis of the inverted terminal repetition in the genome of the parapoxvirus, orf virus, *Virology,* 176, 379, 1990.
180. **Mercer, A. A., Fraser, K., Stockwell, P. A., and Robinson, A. J.,** A homologue of retroviral pseudoproteases in the parapoxvirus, orf virus, *Virology,* 172, 665, 1989.
181. **Fleming, S. B., Fraser, K. M., Mercer, A. A., and Robinson, A. J.,** Vaccinia virus-like early transcriptional control sequences flank an early gene in orf virus, *Gene,* 97, 207, 1991.
182. **Moss, B.,** Poxviridae and their replication, in *Virology,* Vol. 2, 2nd ed., Fields, B. N. and Knipe, D. M., Eds., Raven Press, New York, 1990, chap. 74.
183. **Fleming, S. B., Mercer, A. A., Fraser, K. M., Lyttle, D. J., and Robinson, A. J.,** *In vivo* recognition of orf virus early transcriptional promoters in a vaccinia virus recombinant, in press.
184. **Reid, H. W., Norval, M., Yirrell, D. L., Mercer, A. A., and Robinson, A. J.,** unpublished data, 1989.
185. **Slabaugh, M. B. and Roseman, N. A.,** Retroviral protease-like gene in the vaccinia virus genome, *Proc. Natl. Acad. Sci. U.S.A.,* 86, 4152, 1989.
186. **McGeoch, D. J.,** Protein sequence comparisons show that the 'pseudoproteases' encoded by poxviruses and certain retroviruses belong to the deoxyuridine triphosphatase family, *Nucleic Acids Res.,* 18, 4105, 1990.
187. **Rodriguez, J. F. and Esteban, M.,** Mapping and nucleotide sequence of the vaccinia virus gene that encodes a 14-kilodalton fusion protein, *J. Virol.,* 61, 3550, 1987.
188. **Naase, M., Nicholson, B. H., and Robinson, A. J.,** unpublished data, 1990.
189. **Rosel, J. L., Earl, P. L., Weir, J. P., and Moss, B.,** Conserved TAAATG sequence at the transcriptional and translational initiation sites of vaccinia virus late genes deduced by structural and functional analysis of the *Hin*dIII H genome fragment, *J. Virol.,* 60, 436, 1986.
190. **Fleming, S. B., Blok, J., Mercer, A. A., Fraser, K. M., and Robinson, A. J.,** in preparation.
191. **Goebel, S. J., Johnson, G. P., Perkus, M. E., Davis, S. W., Winslow, J. P., and Paoletti, E.,** The complete DNA sequence of vaccinia virus, *Virology,* 179, 247, 1990.
192. **Drillien, R., Spehner, D., Villeval, D., and Lecocq, J.-P.,** Similar genetic organization between a region of fowlpoxvirus DNA and the vaccinia virus HindIII J fragment despite divergent location of the thymidine kinase gene, *Virology,* 160, 203, 1987.
193. **Gershon, P. D., Ansell, D. M., and Black, D. N.,** A comparison of the genome organization of capripoxvirus with that of the orthopoxviruses, *J. Virol.,* 63, 4703, 1989.
194. **Panicali, D. and Paoletti, E.,** Construction of poxviruses as cloning vectors: insertion of the thymidine kinase gene from herpes simplex virus into the DNA of infectious vaccinia virus, *Proc. Natl. Acad. Sci. U.S.A.,* 79, 4927, 1982.
195. **Mackett, M., Smith, G. L., and Moss, B.,** Vaccinia virus: a selectable eukaryotic cloning and expression vector, *Proc. Natl. Acad. Sci. U.S.A.,* 79, 7415, 1982.
196. **MacDonald, A. and Bell, T. M.,** Growth of contagious pustular dermatitis virus in human tissue cultures, *Nature (London),* 192, 91, 1961.
197. **Sawhney, A. N. and Toschkov, Al.,** Propagation of contagious pustular dermatitis virus in HeLa cell cultures, *Indian J. Exp. Biol.,* 9, 512, 1971.

198. **Salzman, N. P.,** The rate of formation of vaccinia deoxyribonucleic acid and vaccinia virus, *Virology,* 10, 150, 1960.
199. **Hessami, M., Keney, D. A., Pearson, L. D., and Storz, J.,** Isolation of parapox viruses from man and animals: cultivation and cellular changes in bovine fetal spleen cells, *Comp. Immunol. Microbiol. Infect. Dis.,* 2, 1, 1979.
200. **Felgner, P. L., Gadek, T. R., Holm, M., Roman, R., Chan, H. W., Wenz, M., Northrop, J. P., Ringold, G. M., and Danielsen, M.,** Lipofection: a highly efficient, lipid-mediated DNA-transfection procedure, *Proc. Natl. Acad. Sci. U.S.A.,* 84, 7413, 1987.
201. **Graham, F. L. and Van der Eb, A. J.,** A new technique for the assay of infectivity of human adenovirus 5 DNA, *Virology,* 52, 456, 1973.
202. **Mackett, M., Smith, G. L., and Moss, B.,** The construction and characterisation of vaccinia virus recombinants expressing foreign genes, in *DNA Cloning: a Practical Approach,* Vol. 2, Glover, D. M., Ed., IRL Press, Oxford, 1985, chap. 7.
203. **Mercer, A. A. and Robinson, A. J.,** unpublished data, 1990.
204. **Fraser, K. M. and Robinson, A. J.,** unpublished data, 1990.
205. **Mulligan, R. and Berg, P.,** Selection for animal cells that express the *E. coli* gene coding for xanthine-guanine phosphoribosyl transferase, *Proc. Natl. Acad. Sci. U.S.A.,* 78, 2072, 1981.
206. **Falkner, F. G. and Moss, B.,** *Escherichia coli gpt* gene provides dominant selection for vaccinia virus open reading frame expression vectors, *J. Virol.,* 62, 1849, 1988.
207. **Boyle, D. B. and Couper, B. E. H.,** Construction of recombinant fowlpox viruses as vectors for poultry vaccines, *Virus Res.,* 10, 343, 1988.
208. **Lyttle, D. J. and Robinson, A. J.,** unpublished data, 1990.
209. **Mackett, M. and Smith, G. L.,** Vaccinia virus expression vectors, *J. Gen. Virol.,* 67, 2067, 1986.
210. **Moss, B. and Flexner, C.,** Vaccinia virus expression vectors, *Annu. Rev. Immunol.,* 5, 305, 1987.
211. **Tartaglia, J., Pincus, S., and Paoletti, E.,** Poxvirus-based vectors as vaccine candidates, *Crit. Rev. Immunol.,* 10, 13, 1990.
212. **Kaplan, C.,** Vaccinia virus: a suitable vehicle for recombinant vaccines?, *Arch. Virol.,* 106, 127, 1989.
213. **Buller, R. M. L., Smith, G. L., Cremer, K., Notkins, A. L., and Moss, B.,** Decreased virulence of recombinant vaccinia virus expression vectors is associated with a thymidine kinase-negative phenotype, *Nature (London),* 317, 813, 1985.
214. **Gillespie, J. H., Geissinger, C., Scott, F. W., Higgins, W. P., Holmes, D. F., Perkus, M., Mercer, S., and Paoletti, E.,** Response of dairy calves to vaccinia viruses that express foreign genes, *J. Clin. Microbiol.,* 23, 283, 1986.
215. **Appel, M. J. G. and Paoletti, E.,** Immune response to vaccinia virus and recombinant virus products in dogs, *Am. J. Vet. Res.,* 49, 1932, 1988.
216. **Perkus, M. E., Piccini, A., Lipinskas, B. R., and Paoletti, E.,** Recombinant vaccinia virus: immunization against multiple pathogens, *Science,* 229, 981, 1985.
217. **Blakemore, F., Abdussalam, M., and Goldsmith, W. N.,** A case of orf (contagious pustular dermatitis): identification of the virus, *Br. J. Dermatol.,* 60, 404, 1948.
218. **Fastier, L. B.,** Human infections with the virus of ovine contagious pustular dermatitis (scabby mouth), *N.Z. Med. J.,* 56, 121, 1957.
219. **Agger, W. A. and Webster, S. B.,** Human orf infection complicated by erythma multiforme, *Cutis,* 31, 334, 1983.
220. **Wilkinson, J. D.,** Orf: a family with unusual complications, *Br. J. Dermatol.,* 97, 447, 1977.
221. **Erichson, G. A., Cabrey, E. A.,and Gustafson, G. A.,** Generalized contagious ecthyma in a sheep rancher: diagnostic considerations, *J. Am. Vet. Med. Assoc.,* 166, 262, 1957.
222. **Stringer, H.,** personal communication, 1983.

223. **Hooser, S. B., Scherba, G., Morin, D. E., and Whiteley, H. E.,** Atypical contagious ecthyma in a sheep after extensive cutaneous thermal injury, *J. Am. Vet. Med. Assoc.*, 195, 1255, 1989.
224. **Savage, J. and Black, M. M.,** "Giant orf" of finger in a patient with a lymphoma, *Proc. R. Soc. Med.*, 64, 766, 1972.
225. **Hunskaar, S.,** Giant orf in a patient with chronic lymphocytic leukaemia, *Br. J. Dermatol.*, 114, 631, 1986.
226. **Dupre, A., Christol, B., Bonafe, J. L., and Lassere, J.,** Orf and atopic dermatitis, *Br. J. Dermatol.*, 105, 103, 1981.
227. **Taylor, J., Weinberg, R., Languet, B., Desmettre, P., and Paoletti, E.,** Recombinant fowlpox virus inducing protective immunity in non-avian species, *Vaccine*, 6, 497, 1988.
228. **Buddle, B. M.,** personal communication, 1988.
229. **Buddle, B. M., Dellers, R. W., and Schurig, G. G.,** Contagious ecthyma virus-vaccination failures, *Am. J. Vet. Res.*, 45, 263, 1984.
230. Council of the Pharmaceutical Society of Great Britain, *British Veterinary Codex*, Pharmaceutical Press, London, 1953, 453.
231. **Gillard, S., Spehner, D., Drillien, R., and Kirn, A.,** Localization and sequence of a vaccinia virus gene required for multiplication in human cells, *Proc. Natl. Acad. Sci. U.S.A.*, 83, 5573, 1986.
232. **Percus, M. E., Goebel, S. J., Davis, S. W., Johnson, G. P., Limbach, K., Norton, E. K., and Paoletti, E.,** Vaccinia virus host range genes, *Virology*, 179, 276, 1990.
233. **Mayr, A., Herlyn, M., Mahnel, H., Danco, A., Zach, A., and Bostedt, H.,** Bekampfung des ecthyma contagiosum (pustulardermatitis) der schafe mit einem neuen parenteral-zellkultur-lebendimpstoff, *Zentralbl. Veterinaermed. Reihe B*, 28, 535, 1981.
234. **McAleer, W. J., Buynak, E. B., Maigetter, R. Z., Wampler, D. E., Miller, W. J., and Hilleman, M. R.,** Human hepatitis B vaccine from recombinant yeast, *Nature (London)*, 307, 178, 1984.
235. **Smith, G. L. and Moss, B.,** Infectious poxvirus vectors have capacity for at least 25000 base pairs of foreign DNA, *Gene*, 25, 21, 1983.
236. **Murphy, B. R. and Chanock, R. M.,** Immunization against viruses, in *Virology*, Vol. 1, 2nd ed., Fields, B. N. and Knipe, D. M., Eds., Raven Press, New York, 1990, chap. 19.
237. **Samuel, W. M., Chalmers, G. A., Stelfox, J. G., Lowen, A., and Thomsen, J. J.,** Contagious ecthyma in bighorn sheep and mountain goat in western Canada, *J. Wildl. Dis.*, 11, 26, 1975.
238. **Lance, W., Adrian, W., and Widhalm, B.,** An epizootic of contagious ecthyma in Rocky Mountain bighorn sheep in Colorado, *J. Wildl. Dis.*, 17, 601, 1981.
239. **Dieterich, R. A., Spencer, G. R., Burger, D., Gallina, A. M., and VanderSchalie, J.,** Contagious ecthyma in Alaskan musk-oxen and Dall sheep, *J. Am. Vet. Med. Assoc.*, 179, 1140, 1981.
240. **Smith, T. C., Heimer, W. E., and Foreyt, W. J.,** Contagious ecthyma in an adult Dall sheep (Ovis dalli dalli) in Alaska, *J. Wildl. Dis.*, 18, 111, 1982.
241. **Blanc, G., Melanidi, C., and Caminopteris, J.,** Recherches experimentales sur une maladie eruptive de la chevre observee en Grece, *Ann. Inst. Pasteur (Paris)*, 36, 614, 1922.
242. **Guarda, F.** *Ann. Fac. Med. Vet. Univ. Torino*, 9, 37, 1959; cited in Reference 243.
243. **Kater, J. C. and Hansen, N. F.,** Contagious ecthyma in wild thar in the South Island, *N.Z. Vet. J.*, 10, 116, 1962.
244. **Kummeneje, K. and Krogsrud, J.,** Contagious ecthyma (orf) in musk ox (Ovibos moschatos), *Acta Vet. Scand.*, 19, 461, 1978.
245. **Mathiesen, S. D., Jorgensen, T., Traavik, T., and Blix, A. S.,** On contagious ecthyma and its treatment in muskoxen (Ovibos moschatus), *Acta Vet. Scand.*, 26, 120, 1985.
246. **Kummeneje, K. and Krogsrud, J.,** Contagious ecthyma (orf) in reindeer (Rangifer t tarandus), *Vet. Rec.*, 105, 60, 1979.

247. **Moro, M.,** Enfermedades de los auquenidos, in *Bol. Inst. Invest. Trop. Altura (IVTA, Lima, Peru),* 1968, 61; Cited in **Riviera, H., Madewell, B. R., and Ameghino, E.,** Serological survey of viral antibodies in the Peruvian alpaca (Lama pacos), *Am. J. Vet. Res.,* 48, 189, 1987.
248. **Wilkinson, G. T., Prydie, J., and Scarnell, J.,** Possible "orf" (contagious pustular dermatitis, contagious ecthyma of sheep) infection in the dog, *Vet. Rec.,* 87, 766, 1970.
249. **Greig, A. S.,** Contagious ecthyma of the sheep. I. Attempts to infect other hosts, *Can. J. Comp. Med.,* 20, 448, 1956.
250. **Abdussalam, M.,** Contagious pustular dermatitis. III. Experimental infection, *J. Comp. Pathol.,* 67, 305, 1957.
251. **Newsom, I. E. and Cross, F.,** Sore mouth in feeder lambs due to a filterable virus, *J. Am. Vet. Med. Assoc.,* 84, 233, 1934.
252. **Plowright, W., Witcomb, M. A., and Ferris, R. D.,** Studies with a strain of contagious pustular dermatitis virus in tissue culture, *Arch. Gesammte Virusforsch.,* 9, 214, 1959.
253. **Wheeler, C. E. and Cawley, E. P.,** The microscopic appearance of ecthyma contagiosum (orf) in sheep, rabbits and man, *Am. J. Pathol.,* 32, 535, 1956.
254. **Selbie, F. R.,** Experiments on the transmission to the rabbit of the virus of contagious pustular dermatitis of the sheep, *J. Comp. Pathol.,* 54, 161, 1944.
255. **Sinha, O. P., Soman, J. P., and Verma, B. B.,** Isolation and experimental host range studies of contagious ecthyma virus, *Indian J. Anim. Sci.,* 56, 937, 1986.
256. **Lyell, A. and Miles, J. A. R.,** Orf in man, *Br. Med. J.,* ii, 1119, 1950.
257. **Sawhney, A. N.,** Studies on the virus of ecthyma contagiosum, *Izv. Mikrobiol. Inst. Bulg. Akad. Nauk.,* 18, 163, 1966.
258. **Bennett, S. C. J., Horgan, E. S., and Hasseb, M. A.,** The pox diseases of sheep and goats, *J. Comp. Pathol.,* 54, 131, 1944.
259. **Lance, W. R., Hibler, C. P., and DeMartini, J.,** Experimental contagious ecthyma in mule deer, white-tailed deer, pronghorn and wapiti, *J. Wildl. Dis.,* 19, 165, 1983.
260. **Zarnke, R. L., Dieterich, R. A., Neiland, K. A., and Ranglack, G.,** Serological and experimental investigations of contagious ecthyma in Alaska, *J. Wildl. Dis.,* 19, 170, 1983.
261. **Pask, V. M., Mackerras, I. M., Sutherland, A. K., and Simmons, G. C.,** Transmission of contagious ecthyma from sheep to man, *Med. J. Aust.,* 2, 628, 1951.
262. **Cheville, N. F. and Shey, D. J.,** Pseudocowpox in dairy cattle, *J. Am. Vet. Med. Assoc.,* 150, 282, 1974.
263. **Lauder, I. M., Martin, B., Martin, W. B., and Nagington, J.,** Milkers' nodule virus infection and its resemblance to orf, *Vet. Rec.,* 78, 926, 1966.

INDEX

D

E

J

K

L

R

W

X

Y